Europe and Global Climate Change

Politics, Foreign Policy and Regional Cooperation

Edited by

Paul G. Harris

Director, Project on Environmental Change and Foreign Policy
Professor of International and Environmental Studies,
Lingnan University, Hong Kong

Edward Elgar

Cheltenham, UK • Northampton, MA, USA

Published by
Edward Elgar Publishing Limited
Glensanda House
Montpellier Parade
Cheltenham
Glos GL50 1UA
UK

Edward Elgar Publishing, Inc.
William Pratt House
9 Dewey Court
Northampton
Massachusetts 01060
USA

A catalogue record for this book
is available from the British Library

Library of Congress Control Number: 2006934137

ISBN 978 1 84542 944 7

Printed and bound in Great Britain by MPG Books Ltd, Bodmin, Cornwall

Contents

List of figures vii
List of tables viii
List of contributors ix
Preface xi
List of abbreviations and acronyms xiii

INTRODUCTION

1 Europe and the politics and foreign policy of global climate change 3
Paul G Harris

PART I GLOBAL CLIMATE CHANGE AND POLICYMAKING IN EUROPE

2 The politics of climate change in Germany: domestic sources of environmental foreign policy 41
Michael T Hatch
3 The indispensable awkward partner: the United Kingdom in European climate policy 63
Loren R Cass
4 Middle power leadership in the climate change negotiations: foreign policy of the Netherlands 87
Norichika Kanie
5 Climatic issues in Polish foreign policy 113
Anita Bokwa
6 Sweden, climate change and the EU context 139
Kate E Marshall
7 A new climate for Spain: accommodating environmental foreign policy in a federal state 161
J David Tàbara
8 Shaming and framing: Norwegian NGOs in the climate change negotiations 185
Andreas Tjernshaugen and Ho-Ching Lee

PART II THE EUROPEAN UNION AND GLOBAL CLIMATE CHANGE

9 Articulating a consensus: the EU's position on climate change 211
Nuno S Lacasta, Suraje Dessai, Eva Kracht and Katharine Vincent
10 Common policy on climate change: land use, domestic stakeholders and EU foreign policy 233
Martina Jung, Axel Michaelowa, Ingrid Nestle, Sandra Greiner and Michael Dutschke
11 Climate policy in the EU: international regimes and policy diffusion 255
Sebastian Oberthür and Dennis Tänzler
12 Meeting Kyoto commitments: EU influence on Norway and Germany 279
Guri Bang, Jonas Vevatne and Michelle Twena
13 Climate change policy and the enlargement of the EU 305
Leonardo Massai
14 The reflexivity of ideas in climate change policy: German, European and international politics 323
Lyn Jaggard
15 Sharing the burdens of global climate change: international equity and justice in European policy 349
Paul G Harris

CONCLUSION

16 Explaining European responses to global climate change: power, interests and ideas in domestic and international politics 393
Paul G Harris

Index 407

List of figures

1.1 Typology of environmental foreign policy 19
7.1 Change in total aggregate GHG emissions in individual Annex I parties, 1990–2002 167
8.1 Channels of NGO influence in international negotiations 196
10.1 Three-level game in climate change negotiations 240
11.1 The adoption of climate protection plans in the EU and its member states 259
11.2 Introduction of CO_2/energy taxes in the EU and its member states 268
12.1 Emissions of GHGs in Germany and Norway (1990–2002) 287
12.2 Relative GHG emissions in the EU, Norway and Germany, by source (2002) 288
16.1 Theoretical explanations for Europe's responses to global climate change 404

List of tables

4.1	Typologies of leadership modes	90
4.2	Changes in the Dutch CO_2 emissions reduction target (1989–97)	95
4.3	Greenhouse gas reduction targets for 2010 (relative to 1990)	101
6.1	Kronsell's avenues of normative influence	142
7.1	Evolution of GHG emissions in Spain (1990–2002)	164
7.2	Distribution of GHG emissions per sector in Spain (2002)	165
7.3	GHG evolution in Spain 1990–2002: relation to population and selected economic indicators	166
7.4	Key dimensions of climate policy in Spain, 1992–2005	173
9.1	EU 1997 and 1998 burden sharing agreements	225
10.1	Important negotiating coalitions in the climate change regime	242
11.1	Climate protection plans, emissions trading and CO_2/energy taxes: status within the EU (2005)	261
12.1	The EU and emissions trading	283
12.2	Emissions commitments and trends: Germany and Norway	286

List of contributors

Guri Bang is a senior research fellow at the Center for International Climate and Environmental Research in Oslo (CICERO).

Anita Bokwa is a lecturer and researcher in climatology at the Institute of Geography and Spatial Management, Jagiellonian University, Cracow, Poland.

Loren R Cass is an associate professor of political science at the College of the Holy Cross, Worcester, Massachusetts, USA.

Suraje Dessai is a Senior Research Associate at the Tyndall Centre for Climate Change Research and the School of Environmental Sciences at the University of East Anglia, UK.

Michael Dutschke works on international climate policy, specializing in land use, land-use change and forestry, with the Hamburg Institute of International Economics, Germany.

Sandra Greiner, formerly a research fellow at the Hamburg Institute of International Economics, works in the East Asia department of the World Bank.

Paul G Harris is director of the Project on Environmental Change and Foreign Policy and a professor of politics at Lingnan University, Hong Kong.

Michael T Hatch is a professor in the Department of Political Science at the University of the Pacific, Stockton, California, USA.

Lyn Jaggard lectures at Oxford Brookes University and at the University of Greenwich's Summer University, UK.

Martina Jung is a climate policy consultant and former research fellow at the Hamburg Institute of International Economics, Germany.

Norichika Kanie is an associate professor at the Graduate School of Decision Science and Technology, Tokyo Institute of Technology.

Eva Kracht is a lawyer in the German Federal Ministry for the Environment, Nature Conservation and Nuclear Safety.

Nuno S Lacasta is an adjunct professor in the Faculty of Science and Technology at the New University of Lisbon and director of the International Affairs Department of the Portuguese Ministry for the Environment, Spatial Planning and Regional Development.

Ho-Ching Lee is director of the Center of International Programs, Chung-Yuan Christian University, Taiwan.

Kate E Marshall is a doctoral candidate researching environmental dimensions of European integration at the University of Washington.

Leonardo Massai is a researcher in EU Law at the T.M.C. Asser Institute, The Hague, Holland, and a doctoral candidate at the University of Frankfurt am Main, Germany.

Axel Michaelowa, a participant in UN climate negotiations since 1995, heads the Program on International Climate Policy, Hamburg Institute of International Economics, Germany.

Ingrid Nestle, a former member of the German carbon sinks expert team, is a scientific assistant in energy economics at the University of Flensburg, Germany.

Sebastian Oberthür is the academic director at the Institute for European Studies, Vrije Universiteit, Brussels.

J David Tàbara is a senior researcher at the Institute of Environmental Science and Technology at the Autonomous University of Barcelona, Spain.

Dennis Tänzler is a research fellow at Adelphi Research in Berlin.

Andreas Tjernshaugen is a research fellow at the Center for International Climate and Environmental Research, Oslo (CICERO) and a doctoral student at the University of Oslo.

Michelle Twena is a doctoral candidate researching European environmental policymaking at the University of East Anglia, UK.

Jonas Vevatne is a research fellow at the Center for International Climate and Environmental Research, Oslo (CICERO).

Katharine Vincent is a doctoral researcher at the Tyndall Center for Climate Change Research and School of Environmental Sciences, University of East Anglia, UK.

Preface

This is the seventh book from the Project on Environmental Change and Foreign Policy, which I began while on the faculty of London Metropolitan University (formerly London Guildhall University) and which relocated to Hong Kong when I joined Lingnan University. Other books from the project, published between 2000 and 2005, were (in order of publication) *Climate Change and American Foreign Policy* (St Martin's Press/Palgrave Macmillan), *The Environment, International Relations and US Foreign Policy* (Georgetown University Press), *International Equity and Global Environmental Politics: Power and Principles in US Foreign Policy* (Ashgate), *International Environmental Cooperation: Politics and Diplomacy in Pacific Asia* (University Press of Colorado), *Global Warming and East Asia: The Domestic and International Politics of Climate Change* (Routledge) and *Confronting Environmental Change in East and Southeast Asia: Eco-Politics, Foreign Policy and Sustainable Development* (United Nations University Press/Earthscan).

The core objective of the project is to better understand the role of *foreign policy* – the crossovers and interactions between domestic and international politics and policies – in efforts to preserve the environment and natural resources. Underlying this objective is the belief that it is not enough to analyze domestic or international political actors, institutions and processes by themselves. We need to understand the interactions among them, something that explicit thought about foreign policy can help us do.

More specifically, the project has sought to understand foreign policy processes in international efforts to address human-induced changes to the natural environment; to analyze the actors and institutions – domestic and international, governmental and nongovernmental – that constrain and shape actions on environmental issues; to show how environmental changes influence foreign policy processes; and to critically assess environmental foreign policies. Other objectives of the project are to test the waters of research in this field; to showcase research from established *and* new scholars of different backgrounds that has not been forced into traditional empirical, epistemological or ontological boxes; to give insight to governmental and nongovernmental practitioners and activists, which might help improve their understanding of the politics and foreign policy of environmental issues; to disseminate these ideas to government officials and scholars so that they might have some positive effect on policy making and scholarship; and to enlighten students and laypersons interested in environmental protection, sustainable development, international affairs and foreign policy.

I wish to thank the contributors for answering my call to contribute to this project and for persevering through the very long process of research, writing, peer review and revision. Anonymous readers gave valuable comments that forced us to strengthen our thinking and writing. Lingnan University and its Research and Postgraduate Studies Committee provided important financial assistance. KK Chan has given me inspiration and support during this and many other projects, for which I am immensely grateful.

Paul G Harris
Hong Kong

List of abbreviations and acronyms

ACs	autonomous communities (regional)
AGBM	Ad hoc Group on the Berlin Mandate
AGE	Arbeidtsgruppe Emissionshandel zur Bekämpfung des Treibhauseeffektes (German Emissions Trading Group)
AIJ	activities implemented jointly
AND	Autoridad Nacional Designada (Designated National Authority) (Spain)
AOSIS	Alliance of Small Island States
AP	Allocation Plan
ARD	afforestation, reforestation and deforestation
BATNEEC	best available technology not entailing excessive costs
BMU	German Ministry of the Environment (Bundesministerium für Umwelt, Naturschutz und Reaktorsicherheit)
BMWi	Federal Ministry for Economics
BMZ	German Federal Ministry for Cooperation and Development
BPEO	best practicable environmental option
BSA	burden sharing agreement
BuZa	Ministerie van Buitenlandse Zaken
CAN	Climate Action Network
CCAP	Center for Clean Air Policy
CDM	Clean Development Mechanism
CDU	Christian Democratic Union
CG11	Central Group 11
CEEC	Central and Eastern European Countries
CEPI	Confederation of European Paper Industries
CERs	certified emission reductions
CFC	chlorofluorocarbon
CG11	Central Group 11
CGIAR	Consultative Group of International Agricultural Research

CH_4	methane
CHF	Swiss francs
CHP	combined heat and power
CIPE	Committee for Economic Planning (Italy)
CO_2	carbon dioxide
COP	Conference of the Parties (to the FCCC)
COP/MOP	Conference of the Parties serving as the Meeting of the Parties to the Kyoto Protocol
CSU	Christian Social Union
DG	Directorate General (European Commission)
DGXI	Directorate General XI
EATD	European Allowance Trading Directive
EBCSE	European Business Council for Sustainable Energy
EC	European Community
ECCP	European Climate Change Program
EDF	Environmental Defense
EEA	European Economic Area
EEA	European Environment Agency
EEG	Renewable Energy Sources Act
EFP	environmental foreign policy
EFPA	environmental foreign policy analysis
EIT	Economy in Transition
ELO	European Landowners Organization
EMEP	Cooperative Program for Monitoring and Evaluation of the Long-range Transmission of Air Pollutants in Europe
ENDS	Environmental Data Services
ENGO	environmental nongovernmental organization
Enquete Commission	Enquete Commission of the German Bundestag
EP	European Parliament
ERT	European Roundtable of Industrialists
ERUs	emissions reduction units
ET	emissions trading
ETS	emissions trading system
EU	European Union
EU-15	15 Member States of the EU prior to enlargement on 1 May 2004
EZ	Ministerie van Economische Zaken
FAO	Food and Agricultural Organization
FCCC	Framework Convention on Climate Change
FDP	Free Democratic Party

FERN	Forests and the European Union Resource Network
FIVH	Framtiden i Våre Hender (The Future in Our Hands)
FMG	Fellesaksjonen mot gasskraftverk (joint action against gas-fired power plants)
FoEI	Friends of the Earth International
FPA	foreign policy analysis
FSC	Forest Stewardship Council
G77	Group of 77 developing countries
GCC	global climate change
GDP	gross domestic product
GDR	German Democratic Republic
GEF	Global Environment Facility
GHG	greenhouse gas
GNP	gross national product
GPN	Greenpeace Norway
H-CFC	hydro-chlorofluorocarbon
HFC	hydro-fluorocarbon
IAE	International Academy of the Environment
ICC	International Chamber of Commerce
IEA	International Energy Agency
IEE	International Environmental Equity
IET	international emissions trading
IFAD	International Fund for Agricultural Development
IFIR	International Forest Industry Roundtable
IIASA	International Institute of Applied Systems Analysis
INC	Intergovernmental Negotiating Committee
IPCC	Intergovernmental Panel on Climate Change
IPPC	integrated pollution and prevention control
ISPA	Instrument for Structural Policies for Pre-accession
IUC	UNEP Information Unit for Conventions
IUCN	World Conservation Union
IUFRO	International Union of Forest Research Organizations
IWECS	Swiss Intergovernmental Working Group on the Evolution of the Climate System
JI	joint implementation
kWh	kilowatt hour
LCPD	Large Combustion Plant Directive

LDC	least developed country
LNV	Ministerie van Landbouiw Natuurbeheer, en Visserij
LRTAP	Convention on Long-range Transboundary Air Pollution
LTA	Long-Term Agreements
LUCF	land-use change and forestry
LULUCF	land use, land-use change and forestry
MDGs	Millenium Development Goals
MEA	multilateral environmental agreement
MOP	Meeting of the Parties (to the Kyoto Protocol)
mt	million tons
MW	megawatt
N_2O	nitrogen dioxide
NCCC	National Council on Climate Change (Consejo Nacionel del Clima) (Spain)
NCPC	National Cleaner Production Center
NEPP	National Environmentl Policy plan (the Netherlands)
NEPP 2	National Environmental Policy Plan 2 (the Netherlands)
NGO	nongovernmental organization
NHO	Norwegian Confederation of Business and Industry
NNV	Naturvernforbundet (Friends of the Earth Norway/Norwegian Society for the Conservation of Nature)
NO_x	nitrogen oxide
NPAA	National Programs for Adoption of the *Acquis*
NTB	Norsk telegrambyrå (Norwegian news agency)
NU	Natur og Ungdom (Nature and Youth)
ODA	official/overseas development assistance
ODS	ozone-depleting substance
OECD	Organization for Economic Cooperation and Development
OPEC	Organization of Petroleum Exporting Countries
Phare	Pologne, Hongrie Assistance à la Reconstruction Economique
PFC	perfluorocarbon
PSOE	Partido Socialista Obrero Español (Spanish Socialist Party)

QELROs	Quantified Emission Limitation and Reduction Objectives
RFF	Resources for the Future
RMUs	removal units
SAEFL	Swiss Federal Agency for the Environment, Forests and Landscape
SAR	Second Assessment Report
SBSTA	Subsidiary Body for Scientific and Technological Advice
SCOPE	Scientific Committee on Problems of the Environment
SDC	Swiss Development Cooperation Agency
SECO	Swiss State Secretariat for Economic Affairs
SF_6	Sulfur hexaflouride
SFT	Statens Forurensningstilsyn (Norwegian Pollution Control Agency)
SO_2	sulfur dioxide
SPD	Social Democratic Party
SWAPP	Swiss Pilot Program for Activities Implemented Jointly
SWCC	Second World Climate Conference
TAR	Third Assessment Report
TPES	total primary energy supply
UK	United Kingdom
UN	United Nations
UNCED	United Nations Conference on Environment and Development (Earth Summit)
UNECE	United Nations Economic Commission for Europe
UNED	United Nations Environment and Development
UNEP	United Nations Environment Program
UNFCCC	United Nations Framework Convention on Climate Change
UNGA	United Nations General Assembly
UNICE	Union of Industrial and Employer's Confederations of Europe
UNITAR	United Nations Institute for Training and Research
USA	United States of America
USSR	Union of Soviet Socialist Republics
V&W	Ministerie van Verkeer en Waterstaat
VCI	Association of German Chemical Industry

VNO-NCw	het Verbond van Nederlandse Ondernemingen (VNO) en het Nederlands Christelijk Werkgeversverbond (NCW)
VOC	volatile organic compounds
VROM	Ministerie can Volkshuisvesting, Ruimtelijke Ordeningen Milieubeheer
WBCSD	World Business Council for Sustainable Development
WBGU	German Advisory Council on Global Environmental Change
WCED	World Commission on Environment and Development
WGCCD	Working Group on Climate Change and Development
WMO	World Meteorological Organization
WRI	World Resources Institute
WSIS	World Summit on the Information Society
WSSD	World Summit on Sustainable Development
WWF	World Wide Fund for Nature

INTRODUCTION

1. Europe and the politics and foreign policy of global climate change

Paul G Harris

INTRODUCTION

The Earth is experiencing unnatural atmospheric warming that is leading to changes in climate and a host of mostly adverse side effects for humankind and the ecosystems upon which we depend for our wellbeing (see Houghton et al. 2001; Houghton 2004). Of this there is now little dispute.[1] The world's governments, as well as many nongovernmental actors, have started to address the problems of global climate change (GCC) – global warming and resulting climatic changes. Alas, their responses to these problems have been modest at best relative to the scale of painful future changes to environmental and socio-economic systems that are anticipated by scientists. Europe is a crucial actor in the GCC debate and related diplomacy and policy responses. Like the United States (US), as a group the countries of Europe are a primary source of 'greenhouse' gases (GHGs) causing global warming, meaning that the extent to which they limit those emissions will be important for future climate change.[2] They also possess technological and financial resources that are necessary to reduce GHG emissions globally and to assist those countries most vulnerable to GCC so that they can better adapt to its undesirable effects. While the European reaction, like that of the US, has not fully met the challenge of GCC, nowhere has the response to this problem been greater than among the member states of the European Union (EU).[3] As a group they have taken greater steps at the national, regional and international levels to reduce their own and other countries' emissions of GHGs, and they, along with the EU itself, have done the most to at least begin assisting developing countries that will suffer inordinately from the effects of GCC (see Chapter 15).[4]

In this book we explore European and EU responses to global climate change. Our main aims are (1) to analyze the 'what, how and why' of GCC policies of several EU member states and of the EU itself (e.g., the European Commission and its agencies), and (2) to supplement existing knowledge of GCC policies by explicitly adding environmental foreign policy analysis (FPA) to the mix of tools used to understand those policies. While many important arguments are made throughout the chapters that follow, the main one is this: analyzing and thinking about European GCC policies from the perspective of *foreign policy* – the crossover and interaction between domestic and international politics – will help us to better understand how and why Europe has responded to this issue and how (or whether) it will do so in the future. Much research has already been done on the GCC negotiations and related policies of European states and the EU. Our slant toward FPA, often in combination with or supplemental to other analytical approaches, can provide new insights not fully revealed by that of other work.

Toward that end, the chapters in this book describe the responses of individual European countries and the EU to GCC, discuss major issues of domestic and international politics and policymaking underlying those responses, frame the problem of global warming in terms of foreign policy, and place GCC in the context of increasing European regional cooperation, integration and development of a common European foreign policy.[5] (The chapters are summarized below.)

We seek to answer several related questions: How has Europe responded to GCC? What explains Europe's response to what arguably is the world's most pressing environmental problem, and possibly the most important long-term threat to the world yet foreseen? Why have the EU and some of its member states been more proactive than other countries, notably the US? Why have they not done much more, especially given their major contribution to the problem and their ethical obligation to stop polluting the global atmosphere and to do much more to help those most affected by climate change? How can considerations of foreign policy processes – the use of FPA – add to our understanding of European GCC policies? Which theories of foreign policy and international relations can best help us to understand Europe's responses to global climate change?

Undertaking such analyses is important simply because they address such an important topic. Global warming and climate change will have profound impacts on individuals, societies and states in the future. Indeed, some countries, such as the small island states, argue that they are *already* suffering from climate change, notably in the form of sea-level rise. It is likely that the world's rich countries, including those of Europe, will be able to adapt to many of the changes expected for this century, but this will not protect them from unwanted changes in the environment that could make life less enjoyable, might threaten national traditions and cultures dependent on particular local and national environmental conditions, and will inevitably require spending vast sums of money that could otherwise be directed at alternative national priorities. Of course, the adaptability of the world's rich countries, tied directly to their wealth, does not obtain in most of the world; most poor countries will be unable to avoid the many unwanted effects of GCC, and they will require costly assistance from the world's wealthy countries if their peoples are to prosper – or in some cases survive – amidst climate change. Furthermore, GCC poses challenges to national and human security. As droughts become more frequent in parts of the world already suffering shortages of water, food production becomes more difficult and increasing heat and pests take their toll, domestic communities may experience civil unrest that could conceivably spread across borders. While Europe might not be directly affected by most of these problems, its overseas interests will be, and it will have little choice but to respond with aid. For these and other reasons, it is important for us to understand as best as we can Europe's responses to global climate change. By better understanding Europe's GCC politics and policy at all levels – local, national, regional and international – policymakers, scholars, activists and interested laypersons will not only be enlightened; they might also find ways of effecting more positive policy change in the future as the problems of global warming, climate change and related environmental issues grow worse.

The remainder of this chapter first summarizes the problems of global warming and climate change before briefly describing how the international community has responded to GCC, particularly with regard to efforts to foster international cooperation to reduce GHG emissions. The chapter then goes on to show how GCC policy can be thought of in terms of *foreign policy*. That is, politics and policies related to GCC cannot be fully understood by looking solely at domestic politics and policy making, on the one hand, or by examining international politics and diplomacy, on the other. Climate change is a problem, like many other environmental issues and indeed other challenges facing an increasingly globalized world, that by its very nature *crosses over between* the domestic and international arenas of politics and policymaking. As Foreign policy analysis focuses on the crossover between these levels of analysis, and indeed includes each of them, it is a potentially productive way of viewing the problem and its causes and potential solutions. This chapter concludes with a summary of the chapters to follow. Those chapters do much more to detail European and EU responses to GCC.

GLOBAL CLIMATE CHANGE

What are global warming and climate change, and what causes them?[6] Particularly over the last decade or two, scientists have radically improved their understanding of the causes and consequence of global warming – the warming of the Earth as a consequence of GHGs building up in the atmosphere. Many of these GHGs, such as carbon dioxide (CO_2) and methane, while having natural sources, are also the products of human activities and industrialization. Carbon dioxide – the most influential GHG in aggregate – is created by the burning of fossil fuels (e.g., coal, oil, natural gas) for industry, transportation and other purposes, and when trees are felled and subsequently decay or are burned. Other GHGs, such as methane, are the result of agriculture, and yet others (e.g., chlorofluorocarbons, which also deplete the stratospheric ozone layer) are released mostly by industrial activities. The Intergovernmental Panel on Climate Change (IPCC) has concluded that human activities are adding GHGs to the atmosphere, and that this additional contribution over natural sources is having a discernable impact by increasing global temperatures (Houghton et al. 1996; Albritton et al. 2001; Houghton et al. 2001).

Climate change refers to the climatic changes and their consequences resulting from global warming, with the United Nations (UN) Framework Convention on Climate Change (FCCC) including under this rubric atmospheric changes connected directly or indirectly to human activities. The impacts of climate change on natural ecosystems and on human society and economies are potentially severe, ranging from sea level rise and melting ice at higher latitudes (the Arctic and Antarctic) and altitudes (mountain glaciers), to changing weather patterns characterized by increasingly severe storms, floods and droughts, and the attendant impacts of these changes, such as the spread of pests to newly warmed regions. Some areas could experience positive effects of climate change (e.g., an extended growing season in high latitudes), but these will likely be accompanied by adverse impacts and unpredictability at best. Overall, predictions point to adverse impacts, particularly in parts of the world where geographic vulnerability and poverty make adaptation difficult or impossible (see AfDB et al. 2002).

Global Warming and Climate Change: Causes and Consequences

The most authoritative reports on the causes and consequences of GCC come from the IPCC, particularly its 1995 'Second Assessment Report' (SAR) (Houghton et al. 1996; Watson et al. 1996; Bruce et al. 1996) and its 2001 'Third Assessment Report' (TAR) (IPCC 2001; see especially Houghton et al. 2001; Albritton et al. 2001 and McCarthy et al. 2001). The TAR refined the findings of the first assessment, pointing out that climate change is likely to be worse and occur more rapidly than initially predicted (see NRC 2002). The IPCC and many other international and scientific organizations continue to study and refine our understanding of global warming and climate change, but they all tend toward refining what we already know. These trends are quite clear and extremely worrying. This section introduces the IPCC's findings on global warming and summarizes some of the worldwide effects of climate change.

According to the IPCC's TAR (Houghton et al. 2001), there is now a collective picture, derived from an increasing body of observations, of a warming world and other changes in the Earth's climate system. The global average surface temperature increased during the twentieth century (WMO 2003). The ten hottest years since recording began in 1861 have been since 1989, with the 1990s as the warmest decade ever recorded (1998 being the warmest and 2004 [the most recent data] the fourth warmest) (Met Office 2004; WHRC 2004). Snow and ice cover have decreased, the global average sea level has risen, and the heat content of the oceans has increased. Other aspects of climate changed during the twentieth century, including changes in precipitation (e.g., increased heavy precipitation events) and cloud cover; fewer extreme low-temperature periods and more high-temperature periods; more frequent, persistent and intense episodes of the El Niño ocean-warming event (and related adverse effects on weather in many areas); and an increase in the number of areas experiencing drought and severe wet periods. Some climate related events, such as tornadoes or tropical storms, do not appear to have changed based on IPCC data, although the evidence is conflicting and there are suggestions that that is precisely what will happen (if it is not already) (see WMO 2003).

The TAR reports that human activities have increased the amount of GHGs (e.g., CO_2, methane, nitrous oxide, halocarbons) in the atmosphere, as well as their warming potential: 'atmospheric concentration of carbon dioxide (CO_2) has increased 31 per cent since 1750. The present CO_2 concentration has not been exceeded during the past 420,000 years and likely not during the past 20 million years. The current rate of increase is unprecedented during at least the past 20,000 years' (Albritton et al. 2001: 7). Three-quarters of human-induced emissions of CO_2 over the last two decades have come from the burning of fossil fuels, with most of the remainder a consequence of land-use changes, particularly deforestation. Natural causes of GCC have been relatively small.

According to the TAR, while uncertainties remain, understanding of climate processes and predicted effects has improved and models for predicting future climate are increasingly accurate and precise. New and stronger evidence points to human activities as the sources of observed global warming over the last 50 years, further strengthening the SAR's conclusion that the 'balance of evidence suggests a discernible human influence on global climate' (IPCC 1995; reaffirmed in IPCC 2001).[7] Warming over the last 100 years is unlikely to have been natural, with studies showing that global warming, particularly during the last 35–50 years, most likely resulted from human activities. Thus the TAR concludes: 'In light of new evidence and taking into account the remaining uncertainties, most observed warming over the last 50 years is likely to have been due to the increase in greenhouse gas concentrations. Furthermore, it is very likely that the twentieth century warming has contributed significantly to the observed sea level rise … and widespread loss of land ice' (Albritton et al. 2001: 10).

The TAR determined that human activities will continue to shape the Earth's atmosphere throughout this century and beyond, and average global temperatures and sea levels are projected to rise. Emissions from burning fossil fuels will be the dominant source of atmospheric CO_2 during this century. These emissions and those of other GHGs would have to be reduced to 'a very small fraction of current emissions' to stabilize climate (Albritton et al. 2001: 12). Global average temperature is projected by the IPCC to increase by 1.4 to 5.8 degrees Celsius during this century (more than anticipated in the SAR). This warming will occur at a rate faster than that observed in the twentieth century, 'very likely to be without precedent during at least the last 10,000 years' (Albritton et al. 2001: 13). During this century, warming is expected to occur in most areas, but it should be particularly pronounced at northern high latitudes during winter. The global mean sea level is expected to rise .09 to .88 meters in this century, with other likely changes to include higher maximum temperatures and more hot days over most land areas, higher minimum temperatures and fewer cold days over most land areas, more intense precipitation events over many areas, increased summertime continental drying and drought over mid-latitude continental interiors, and more severe storms in some regions.

The ecological and socio-economic impacts of GCC are likely to be very significant and frequently very painful. The TAR's findings on these impacts include the following (IPCC Working Group II 2001; McCarthy et al. 2001): Regional changes in climate have already affected many physical and biological systems, with temperature increases being the proximate cause. Observed changes in regional climate have occurred in terrestrial, aquatic and marine environments, and effects have included shrinking glaciers, thawing permafrost, reduced periods in which lakes and rivers are frozen, longer mid- and high-latitude growing seasons, shifts in animal and plant ranges to higher latitudes and altitudes, declines in populations of some animals and plants and reduced egg-laying in some birds, and insects populating new areas. It appears that increased floods and drought have already affected some social and economic systems. However, separating these ecological events from socioeconomic factors is difficult.

The TAR shows how many human systems are sensitive to GCC. These include water resources, agriculture, coastal zones and marine fisheries, settlements, energy, industry, financial services (e.g., insurance industries affected by increased claims) and human health. Adverse impacts of GCC include reduced crop yields in most tropical and sub-tropical regions; decreased water availability in many water-scarce areas, especially the sub-tropics; more people exposed to increased mortality from heat stress and vector-borne diseases like cholera; widespread increase in flood risk from rising sea levels; and increasing demand for energy to cool areas affected by higher summer temperatures. Some impacts may be positive, such as increased crop yields in some mid-latitude areas; potentially more timber if forests are managed appropriately (although increased pests could more than offset this); increased water availability for some water scarce areas (e.g., Southeast Asia, at least until glaciers disappear); lower winter mortality in traditionally cold areas; and reduced winter demand for energy due to higher winter temperatures. Many of the risks are unclear, and there is substantial potential for 'large-scale and possibly irreversible impacts' from changing ocean currents, melting ice sheets, accelerated global warming due to atmospheric feedback effects, and so forth (IPCC Working Group II 2001: 4–5; see McCarthy et al. 2001).

What are the predicted consequences of GCC for Europe? According to the IPCC's assessment of regional impacts (Watson et. al 1998), Europe's ecosystems are particularly sensitive. Grasslands may shift and northern forests will likely spread to regions that are now tundra. The migration of local climate zones may be too fast for some species and types of forest to keep pace, making them potentially endangered – something that is almost certain for high-altitude species, which have nowhere to migrate. Wetlands and the water table will be degraded with warming, floods may increase in northern Europe, and water in southern Europe will likely become scarcer. Of particular concern are the anticipated changes to snow and ice cover. According to the report, as much as '95 per cent of Alpine glacier mass could disappear by 2100, with subsequent consequences for the water flow regime – affecting, for example, summer water supply, shipping, and hydropower'. Winter tourism would of course be adversely affected. Yields of winter crops may increase, 'assuming that neither precipitation nor irrigation are limiting and that water-use efficiency increases with the ambient atmospheric concentration of CO_2'. However, there may be decreased crop yields in southern Europe due to reduced water supplies, making irrigation 'an even larger competitor to domestic and industrial water use'. Coastal areas are already vulnerable to storm surges and will be susceptible to sea level rise and other changes associated with GCC. Without intervention, human health in Europe may suffer from increased heat-related mortality, more urban air pollution and expansion of vector-borne diseases, although there may be fewer cold-related deaths. The IPCC report concludes that the greatest impacts for Europe 'are likely to be felt through changes in the frequency of extreme events and precipitation, causing more droughts in some areas and more river floods elsewhere' (Watson et al. 1998).[8]

In addition to efforts to mitigate GCC through limiting GHG emissions, the TAR argues that adaptation is a necessary strategy.[9] Europe is relatively wealthy and can, with significant investment, adapt to many adverse environmental changes resulting from GCC. However, those people and societies with the fewest resources are most vulnerable because they are least able to adapt. Projected warming may result in a mixture of economic gains and losses for developed countries, but developing countries can expect mostly losses: 'the projected distribution of economic impacts is such that it would increase the disparity in well-being between developed countries and developing countries,' with the disparities increasing the more the temperature increases (IPCC Working Group II 2001: 6). A report from the Working Group on Climate Change and Development (WGCCD) warns that GCC could 'threaten attainment' of the Millennium Development Goals (MDGs) for substantially reducing global poverty, disease and environmental destruction by 2015, and it 'threatens to reverse human progress, making the MDGs for poverty reduction unachievable' (WGCCD 2005: 2). The upshot is that, around the world, 'more people are projected to be harmed than benefited by climate change', even if temperature increases are somehow limited (IPCC Working Group II 2001: 6).

THE INTERNATIONAL POLITICS AND DIPLOMACY OF GLOBAL CLIMATE CHANGE

Europe's actions on GCC have taken place within the milieu of international negotiations that began in earnest in the 1980s and grew in intensity in the 1990s.[10] The initial stimulus for these efforts was science, but – as is true in other international environmental deliberations – diplomatic efforts eventually took on a life of their own that was partly separated from science. International and domestic politics joined the science as ongoing stimuli for efforts to address GCC. The First World Climate Conference, held in 1979, was a gathering of scientists interested in GCC and its relationship with human activities. The conference issued a statement calling on countries to 'foresee and prevent potential man-made changes in climate that might be adverse to the well-being of humanity' (UNFCCC 2002). From that conference a program of scientific research was established that would contribute to the creation of the IPCC in 1988 by the World Meteorological Organization and the United Nations (UN) Environment Program. The IPCC's 1990 first assessment report (see the previous section) and the Second World Climate Conference in 1990 added stimulus to initial concerns about climate change among diplomats. In December 1990, therefore, the UN General Assembly established the Intergovernmental Negotiating Committee (INC) for a Framework Convention on Climate Change (FCCC). The goal of the climate change INC was to negotiate a *framework* convention that would be the basis for subsequent international protocols dealing with GCC.[11]

From that point until the 1992 UN Conference on Environment and Development – the 'Earth Summit', held in Rio de Janeiro – representatives of over 150 countries negotiated the FCCC. The FCCC called on the world's most economically developed countries to reduce their emissions of GHGs to 1990 levels by 2000. None did so.[12] Particular responsibility is laid on the developed states (listed in the FCCC as 'Annex I' countries) to provide 'new and additional' resources to developing countries to help them with their efforts to limit GHG emissions. The FCCC came into force in 1994, after ratification by 50 countries. In 1995, the INC was replaced by the Conference of the Parties (COP), which became the FCCC's overriding authority. Almost all of the world's countries have now ratified the Convention. In the years following its adoption, several COP meetings were held to negotiate the details of how GHG emissions limitations would be achieved. While the negotiations leading to the FCCC were fraught with differences, particularly between developed and developing countries, it was the negotiations after 1992 that were most contentious. They were noteworthy for several milestones, particularly the Berlin Mandate, the Kyoto Protocol, the Buenos Aires Plan of Action, the Marrakech Accords and the New Delhi Declaration.

The first COP was held in Berlin in 1995. At this meeting the developed countries acknowledged that they had a greater share of the responsibility for climate change and would act first. The conference established the Ad Hoc Group on the Berlin Mandate, which subsequently negotiated details for implementing the objectives of the FCCC. Central to the Berlin Mandate was the demand by developing countries that the industrialized countries take on greater commitments to reduce their GHG emissions and to assist the developing countries, which were excluded from commitments to make emissions cuts. Thus the first COP affirmed the notion of 'common but differentiated responsibilities', meaning that, while all states have a common responsibility to address GCC, the developed countries have a greater 'differentiated' obligation to do so (see Chapter 15).

At the second COP, held in Geneva in 1996, countries endorsed the IPCC's second assessment report (which, again, concluded that 'the balance of evidence suggests that there is a discernible human influence on global climate') and called for a legally binding protocol with specific targets and timetables for reductions of GHG emissions by developed countries (Houghton et al. 1996). The resulting Geneva Declaration served as the negotiating basis for the Kyoto Protocol, which was agreed in December 1997 at the third COP in Kyoto, Japan (see UNCOP 1996).

The Kyoto conference proved to be especially contentious, not least because the US seemed to be reneging on the Berlin Mandate when President Bill Clinton called for 'meaningful participation' of the developing countries. Nevertheless, diplomats at the conference managed to agree to the Kyoto Protocol, which established specific emissions goals for developed countries, but which did not require significant commitments by developing countries (see Grubb et al. 1999). The developed countries agreed to a goal of reducing their aggregate GHG emissions by 5.2 per cent below 1990 levels between 2008 and 2012. The EU member states agreed to reduce their collective emissions (what is called the EU 'bubble') by 8 per cent, with Germany and Britain making the largest cuts. The Kyoto Protocol also catered to some US demands by endorsing emissions-trading programs that would allow developed countries to buy and sell emissions credits among themselves. Other so-called flexible mechanisms adopted in the context of the protocol were joint implementation (JI), whereby developed countries could earn emissions credits when investing in one another's emissions-reduction projects, and the Clean Development Mechanism (CDM), which would manage programs of joint emissions efforts between developed and developing countries. The CDM was intended to allow developed countries to pay for – and receive emissions credits for – emissions-reduction projects in developing countries.

Some of the means by which the Kyoto Protocol's 5 per cent goal would be reached were codified at the fourth COP, held in Buenos Aires in November 1998. The Buenos Aires Plan of Action set a two-year deadline for finalizing many of the details for implementing the Kyoto Protocol. Parties agreed to reach decisions by the end of 2000 on several key issues, including guidelines for emissions trading, JI, the CDM and technology transfers (FCCC 1998). At the fifth COP in Bonn during October 1999, parties agreed to a timetable for completing outstanding details of the Kyoto Protocol by the sixth COP and, in an effort to speed up negotiations, gave the conference president the power to 'take all necessary steps to intensify the negotiating process on all issues during the coming year' (FCCC 1999).

The sixth COP began in November 2000 in The Hague, but the talks broke down due to disagreements among delegates, particularly on the question of carbon sinks, which are processes, such as planting trees (afforestation) that can remove GHGs from the atmosphere (Betsill 2005: 113). The Kyoto Protocol's likelihood of ratification was put into doubt with the advent of President George W Bush in the US. Upon taking office he declared it 'fatally flawed' and withdrew all US support (which still obtains as this book goes to press in Autumn 2006). The sixth COP resumed in Bonn during July 2001. The resulting Bonn Agreement clarified plans for emissions trading, sinks, compliance mechanisms and aid to developing countries. At the seventh COP, held in Marrakech in late 2001, parties to the FCC agreed to a long list of ways to meet the Kyoto commitments. The result was the Marrakech Accords, a complicated mix of measures for implementing the Kyoto protocol, largely designed to garner ratification from enough states to allow the protocol to enter into force. Parties agreed to increase funding for the FCCC's financial mechanism, the Global Environmental Facility (GEF), as well as to establish three new funds that would provide more aid to poor countries: the Least Developed Countries Fund, the Special Climate Change Fund and the Adaptation Fund.

A milestone of sorts (arguably a dubious one) was reached at the October 2002 eighth COP in New Delhi. At that conference there was tacit agreement between the US, a few other developed countries (but not Europe) and several large developing countries, notably China and India, that shifted much of the focus away from mitigating GCC and toward adaptation (i.e., wealthy countries agreeing to help developing countries adapt to GCC, rather than the former countries having to reduce their GHG pollution) (Harris 2003: 32). Most European countries, while agreeing that aid to developing countries was an important obligation – and being far more willing than the US to act on that obligation – nevertheless actually started, more or less unilaterally, to limit and reduce their GHG emissions. It was at this COP, as well as the ninth one, held in Milan, Italy, that diplomats discussed ways to implement the Marrakech Accords and prepare for ratification of the Kyoto Protocol. The tenth COP was held in Buenos Aires in December 2004. It was dubbed the 'Adaptation COP' because more of the discussion was about adaptation to GCC than the more common COP discussions about efforts to mitigate it through emissions limitations (ENB 2004). In the end, there were pledges for more assistance to aid poor countries most affected by GCC, but there were no firm commitments to make access to adaptation funds easier, and observers were left skeptical of how much the poor would benefit (ENB 2004). Importantly, it was also in 2004 that Russia ratified the Kyoto Protocol, allowing the agreement to finally enter into force in February 2005.

One visible aspect of the GCC negotiations has been the acrimony between the developed countries – particularly the US – and the developing world. Until the last few years, when Europe clearly diverged from US positions, the developed countries generally sought *global* restrictions on emissions reductions with flexible mechanisms for their implementation. What this would mean in practice is that the developing countries would be required to manage and eventually reduce their emissions of GHGs. As GHGs originate all over the world, the developed countries have argued that all countries, at least all the large ones, need to be part of emissions reduction efforts. In contrast, the developing countries have pointed out that their per capita emissions of GHGs, particularly CO_2, remain very low relative to their developed counterparts. What is more, it is the industrialized developed countries that have benefited from past emissions of GHGs since the Industrial Revolution. It is the responsibility of the developed countries, therefore, to reduce their emissions of GHGs, while they allow the countries of the global South to focus on economic development. The developing countries also, by definition, have far weaker economies and often-widespread poverty; they argue that they ought to be allowed to raise the living standards of their citizens without being constrained by costly measures to reduce their GHG emissions.

Thus, the international negotiations have been plagued by the efforts of developed countries to persuade developing countries to commit to emissions limitations, on the one hand, and developing country efforts to avoid such commitments on the other. These differences were briefly 'solved' by the Berlin Mandate and the affirmation of the principle of common but differentiated responsibility. Although the US challenged this principle at Kyoto, it failed to persuade developing countries to undertake GHG emissions limitations. This outcome was consistent with the policies of most European countries and the European Commission, which – while recognizing the need to get developing country involvement – adhered to the Berlin Mandate's requirement that the wealthy countries of the world reduce their GHG emissions before asking the rest of the world to do so (see Chapter 15; Harris 2003).

It would be wrong to assume, however, that there were no differences among the developing countries themselves. Indeed, the goals of some groups of developing countries have differed quite sharply. At one extreme are the oil-producing countries, notably members of the Organization of Petroleum Exporting Countries (OPEC). They agreed until recently with many politicians and industry interest groups in the US that the science of GCC was uncertain. Therefore, they argued, action on GCC should be postponed until this uncertainty is reduced or eliminated. They joined with the US and some other developed countries in attempting to water down proposals for substantive reduction of CO_2 emissions in particular. China, often in collaboration with other members of the Group of 77 developing countries, consistently sought to prevent wording in international agreements that would require developing countries to take action, even 'voluntary' action. However, at COP4 in Buenos Aires, Kazakhstan and Argentina agreed in principle to voluntary limitations on their GHG emissions. China's position and that of OPEC also contrasted with the goals of many very poor coastal states and members of the Alliance of Small Island States (AOSIS), which fear that they will be (or already are) among those countries most severely affected by GCC impacts, notably by rising sea levels. Indeed, while OPEC countries were calling for no action, AOSIS countries consistently called for far more action than almost all other countries were willing to accept. Developed countries wishing to avoid obligations to cut GHGs were able to exploit these and other differences within the developing world.

Similarly, the developed countries were not always in agreement. In the run up to the Kyoto conference, for example, the EU was calling for firm targets and timetables that would require emissions reductions to be much higher than those finally agreed upon. The Americans were calling for reductions lower than those agreed, and the Australians were demanding that they be allowed to increase their emissions (a concession they were granted in the Kyoto Protocol). The Europeans were also more willing to meet the demands of developing countries for special treatment in the form of new and additional financial and technological assistance, and they wanted to live by the understandings of the Berlin Mandate. The US and a few other developed countries have continued to call for firm commitments from the largest and most well off developing countries.

These many ongoing differences among countries were manifested during the late 2005 conference of the parties to the FCCC, COP11, and the First Conference of the Parties Serving as the Meeting of the Parties to the Kyoto Protocol (COP/MOP-1) to the Kyoto Protocol, which were held simultaneously in Montreal (ENB 2005). At this gathering, European states were noteworthy for countering attempts by the US to derail the meeting and to prevent the robust implementation of the Kyoto Protocol. The meeting adopted the Marrakech Accords and formalized rules for implementing the protocol (e.g., rules for emissions trading, JI, crediting of emissions sinks and penalties for non-compliance), streamlined and strengthened the CDM, began negotiations for further commitments by developed country parties to the protocol beyond 2012 (when the Kyoto commitments expire), set out guidelines for the Adaptation Fund, and initiated a process for negotiating long-term action to combat climate change. Several developing countries, while still opposed to binding obligations, showed new interest in undertaking voluntary measures, in keeping with the principle of common but differentiated responsibility (see Pew Center 2005).

In the chapters that follow, the contributors look at different aspects of GCC diplomacy in much greater detail.[13] Their objective is to explain how and why individual European states and the EU have responded the way they have to this problem. They are particularly interested in the politics underlying European national and international policies on GCC. Climate change is, of course, fundamentally caused by activities within countries. However, its consequences are felt much more broadly and, very importantly, it is simultaneously an issue for domestic and international politics and policymaking. One way of looking at these politics and the policies that result from them is to frame the issue of GCC in terms of environmental *foreign policy*, and to look in some detail at the actors and processes of foreign policy that, by definition, operate in the analytical space that crosses over between the domestic and the international.

GLOBAL CLIMATE CHANGE AND FOREIGN POLICY

Environmental foreign policy (EFP) can be conceived of as the interplay between (1) domestic forces, institutions and actors involved in environmental decision making and the implementation of environmental policies, and (2) international forces, institutions and actors, such as environmental changes themselves and their interaction with other forces (e.g., democratization, globalization), international environmental organizations and regimes, and powerful countries, corporations and non-governmental organizations (NGOs) with a role in shaping human responses to environmental changes.[14] From a policy perspective, EFP is about the international *environmental* objectives that officials of national governments seek to attain; the values and principles – including but not restricted to environmental ones – underlying those objectives; the methods by which the objectives are sought; the processes by which these objectives, values and principles, and methods are developed and implemented; and the domestic and international actors and forces – including but not exclusively environmental ones – shaping environmental policies and actions both at home and abroad, but which have some international or external character.

Foreign policy objectives, actors and processes can be central in determining whether countries cooperate to address GCC and other environmental problems. What is particularly important about foreign policy is that it involves the *crossover and interaction* between domestic politics and processes, on the one hand, and international relations and institutions, on the other. Many environmental policy officials are simultaneously pressured to follow international norms and promote national interests and ideals. That is, they are buffeted by *both* domestic and international forces. They are part of EFP. Thus, looking at purely local or international variables seldom explains environmental policy within and among countries.

Environmental issues are often distinctive in the manner in which they ignore state borders; problems in one country affect others and problems restricted to one country require the involvement of others (e.g., financial assistance and technology) if they are to be resolved or remain local. Many issues, actors and forces acting domestically and internationally affect and influence countries' national environmental regulations and their environmental foreign policies, and hence they impact international environmental cooperation. Yet, despite obvious (albeit not fully comprehended) connections between local and international policy processes, many studies do not fully account for the foreign policy aspects of environmental protection efforts. They tend to focus on international regimes or domestic politics as drivers of international environmental relations – which is indeed often the case – without also looking at the foreign policy processes and actors that can impact on responses to environmental issues.

Thinking about foreign policy focuses our attention on interactions among domestic political preferences and positions governments take in negotiations, the balancing of economic growth and popular demands for development with foreign pressures to join environmental regimes, and the rivalries and alliances between Foreign policy agencies and the individuals working in them (among many other considerations). A good reason for also looking at foreign policy processes more systematically is that they can reveal important national and institutional characteristics shaping state environmental behavior, both domestically and internationally.

Foreign policy is, to be sure, about pursuing and promoting national interests. Already complexities arise, however. It is not always clear what a country's national environmental interests are – or what a regional organization's environmental interests are – or ought to be, particularly with regard to complex environmental issues, and it is almost always debatable how best to promote them (Webber and Smith 2002: 43–44). As Roy argues, policy-making elites will disagree over national goals and how to achieve them:

> Beyond its most basic formulation, the national interest is not a monolithic, objective concept, but rather a dynamic and unsettled one, subject to constant debate. [Moreover,] powerful groups and individuals are subject to self-interested behavior, and may support the policy option they calculate will enhance their power and prestige, even if it is not necessarily the best option for the nation as a whole. (Roy 1998: 137–38)

Thus, defining national interests, including the interests of states that have joined together rather tightly in a regional organization, and the ways to promote them, is a problematic and complex undertaking, involving actors and institutions seemingly unimportant to the casual observer, even when issues and associated interests are better understood than they usually are in environmental cases.[15]

We cannot of course completely abstract out the forces of foreign policy, particularly if foreign policy is broadly defined. Foreign policy cannot be separated from, for example, domestic politics and institutions, at one end of a spectrum, and global regimes and international power balances, at the other end. What is useful, perhaps, is to go beyond thinking in terms of domestic and international levels of analysis to a 'two-levels-plus' game (cf. Putnam 1988; Evans et al. 1993). That is, we ought to consider international political dynamics and domestic politics, but we can also think *explicitly* about the additional 'level' of foreign policy processes, which almost always falls between and affects the international and domestic level factors.

Foreign policy analysis is well suited to studying responses to environmental changes because it considers the 'continuing erosion of the distinction between domestic and foreign issues, between the sociopolitical and economic processes that unfold at home and those that transpire abroad' (Rosenau 1987: 3). Applying FPA techniques to our study of international environmental issues, or at least considering them along with other analytical approaches, can result in interesting findings simply because environmental issues like GCC are quite distinct from many others – given their often high level of uncertainty, often temporally distant impacts and the great number of, and disparity in, stakeholders actively involved in the issue area.[16]

How might we analyze and thereby better understand and explain (and help policymakers and activists possibly manipulate) EFP in Europe? We can start by applying FPA to European foreign policy generally (White 1999) and by using FPA approaches to help us think systematically about the relationship between foreign policy and environmental change in particular. Barkdull and Harris (2002: 63–91) have shown how a number of theories and approaches to FPA might be deployed to better understand *environmental* foreign policy. They propose a typology (see Figure 1.1) that highlights a variety of potentially important variables in the shaping of foreign polices in particular circumstances related to addressing international environmental issues.[17]

Barkdull and Harris note that most theoretical approaches to foreign policy are of three major types based on the explanatory forces they emphasize: systemic, societal or state-centric (the top row of Figure 1.1; cf. Ikenberry et al., 1988). Systemic approaches argue that foreign policy stems from the role, identity or interests given to the state by systemic factors (e.g., regional or global configurations of power, hegemonic ideas). They direct our attention to the structural characteristics of international relations, showing that states may 'arrive at their roles, identities, and national interests as a consequence of the regional or global configuration of power … or as a consequence of ideas. Systemic theory is distinct in that it does not attribute outcomes to factors such as domestic politics and institutions' (Barkdull

	Systemic (International System)	**Societal** (Domestic Politics)	**State-centric** (Government)
Power	Structural Realism; Hegemonic Model	Elite Theory; Class Analysis	Executive-Legislature Relations
Interests	National Interest Model	Interest Group Politics; Pluralist Models	Bureaucratic Politics Model
Ideas	Sociological Approach to International Regimes	Media Studies; Social Movements; Public Opinion	Foreign Policy Executive; Psychology

Source: Adapted from Barkdull and Harris (2002: 69). Cells in the matrix show examples of illustrative foreign policy studies.

Figure 1.1 Typology of environmental foreign policy

and Harris 2002: 68). Consequently, this type of approach to EFP would divert our focus away from bureaucracy, public opinion and interest groups, pointing instead to, for example, 'environmental hegemons' like the US in largely determining GCC policies, even in Europe.

Societal theories point to the preferences of domestic actors, which are translated into policies adopted and implemented by the government. From this perspective, explanations for foreign policy 'are found in the ongoing struggle for influence among domestic social forces or political groups' (Ikenberry et al., 1988: 7). This approach might suggest that governments do not independently decide EFP; they are instead neutral or passive arbiters of policy struggles, or perhaps merely fragmented arenas for bargaining over policy. The international arena might be viewed as merely a venue for the expression of policies determined by society (or societies). Societal theories suggest that the forces shaping Europe's GCC policies are found within European societies – among elites, interest groups, social movements and public sentiments – not within the governments of member states or in international distributions of power (Barkdull and Harris 2002: 74–5).

Alternatively, state-centric approaches suggest that foreign policy is shaped by the structure of government and the individuals and agencies that promulgate and implement foreign policies on its behalf, often with an emphasis 'on the goal-oriented behavior of politicians and civil servants as they respond to internal and external constraints in an effort to manipulate policy outcomes in accordance with their preferences' (Ikenberry et al. 1988: 10). State-centric approaches often discount the importance of societal actors and forces in shaping foreign policy, instead placing a premium on the influence of institutions or the focus of top policymakers on promoting the national interest. 'The general message of this perspective', according to Barkdull and Harris (2002: 79), 'is that the state can act independently of societal interests', and 'foreign policy outcomes cannot be read off from the structure of the international system, however defined'. Alternatively, the structure of the state, such as the distribution of power between executive and legislative branches of government, or the workings and influence of bureaucratic agencies, largely determine foreign policy. Thus we might say that European GCC policies derive not from public pressure or international forces but instead from perceptions of threats to national interests by leaders, pressure from legislatures on policymakers, or the degree to which GCC is important to bureaucratic actors. In the latter case, the influence of environmental and foreign ministries, and perhaps the European Commission, deserve attention.

These three broad approaches to understanding foreign policy – focusing on the international system, domestic society or the state – can be refined by simultaneously considering, in each case, the role of power, interests and ideas (the left-hand column of Figure 1.1) (Barkdull and Harris: 67–8; cf. Hasenclever et al., 1997). According to some power-based approaches, often characterized as 'realism', countries join international regimes due to hegemonic or oligopoly distributions of power in the international system. Hegemons (or small groups of leading powers) create regimes that serve their interests, and then force them upon other countries. Alternatively, interest-based theories, often associated with 'liberal institutionalism', posit that international cooperation stems from the desires of states to promote their interests in a given issue area. According to this perspective, hegemonic power is not essential because rational state actors will cooperate to achieve joint gains. Yet another set of theories focus on ideas and what Smith calls the 'social construction of foreign policy' (Smith 2004: 123). Ideas can direct international actors toward new ways to pursue their interests, whether unilaterally or multilaterally. From this 'constructivist' perspective, material interests and power may have limited influence compared to even more influential identities that the international system generates for global actors.

No single theory, or even type of theory, is necessarily best for explaining and understanding the EFPs of all states in all circumstances. Indeed, approaches might suitably be combined to arrive at the richest explanations for policy (although theoretical purists might not like this). For example, power in the international system is seldom going to adequately explain GCC policy. Certainly, the distribution of 'environmental power' of the US and China matters greatly – both can substantially diminish the efficacy of the GCC regime and thus wield tremendous influence in this context – and Europe's 'power' in this respect is on the rise and of increasing utility. However, US power is wielded as it is largely due to state and society-based factors (e.g., the influence of special interests and bureaucracy on GCC policy) and is now so far beyond the pale that it has not prevented Europe starting to cut its GHG emissions even as the US does nothing.

Alternatively, principled ideas, such as the obligation of the EU to take on its fair share of the burdens of GCC (see Chapters 14 and 15), have influenced policy in this issue area, but power configurations among European and other states, as well as assessments of national interests in this context, mean that the influence of those ideas in shaping GCC policy are substantially reduced. In any case ascertaining how those ideas are influential requires looking at potential interactions and feedback loops at all levels of analysis.

A number of ways in which single or multiple FPA approaches can be useful in explaining EU and European national policies on GCC are demonstrated in the chapters that follow. The contributors to this volume draw on, or build upon, the typology described by Barkdull and Harris (Figure 1.1) as they explore Europe's responses to this major global problem. Chapter 16 returns to this framework to highlight a number of theoretical lessons that can be learned from the chapter case studies.

EUROPE'S RESPONSES TO GLOBAL CLIMATE CHANGE: CASE STUDIES

How has Europe responded to the problem of GCC? What explains those responses? In the chapters that follow, contributors undertake case studies of European and EU foreign policy to give detailed answers to these questions. The case studies are divided into two parts. In Part I, the contributors examine the politics and foreign policy of GCC in individual European states. It is difficult for European states to devise and implement policies on GCC. Each is a pluralistic democracy, meaning that industries, NGOs and the public play a part in pushing for – and pushing against – each government's policies related to GCC and GHG emissions in particular. Each state is also a member of the EU (or greatly influenced by it, as in the case of Norway, which is examined in Chapter 8), meaning that their policies must be devised and actualized in the context of EU membership and all the requirements and constraints that it inevitably entails. Consequently, Part I examines efforts by European states, including new members of the EU, to maintain policy autonomy (e.g., 'subsidiarity') amidst attempts, some of them successful, by arms of the EU organization to co-ordinate a regional, EU-wide policy on GCC.

Part II of the book examines European regional cooperation on GCC, and especially the EU's role in international climate change negotiations and regime building. Global climate change is an issue that falls under the rubric of subsidiarity within the EU, which 'confines action by the EC to those matters whose objectives cannot be sufficiently achieved by the Member States and that can be better achieved by the EC' (Haigh 1996: 160; see Collier 1997: 48). By its very nature, *global* climate change cries out for European governments (and other actors in the region) to work together to address its causes and consequences both in Europe and internationally. The 'competence' of the EU (more specifically, in legal terms, the European Community [EC]) to be involved in international environmental agreements and negotiations, while not explicitly envisioned in the Treaty of Rome, has nevertheless evolved over time, reinforced by the Maastricht Treaty and, put simply, the need for the EU to complement the environmental foreign policies of the member states (see Chapter 9; Sbragia and Hildebrand 1998: 217). Compared to GCC policies of European states, formulating and implementing EU-level, and EU-wide, GCC policy can be equally, or even more, fraught with challenges for national and European policymakers, as demonstrated by doomed EU efforts in the 1990s to implement a carbon tax throughout the union. Nevertheless, the EU – that is, the European Commission – has almost inevitably taken on a co-ordination and even leadership role on matters related to climate change, despite resistance from some member states. Taken together, case studies from both parts of the book provide rich narratives and thoughtful explanations of GCC politics, foreign policy and regional cooperation in Europe.

Global Climate Change and Policymaking in European States

As Ute Collier points out, it is subjective and 'difficult to make any kind of classification' of European 'leaders' and 'laggards' on climate change (Collier 1997: 185). Nevertheless, the country case studies in Part I are ordered based on commitments made by EU states to contribute to the

Kyoto Protocol commitment of reducing overall EU emissions by 8 percent. Thus we begin by looking at three relative 'leaders' in this respect, namely, Germany, which agreed to cut its GHG emissions by 21 per cent, the United Kingdom (UK) (–12.5 per cent) and the Netherlands (–6 per cent), followed by Poland, which agreed to a 6 per cent cut before joining the EU. We then look at two relative laggards in the GHG emissions-control game, namely Sweden and Spain. Within the EU bubble, Sweden is allowed to increase its emissions by 4 per cent and Spain can increase its GHGs by up to 15 per cent.[18] The final chapter in this section examines a state that has persistently avoided joining the EU, but which plays an important role in Europe's collective efforts to address GCC. That state is Norway, which is permitted under the Kyoto Protocol to increase its emissions up to 1 per cent.

In Chapter 2, Michael T Hatch examines the politics of GCC in Germany. Germany has been an extraordinarily important player in the international negotiations on GCC and is central to the commitments assumed by the EU under the Kyoto Protocol. In the absence of substantial reductions of GHG emissions by Germany, the EU has little chance of meeting its international obligations. Hatch asks why the German government has pursued such a central role. The answer, he argues, is located largely in a domestic policy process that resulted in commitments to reductions in CO_2 emissions that are among the most ambitious in the industrialized world. More specifically, German policy has been shaped by a collaborative approach that enabled a consensus to coalesce around the view that global warming is a serious threat that warrants concerted action. Out of this consensual policy process emerged a commitment to dramatic reductions in GHG emissions. Domestic decisions over the extent of reduction targets and the means to achieve them, in turn, became intimately linked through German foreign policy to negotiations, at regional and international levels, on the shape of the FCCC and the Kyoto Protocol.

In Chapter 3, Loren R Cass analyzes the UK in the context of European GCC policy. The UK's role in the development of a common European response to GCC presents a fascinating case study of the ability of individual member states to pursue their national preferences within the context of the broader European EFP process. Along with the Germans, the British have provided the vast majority of the EU's aggregate CO_2 emission reductions. Without the British cuts it would have been much more difficult for the EU to pursue a relatively aggressive position (compared to, say, the US) on international GHG cuts. The UK played a critical role in shaping European negotiating positions and in acting as an intermediary between the US and the EU in the negotiations for the FCCC, the Kyoto Protocol, and the Hague and Bonn agreements on implementing the protocol. Cass's analysis of the interaction between British and European positions in the

GCC negotiations suggests that EU member states have significant room to manoeuver within common European positions on climate change. Successive British governments supported EU initiatives when they served UK interests, but they were also willing to use the US and international institutions to constrain European initiatives that ran counter to Britain's broader interests. Cass shows that the UK played an indispensable role within the EU's climate policy by providing sizable emission reductions that allowed the EU to seize an international leadership position, while simultaneously preventing European foreign policy on GCC from developing too quickly for British tastes.

In Chapter 4, Norichika Kanie examines 'middle-power' diplomacy in the multilateral environmental institution-making process, focusing on the case of the Netherlands in the early climate talks leading to the Kyoto Protocol. For a middle power, the Netherlands has exerted more than its proportional influence in GCC regime building and the Kyoto Protocol-related process in particular. This can be attributed mainly to two factors. One is the role that 'domestic capacity' (or leadership potential) played in the Netherlands's GCC policy. The term capacity here encompasses domestic policies and associated institutional developments, as well as the foreign policy knowledge base that emerged from policy-oriented research projects. In other words, experience in implementing policies, including growth of policy networks among actors working on GCC and the ideas that fed into negotiations, grew into leadership potential for this middle-power state.

While success in introducing any unilateral policy initiatives and connecting them with external policy strategies depends very much on time and issue specific domestic and external conditions and constraints, the efficient interaction between governmental and non-governmental actors, on the one hand, and an efficient strategy on the side of the non-governmental actors, on the other hand, may be key factors in facilitating unilateral action on GCC. In tackling global issues, the effective role of government may function as a link not only between domestic and global policies, but also between the government and NGOs. Kanie argues that the early introduction of an advance or experimental GCC policy can be advantageous because it can set an example and thus provide a basis for a solid argument against countries that have not introduced a comparable policy. However, potential does not necessarily mean the attainment of a leadership role. This leads to a second reason for the Netherlands' disproportionate influence in shaping the early GCC negotiations and the Kyoto Protocol: the framework provided by a regional organization, namely the EU, functioned as a lever that enabled the Netherlands to actualize leadership in the GCC negotiations.

The EU expanded to 25 members in May 2004. Turning to one of the newest member states, in Chapter 5 Anita Bokwa describes climatic issues in Poland's foreign policy. Political decisions in Poland concerning environmental issues, including those related to GCC, have been pushed by the country's requirements for accession to the EU and its membership in the UN. Bokwa argues that the demands of membership in the EU shaped Poland's state identity, determined its national interests and thereby influenced its behavior. After 1990, joining the EU became the primary foreign policy objective of Poland and the source of its domestic environmental requirements. As she puts it, 'the aim of joining the EU was of such a high priority that other factors were put aside'. However, most improvements in environmental conditions within Poland have been a consequence of the deterioration of its industrial economy after the fall of communism. There is no strong green political party in Poland that could counteract the powerful coal-energy lobby inherited from the country's communist history. The upshot is that domestic policy on the environment has been driven by foreign policy. Bokwa argues that this is largely a consequence of the rigid thinking of bureaucrats who came of age in the communist system: they are accustomed to following diktats – now from Brussels – rather than thinking and acting creatively and proactively. If Poland's GCC policies are to be more than reactive, there will have to be 'a huge change of the social mentality, which most probably can be achieved only by the next generation'. Thus, Bokwa's chapter shows how much EU membership can affect a European state's GCC policies, although perhaps not quite in the manner advocates of the organization would want.

Chapter 6, by Kate E Marshall, uses Barkdull and Harris's (2002) typology of EFP (see above) to explore Swedish actions on GCC. Sweden is usually a lead state in global and European environmental politics, and Marshall notes its past successes in combating air pollution. Sweden has had particular success in engaging in what Marshall calls 'norms-based diplomacy' in European and international negotiating forums, corresponding to the ideational approaches in the Barkdull and Harris typology. However, Sweden has not emerged as a lead state in GCC diplomacy, a position at odds with its past performance. For example, in the late 1990s Sweden stalled intra-EU GCC negotiations due to its concern about the scope of a GHG emissions-trading program. Sweden wanted an unlimited trading scheme, while most EU members preferred one with limits on the quantity of GHG emissions credits that could be traded (meaning that member states would not be able to avoid making domestic GHG cuts). The factors that had historically pushed Sweden toward an environmentally friendly, norms-based foreign policy were present, but such a policy did not materialize. In an effort to explain this discrepancy, Marshall uses a combination of societal, state and systemic levels of analysis. She pays particular attention to the role of nuclear power within Sweden, and a 1980 referendum in which Swedes voted against the construction of new nuclear power plants. The upshot is that in seeking to protect its sovereignty, Sweden did not engage in its usual form of norms-based diplomacy, despite the fact that all the postulated factors for normative influence were present. Consequently, this chapter not only provides empirical material on an interesting incident in GCC politics, but also shows that a norms-based EFP has limits, and that, even in environmental lead states, sovereignty concerns can take priority.

The politics of GCC in Spain is the subject of Chapter 7, written by J David Tàbara. Tàbara's chapter aims to explain the reasons for what he characterizes as the failure of Spain's GCC foreign policy, and for the growing role taken by Spanish regional actors in the implementation of European and international environmental agreements. Tàbara argues that Spain is a long way from fulfilling its international commitments on GCC. He examines institutional and political-cultural factors underlying Spain's GCC politics and the growth of its GHG emissions. His chapter emphasizes the lack of public participation in this area of environmental policymaking. Tàbara argues that Spain will have to buy emission reductions from abroad or find new ways to reduce GHGs in a more decentralized manner in tune with its political organization, which is based on regional Autonomous Communities (ACs). The governments comprising ACs in Spain have substantial environmental responsibilities that are held by centralized state agencies in most other countries. This strongly decentralized political structure, particularly as it relates to environmental, industrial and energy issues, poses special difficulties for co-ordination of the Spain's implementation of the GCC regime. However, this decentralization is not the cause of Spain's failure in this regard. Rather, Tàbara concludes that the governments of Spain's ACs provide differentiated and potentially more flexible opportunities for the successful implementation of international environmental policies.

In Chapter 8, Andreas Tjernshaugen and Ho-Ching Lee examine Norway's role in international climate change negotiations, in so doing focusing on the impact of NGOs. As Tjernshaugen and Lee point out, in countries where governments and the public are inclined to listen to the environmental movement, environmental NGOs (ENGOs) are likely to be one of several groups of actors with some degree of influence over foreign policy regarding climate change. Their chapter discusses the strategies chosen by Norwegian ENGOs and their international allies in seeking to influence Norway's behavior in the UN climate change negotiations. They focus on ENGO activities at major negotiating sessions and find that they mostly chose *not* to try to affect negotiation outcomes through lobbying negotiators directly. Instead, the ENGOs used the negotiating sessions as opportunities to shape the domestic political agenda by transmitting information and arguments to the media. The aim was to indirectly influence Norway's foreign policy and negotiating position by *shaming* its government into adopting policies preferred by the ENGOs and *framing* the domestic climate policy debate in terms favorable to the ENGO's objectives. Tjernshaugen and Lee show how actions taken in the international and domestic political arenas were closely intertwined in this case, and they argue that ENGO influence in international negotiations should be understood with such linkages in mind. Specifically, they emphasize that the ENGO's ability to threaten a government's legitimacy and popular image, as well as their ability to shape domestic actors' interpretations of international commitments, are potentially important vehicles of influence in EFP and thereby in international GCC (and other international environmental) negotiations.

The EU and Global Climate Change

Part II focuses on the politics and foreign policymaking of the EU, including the EU's role in international negotiations on GCC and further analysis of interactions between the EU and its member states in this issue area. Part II begins with a discussion, in Chapter 9 by Nuno S Lacasta, Suraje Dessai, Eva Kracht and Katharine Vincent, of how the EU 'articulated a consensus' on climate change. The authors argue that GCC ranks highly on Europe's political agenda and continues to be a key area of foreign policy for the EU. Their chapter gives an overview of key processes and actors in the evolution of EU climate change policy since the adoption of the FCCC in 1992. They highlight the complexities arising out of the unique multilevel governance structure of the EU, whereby the issue of GCC contains elements of 'exclusive' and 'shared' competences. Major actors involved in shaping GCC policy are introduced in the chapter, including the member states, the European Commission, businesses and industry, ENGOs, the media and the public. Lacasta and colleagues undertake three case studies on several key elements of European GCC policy, namely those regarding policies and measures, burden sharing and the developing countries. They evaluate and offer predictions regarding the future prospects of EU GCC policy. The authors believe that the EU's main priorities should be to lead by example through effective domestic policies on GCC, to reform EU structures so that a collective position emerges more readily, and to attempt to bridge the gap between the US and developing countries so that both participate in future GCC efforts. By doing these things, the EU can lead international negotiations toward more effective management of the global commons.

In Chapter 10, another collaborative group of scholars – Martina Jung, Axel Michaelowa, Ingrid Nestle, Sandra Greiner and Michael Dutschke – look at land-use processes that absorb GHGs from the atmosphere, domestic stakeholders, and the formulation of a common EU foreign policy on GCC. Terrestrial 'land use, land-use change and forestry' – LULUCF in the official lexicon – have been among the most debated and complex issues in the international climate change negotiations. The EU was one of the most prominent opponents of the use of forests as carbon sinks for mitigation of climate change, much as environmentalists have often opposed them because they allow polluters to forgo emissions cuts. The chapter outlines the main land use-related questions under discussion since the Kyoto conference in 1997. It describes the complex structure of country groupings in the international GCC negotiations as well as the different positions of stakeholders, including NGOs and industries within and outside the EU.

The chapter also examines various channels of influence used by these stakeholder groups to influence the European position in the LULUCF negotiation process. In addition to these external factors, the EU has faced a relatively complex internal co-ordination process and conflicting interests of different member countries that slows the organization's reaction to new developments in international GCC negotiations. As the authors of Chapter 10 note, despite EU opposition, a relatively powerful coalition of supporters of carbon sinks and other land-use changes finally managed to include LULUCF activities in the Kyoto framework. However, following the EU's failure to reach its goal of completely excluding carbon sinks, it shifted toward influencing the role of carbon sinks by proposing conservative rules and procedures for their implementation.

In Chapter 11, Sebastian Oberthür and Dennis Tänzler seek to describe and explain how external factors, namely international regimes, influence GCC policies within the EU. Their chapter highlights how the Kyoto Protocol in particular has influenced the adoption and diffusion of policy instruments in the EU and its member states, in the process changing the conditions for the EU's foreign policy on GCC. While the literature on policy diffusion has tended to focus on processes of spatial policy learning, Oberthür and Tänzler argue that the spread of selected climate policy instruments across the EU cannot be understood and explained properly without reference to the role of international institutions, in this case the Kyoto Protocol. The international regime on GCC was an important force driving the increasing acceptance and diffusion of policy instruments, such as climate policy plans and emissions trading, across Europe.[19] Neither the FCCC nor the Kyoto Protocol – the constitutive legal foundations of the international climate change regime – mandates the adoption of such policy instruments. Nonetheless, the international regime played a crucial role in triggering the diffusion of related GCC policies by increasing pressure on governments to take effective measures in order to reach their emissions targets and by providing incentives to implement innovative policy instruments like emissions trading. Importantly, these effects of the Kyoto Protocol occurred prior to its entry into force, arguably reinforcing the EU's leadership role in international climate politics.

Chapter 12 turns more explicitly inward, looking at how the EU has influenced the GCC policies in one member state and one non-member state. Guri Bang, Jonas Vevatne and Michelle Twena seek to understand and explain how the EU has influenced the GCC policymaking processes in Germany and Norway. They ask how the EU has influenced climate instrument choice. To answer this question empirically, Bang, Vevatne and Twena focus on the formulation of a prominent climate policy issue, namely emissions trading (ET). At Kyoto in 1997, parties to the FCCC agreed that trading in GHG emissions would be a cost-effective way to achieve overall emissions reductions. Thus ET schemes became central in foreign policy discourses. However, the specifics of how to implement ET were not agreed at Kyoto, instead being left to subsequent international deliberations. Only at the seventh conference of the parties in Marrakech in 2001 was agreement reached on the issue. The authors of Chapter 12 argue that the EU has been a decisive factor in laying the ground for an emissions trading system (ETS) in Germany, and has been influential in shaping the design of Norway's ETS. They explain how Norway started off being enthusiastic about ET, while Germany was openly against it. When the EU changed its position as a result of what happened at Kyoto, Norway and Germany changed their positions on domestic ET. Underlying Chapter 12's discussion is an analytical framework based on social constructivism, elitism and interest-based theory, which examines the roles of norms, ideas, power and interests in international climate change policy. The authors argue that these perspectives frame policy instrument choice, which results from the interplay between domestic actors and institutions, on the one hand, and international organizations, institutions and regimes, on the other.

In Chapter 13, Leonardo Massai discusses GCC politics and policies in the context of EU enlargement, building on issues raised in Chapter 5's discussion of Poland's GCC policies. The EU's enlargement process reached a milestone in May 2004 with the admission of 10 Central and Eastern European Countries (CEECs). Massai argues that this enlargement will considerably affect European EFP, as well as many EU policy institutions. The accession procedure required that applicant states implement the *Acquis communautaire*, the EU's collection of primary and secondary legislation (and related instruments). After considering the requirements of EU enlargement, Massai analyzes correlations between the environmental *Acquis* and European climate related legislation, and the extent to which representatives of CEECs have been involved in the shaping of new EU instruments on GCC. The CEEC's GHG emissions will fall (compared to pre-accession, business-as-usual policies) as a consequence of joining the EU. The accession requirements have forced these countries to set up new policies and measures that in turn have affected their national strategies for reducing GHGs within the framework of the FCCC and the Kyoto Protocol. However, in terms of reduction obligations, CEECs are considered separately from the 'old' member states, which are jointly committed to fulfilling the Kyoto Protocol under the EU's burden sharing agreement (the emissions bubble). With this in mind, the chapter examines the European Allowance Trading Directive, an important Kyoto Protocol-based mechanism being implemented in the enlarged EU.

In Chapter 14, Lyn Jaggard examines the 'reflexivity' of ideas in national, regional and international policy making on GCC. She uses the case of Germany to illustrate how discourse and particularistic ideas can influence policy. She points out that it is essential for individual countries to implement a variety of policies in order to manage the effects of GCC and to minimize its future adverse effects. However, as we have seen, the transboundary nature of this phenomenon necessitates international cooperation. The way GCC politics are conducted is influenced by the way in which the problem and its possible solutions are perceived. Jaggard highlights the manner in which values, ideas and concerns have been influential in shaping Germany's domestic climate change policies, which have in turn impacted related German foreign policy, and thus international negotiations. She argues that 'reflexivity' is evident among German domestic policy, foreign policy processes and cooperative international negotiations leading to GCC policies. A feature of German politics is the extensive discussion that is often required to reach a consensus, something that is particularly evident in Germany's foreign policies. The acceptability of the terms and inclusiveness of discussions, and hence the perceived justice of the results, must normally be addressed in the policymaking process. This is important because policies are more likely to succeed if everyone agrees with them – that is, if they are reflective of the ideas of the participants who have, through discussion, arrived at a consensual decision. Jaggard argues that analyzing and highlighting the linkages and feedbacks between the various levels of governance and policymaking can aid us in better understanding these processes. According to Jaggard, greater knowledge of these 'reflexive' processes can enable practitioners and policymakers to maximize the effectiveness of their policies.

In Chapter 15, Paul G Harris looks at some of the ethical-normative considerations, notably 'international environmental equity' (IEE) and fairness, which are central to efforts to address global climate change. Most economically developed countries, especially in Europe, have started to recognize and accept the proposition that they should take on their fair share of international burdens associated with GCC. In Chapter 15, Europe's policies and actions on climate change are subject to some (potentially controversial) normative assessment. Harris starts by introducing the notion of IEE in the context of the GCC agreements and shows how it, arguably, should and does apply in this issue area. He points out how Europe has been a leader on IEE in the GCC negotiations over the last decade and more, at least compared to other major actors such as the US, and he points to what European states and the EU have done to take on some of the burden of GCC, such as starting to actually reduce their emissions of GHGs and to help developing countries cope with suffering that will come from climate change. The upshot is that Europe is doing more than any other part of the world to address GCC and to share the burdens associated with it.

However, Harris argues that it is nevertheless not doing nearly enough. Both practical and normative considerations point to the need for much more urgent action by Europe to share the burdens of GCC. While ideas like IEE have helped shape or 'construct' Europe's identity and interests in this issue area, they must still serve as a normative guide for additional action in the future.

Why has Europe done what it has with regard to GCC? Why has it not done more? We return to these questions in the Conclusion (Chapter 16), highlighting a number of key theoretical lessons learned from the chapter case studies.

CONCLUSION

Europe's role in causing and combating GCC cannot be overstated. In this book we attempt to illuminate and explain the politics, foreign policy processes and regional cooperation that underlie European national and EU policies on this profound problem. The analyses that follow show that those policies are almost always a consequence of crossover effects between national and even sub-national politics and policymaking – that is, domestic politics – and international negotiations, diplomacy and regimes – international politics. These crossovers can be understood by thinking in terms of *foreign policy*, which is by definition about actors, issues and forces at both the domestic and international levels of policy action, as well as the interaction among them. This is not to dismiss the importance of what happens within states alone, or the crucial role that international factors have by themselves in shaping outcomes related to GCC. Rather, the research portrayed in this book shows that we can focus our attention on what happens within or among countries, but we cannot do so exclusively – at least not if we want to understand reality. These chapters compel us to question what we mean by systemic, societal and state-centric approaches to international relations, especially the latter. Increasingly, we can think in terms of a European society and a European, or EU, 'state' that exists alongside the individual countries of Europe – despite setbacks associated with attempts to ratify a European constitution. Increasingly, even those Europeans not encompassed by the EU are part of the society shaping European and EU policies on GCC, and they are inevitably subject, at least in part, to the EU 'state'. Likewise, despite their collective wealth and power, neither the European countries nor Europe – the EU – can avoid the global impacts of what happens at home and the influence and implications, both practical and ethical, of what happens abroad. Global climate change, despite being caused by what each person does on a daily basis, is just that – global. As such, it not only tells us about Europe's impact on the planet, but also Europe's role in the world and its interdependence with it.

NOTES

1. This scientific consensus is now well entrenched in Europe and beyond (see Oreskes 2004). Those who believe that global warming and climate change are not happening are among a dwindling number, now mostly restricted to the US. For discussions of the tactics of the 'climate skeptics', see e.g. Leggett (2001) and Gelbspan (2004).
2. For a discussion of US climate change policy and its related obligations, see Harris (2000a, 2001).
3. For an official summary of the EU position on climate change, see DGE (2002).
4. See Vogler and Bretheron (2006) for an investigation of the EU as an actor, and particularly the nature of the EU as a protagonist, in the GCC regime.
5. The integration of European *domestic* environmental policies more generally is beyond the scope of this book, although the manner in which integration has occurred is germane to climate change policy. For a discussion of environmental policy integration and co-ordination among EU countries, see, for example, Jordan (2005); Jordan and Liefferink (2004); Jordan et al. (2003); and Jordan, Schout and Zito (2004). For a summary of EU environmental foreign ('external') policy, see Thieme (2001); Vogler (2004) and Ott (2005). On the common foreign and security policy and convergence in EU external relations generally, see Bale (2000); Keisala (2004); Laursen (2002); Musu (2003) and Smith (2004).
6. See Drake (2000) for an introduction to global warming. This section adapts Harris (2003: 20–23).
7. While the EU has, for a number of years, fully endorsed the IPCC's assessments of GCC, it was a breakthrough of sorts when President George W Bush, after years of refusing to accept that global warming is a reality, said at the 2005 G-8 (group of eight industrialized countries) summit that human activities are 'to some extent' to blame for it (BBC News 2005).

8. For a summary of anticipated climate change impacts throughout Europe, see European Commission (2005: 12–16).
9. For a detailed examination of GCC *mitigation* strategies, see Metz et al. (2001). On the topic of *adaptation* to climate change, see McCarthy et al. (2001) and Lim et al. (2004).
10. For a largely first-hand account of GCC diplomacy, see Leggett (2001), and for complete accounts of all the FCCC-related negotiations, see the archives of *Earth Negotiations Bulletin* (2005) (available at http://www.iisd.ca/ vol12/). This section builds on Harris (2000b: 11–16).
11. The call for a 'framework' convention was based on the success of the framework Vienna Convention on stratospheric ozone depletion that resulted in the 1987 Montreal Protocol on Substances that Deplete the Ozone Layer. See Benedick (1998).
12. There is some debate about possible exceptions, but, insofar as there are any, their rarity proves the point.
13. For an ongoing historical account of the GCC negotiations, see the latest edition of the *Earth Negotiations Bulletin*, available at http://www.iisd.ca/enbvol/enb-background.htm.
14. The interaction between the domestic and the international that defines EFP is analogous to what Kern et al. (2001: iv) characterize as 'diffusion between the poles of national and international environmental policy.' Portions of this section first appeared in Harris (2004: 14–17).
15. The ongoing debate about whether one can even talk of a 'European' (i.e. EU) foreign policy, and from that how one can (or cannot) go about analyzing it, is described succinctly in Carlsnaes (2004). This debate applies particularly to the second part of this volume, where the contributors assume varying degrees of 'actorness' on the part of the EU. See Vogler and Bretheron (2006) and Chapter 9.
16. On the importance of issue-area definition in FPA, see Potter (1980: 405–27).
17. See Barkdull and Harris (2002: 66–84), for a detailed exposition of this typology and elaboration of the examples in Figure 1.1. For a discussion of how major theories of international relations help explain the first decade of GCC negotiations, see Paterson (1996).
18. Actual changes in GHG emissions (percentages) from 1990 to 2001 for those in the former EU-15 are Germany (–32.1), UK (–12.0), Sweden (–3.3) and Spain (+32.1) (EEA 2003: Table 1).
19. The EU was opposed to GHG emissions trading before the Kyoto conference in 1997, subsequently becoming a leader on the issue. See Christiansen and Wettestad (2003) on the political process and Klepper and Peterson (2005) on implementation of the scheme. Heller (1998) describes early EU efforts to co-ordinate European policies on climate change.

REFERENCES

African Development Bank et al. (AfDB) (2002), *Poverty and Climate Change: Reducing the Vulnerability of the Poor*, Consultation Draft (October), Washington, DC: World Bank.

Albritton, Daniel L et al. (2001), *Summary for Policymakers: A Report of Working Group I of the Intergovernmental Panel on Climate Change*, Bonn: Intergovernmental Panel on Climate Change, available at http://www.ipcc.ch/pub/spm22–01.pdf.

Bale, Timothy (2000), 'Field-level CFSP: EU diplomatic cooperation in third countries', *Current Politics and Economics of Europe*, **10**(2), 187–212.

Barkdull, John and Paul G Harris (2002), 'Environmental change and foreign policy: a survey of theory,' *Global Environmental Politics*, **2**(2), (May): 63–91.

BBC News (2005), 'Bush Rejects Kyoto-Style G8 Deal' (4 July), available at http://news.bbc.co.uk/1/hi/world/americas/4647383.stm.

Benedick, Richard E (1998), *Ozone Diplomacy* (enlarged edition), Cambridge, MA: Harvard University Press.

Betsill, Michele M (2005), 'Global climate change policy: making progress or spinning wheels?', in *The Global Environment: Institutions, Law, and Policy*, edited by Regina S Axelrod, David Leonard Downie and Norman J Vig. Washington, DC: CQ Press.

Bruce, J, H Lee and E Haites (eds) (1996), *Climate Change 1995: Economic and Social Dimensions of Climate Change*, Cambridge: Cambridge University Press.

Carlsnaes, Walter (2004), 'Where is the analysis of European foreign policy going?', *European Union Politics* **5**: 495–508.

Christiansen, Atle C and Jorgen Wettestad (2003), 'The EU as a frontrunner on greenhouse gas emissions trading: how did it happen and will the EU succeed?', *Climate Policy* **3**: 3–18.

Collier, Ute (1997), 'Comparative analysis and conclusions', in Ute Collier and Ragnar E Loftsedt (eds), *Cases of Climate Change Policy: Political Reality in the European Union*, London: Earthscan.

Directorate-General for the Environment of the European Commission (DGE) (2002), *EU Focus on Climate Change*, Luxembourg: Office for Official Publications of the European Communities.

Drake, Frances (2000), *Global Warming: The Science of Climate Change*, New York: Oxford University Press.

Earth Negotiations Bulletin (ENB) (2004), 'Summary of the tenth conference of the parties to the UN Framework Convention on Climate Change: 6–18 December 2004', *Earth Negotiations Bulletin* **12**, 260 (20 December), available at www.iisd.ca/vol12/enb12260e.html.

Earth Negotiations Bulletin (ENB) (2005), 'Summary of the eleventh conference of the parties to the UN Framework Convention on Climate Change and first conference of the parties serving as the meeting of the parties to the Kyoto Protocol: 28 November–10 December 2005', *Earth Negotiations Bulletin* **12**, 291 (12 December), available at www.iisd.ca/vol12/enb12291e.html.

European Environment Agency (EEA) (2003), 'EU greenhouse gas emissions rise for second year running', *News Release* (6 May), available at http://org.eea.eu.int/documents/newsreleases/ghg-2003-en.

European Commission (2005), 'Winning the battle against global climate change', Communication from the Commission to the Council, the European Parliament, the European Economic and Social Committee and the Committee of the Regions, Brussels: European Commission.

Evans, Peter B, Harold K Jacobson, and Robert D Putnam (1993), *Double-Edged Diplomacy: International Bargaining and Domestic Politics*, Berkeley: University of California Press.

Framework Convention on Climate Change (FCCC) (1998), 'Climate change meeting adopts Buenos Aires plan of action' (14 November 1998), available at www.unfccc.de/media/p111498.html.

Framework Convention on Climate Change (FCCC) (1999), 'Ministers pledge to finalize Climate Agreement by November 2000', Press release (5 November).

Gelbspan, Ross (2004), *Boiling Point*, New York: Basic Books.

Grubb, Michael, Christian Vrolijk and Duncan Brack (1999), *The Kyoto Protocol: A Guide and Assessment*, London: Royal Institute of International Affairs.

Haigh, Nigel (1996), 'Climate Change Policies and Politics in the European Community', in *Politics of Climate Change*, Tim O'Riordan and Jill Jager (eds). London: Routledge.

Harris, Paul G (ed) (2000a), *Climate Change and American Foreign Policy*, New York: St Martin's Press.

Harris, Paul G (2000b) 'Climate change and foreign policy: an introduction', in Paul G Harris (ed), *Climate Change and American Foreign Policy*, New York: St Martin's Press.

Harris, Paul G (2001), *International Equity and Global Environmental Politics*, Aldershot: Ashgate.

Harris, Paul G (2003), 'Climate change priorities for east asia: socio-economic impacts and international justice', in *Global Warming and East Asia*, Paul G Harris (ed.), London: Routledge.

Harris, Paul G (2004) 'Environmental foreign policy: a new subfield in international studies?', Paper presented at the Annual Convention of the International Studies Association, Montreal, Quebec (17–20 March 2004).

Hasenclever, Andreas, Peter Mayer and Volker Rittberger (1997), *Theories of International Regimes*, Cambridge, UK: Cambridge University Press.

Heller, Thomas (1998), 'The path to EU climate change policy', in Jonathan Golub (ed.), *Global Competition and EU Environmental Policy*, London: Routledge.

Houghton, John (2004), *Global Warming: The Complete Briefing*, Oxford: Oxford University Press.

Houghton, JT, Y Ding, DJ Griggs, M Noguer, PJ van der Linden and D Xiaosu (2001), *Climate Change 2001: The Scientific Basis*, Cambridge: Cambridge University Press.

Houghton, John T, LG Meiro Filho, BA Callander, N Harris, A Kattenburg, and K Maskell (eds) (1996), *Climate Change 1995: The Science of Climate Change*, New York: Cambridge University Press.

Ikenberry, John G, David A Lake and Michael Mastanduno (eds) (1988), *The State and American Foreign Economic Policy*, Ithaca: Cornell University Press.

Intergovernmental Panel on Climate Change (IPCC) (1995), 'Summary for policymakers: the science of climate change – IPCC working group I', available at www.ipcc.ch/pub/sarsum1.htm#four.

Intergovernmental Panel on Climate Change (IPCC) (2001), *Climate Change 2001*, 3 vols. Cambridge: Cambridge University Press.

IPCC Working Group II (2001), *Summary for Policymakers, Climate Change 2001: Impacts, Adaptation, and Vulnerability*, Geneva: IPCC Working Group II (19 February draft).

Jordan, Andrew (ed.) (2005), *Environmental Policy in the European Union: Actors, Institutions and Processes*, London: Earthscan.

Jordan, Andrew, Adriaan Schout and Anthony Zito (2004), 'Coordinating European Union Environmental Policy: Shifting from Passive to Active Coordination?', Paper presented at the workshop on 'Coordinating the EU', Birkbeck College, University of London (London: 28–29 November 2003).

Jordan, Andrew and Duncan Liefferink (2004), *Environmental Policy in Europe: The Europeanization of National Environmental Policy*, London: Routledge.

Jordan, Andrew, Rüdiger Wurzel and Anthony Zito (2003), 'How "new" environmental policy instruments (NEPIs) spread in the European Union: an analysis of the role of the EU in shaping environmental governance', European Union Studies Association Biennial Conference (27–29 March 2003).

Keisala, Katja (2004), *The European Union as an International Actor: Strengths of the European Civilian Power*. Academic Dissertation, Finland, University of Tampere.

Kern, Kristine, Helge Jorgens and Martin Janicke (2001), 'The diffusion of environmental policy innovations: a contribution to the globalization of environmental policy', Discussion Paper FS II 01–302, Berlin: Social Science Research Center, Free University.

Klepper, Gernot and Sonja Peterson (2005), 'Emissions trading, CDM, JI, and more: the climate strategy of Europe', Paper for the Climate Change Modelling and Policy Series, Fondazione Eni Enrico Mattei (April), available at www.feem.it/Feem/Pub/ Publications/WPapers/default.htm.

Laursen, Finn (2002), 'The EU in the global system', Paper for a lecture on 'The European Union in the global system: external relations and the common foreign and security policy (CFSP)', Graduate Institute of European Studies, Tamkang University, Taipei: (March).

Leggett, Jeremy (2001), *The Carbon War: Global Warming and the End of the Oil Era*, New York: Routledge.

Lim, Bo, Ian Burton, Sallemul Huq, Elizabeth Malone and Erika Spanger-Siegfried (2004), *Adaptation Policy Frameworks for Climate Change: Developing Strategies, Policies and Measures*, Cambridge: Cambridge University Press.

McCarthy, James J, Osvaldo F Canziani, Neil A Leary, David J Dokken and Kasey S White (eds) (2001), *Climate Change 2001: Impacts, Adaptation and Vulnerability*, Cambridge: Cambridge University Press.

Met Office (UK), (2004), '2004: another warm year', News Release (16 December), available at www.meto.gov.uk/corporate/pressoffice/2004/ pr20041216.html.

Metz, Bert, Ogunlade Davidson, Rob Swart and Jiahua Pan (eds) (2001), *Climate Change 2001: Mitigation*, Cambridge: Cambridge University Press.

Musu, Costanza (2003), 'European foreign policy: a collective policy or a policy of converging parallels?', *European Foreign Affairs Review* **8**: 35–49.

National Research Council (NRC) (2002), *Abrupt Climate Change: Inevitable Surprises*, Washington, DC: National Academy Press.

Oreskes, Naomi (2004), 'The Scientific Consensus on Climate Change', *Science* **306**, 5702 (December): 1686.

Ott, Hermann E (2005), 'The European Union: a strategic actor for sustainable global governance', in Andreas Rechkemmer (ed.), *UNEO: Towards an International Environment Organization*, Baden-Baden: Nomos.

Paterson, Matthew (1996), *Global Warming and Global Politics*, London: Routledge.

Pew Center on Global Climate Change (2005), 'COP 11 and COP/MOP 1 Montreal' (December), available at www.pewclimate.org/what_s_being_done/in_the_world/cop11/index.cfm.

Potter, William C (1980), 'Issue area and foreign policy analysis', *International Organization* **34**(3) (Summer): 405–27.

Putnam, Robert D (1988), 'Diplomacy and domestic politics: the logic of two-level games', *International Organization*, **19**: 345–66.

Rosenau, James N (1987), 'Introduction: new directions and recurrent questions in the comparative study of foreign policy', in Charles F Hermann, Charles W Kegley Jr and James N Rosenau (eds), *New Directions in the Study of Foreign Policy*, Winchester, MA: Unwin Hyman.

Roy, Denny (1998), 'Restructuring foreign and defense policy: the People's Republic of China', in Anthony McGrew and Christopher Brook (eds), *Asia-Pacific in the New World Order*, London: Routledge.

Sbragia, Alberta M and Philipp M Hildebrand, (1998), 'The European Union and compliance: a story in the making', in Edith Brown Weiss and Harold K. Jacobson (eds), *Engaging Countries: Strengthening Compliance with International Environmental Accords*, Cambridge, MA: MIT Press.

Smith, Michael (2004), 'Institutionalization, Policy adaptation and european foreign policy cooperation', *European Journal of International Relations* **10**(1): 95–136.

Thieme, Dominik (2001), 'European community external relations in the field of the environment', *European Environmental Law Review* **10**, (8–9 August): 252–64.

United Nations Conference of the Parties to the FCCC (UNCOP) (1996), 'Review of the Implementation of the Convention and of Decisions of the First Session of the Conference of the Parties: Ministerial Declaration' (July 18), FCCC/CP/1996/L.17.

United Nations Framework Convention on Climate Change (UNFCCC) (2002), 'The International Response to Climate Change', (July), available at www.klima.ph/resources/UNFCCC/iuckit/fact17.html.

Vogler, John (2004), 'The External Environmental Policy of the European Union', in Olav Schram Stokke and Øystein B Thommessen (eds), *Yearbook of International Co-operation on Environment and Development 2003/04*, London: Earthscan.

Vogler, John and Charlotte Bretherton (2006), 'The European Union as a protagonist to the United States on climate change', *International Studies Perspectives* **7**: 1–22.

Watson, RT, MC Zinyowera and RH Moss (1996), *Climate Change 1995: Impacts, Adaptations and Mitigation of Climate Change – Scientific-Technical Analysis*, New York: Cambridge University Press.

Watson, Robert T, et al., (1998), *The Regional Impacts of Climate Change: An Assessment of Vulnerability*, New York: Cambridge University Press, available at www.grida.no/climate/ipcc/regional/index.htm.

Webber, Mark and Michael Smith (2002), *Foreign Policy in a Transformed World*, Harlow: Prentice Hall.

White, Brian (1999), 'The European challenge to foreign policy', *European Journal of International Relations* **5**(1): 37–66.

Woods Hole Research Center (2004), 'Scientific evidence: increasing temperatures and greenhouse gases', available at www.whrc.org/resources/online_publications/warming_earth/scientific_evidence.htm.

Working Group on Climate Change and Development (WGCCD) (2005), *Up in Smoke?: Threats from, and Responses to, the Impact of Global Warming on Human Development*, London: International Institute for Environment and Development.

World Meteorological Organization (WMO) (2003), 'WMO statement on the status of global climate in 2002', Press Release No. 684 (17 December), available at www.wmo.ch/web/Press/Press.html#pr.

PART I

Global climate change and policymaking in Europe

2. The politics of climate change in Germany: domestic sources of environmental foreign policy

Michael T Hatch

INTRODUCTION

Throughout the series of international negotiations leading to the United Nations (UN) Framework Convention on Climate Change (FCCC) and the Kyoto Protocol, Germany has been a strong proponent of ambitious emissions targets. In October 1990, for example, the European Community (EC) adopted a target of stabilizing CO_2 emissions at 1990 levels by the year 2000 – a position pushed, in part, by the German and like-minded European governments to give them greater influence in the negotiations on an international climate change agreement. In advance of the third conference of the parties (COP3), the European Union (EU) called for a 15 per cent cut in CO_2 emissions by 2010. This EU target was based on a burden-sharing arrangement in which Germany was a major contributor – a 25 per cent reduction in domestic CO_2 emissions, which translated into an estimated 80 per cent of total EU reductions. In the aftermath of the compromize reached at Kyoto, the burden-sharing arrangements negotiated within the EU called for Germany to undertake a 21 per cent domestic cut in emissions of the basket of six greenhouse gases (GHGs) stipulated in the Kyoto Protocol.

Clearly, Germany has been an important player in the global climate change (GCC) negotiations and is central to the commitments assumed by the EU under the Kyoto Protocol – in the absence of substantial reductions of GHG emissions in Germany, the EU has little chance of meeting its international obligations. The core question to be addressed in this chapter is why the German government pursued such a role. The answer, it will be argued, is located largely in a domestic policy process that resulted in commitments to reductions in CO_2 emissions that are among the most ambitious in the industrialized world.[1]

DOMESTIC POLITICS AND FOREIGN POLICY IN GERMANY

Any significant attempt to address the global warming question has profound effects on a broad set of interests spanning society and, hence, have important ramifications both for domestic politics and foreign policy. Given the potential impact on important constituencies ranging from industry and labor to environmentalists, GCC becomes a concern of parliament, politicians engaged in electoral politics, and bureaucratic interests activated as activities coming under their purview become subject to actions. At the same time, it becomes the subject of foreign policy, since effective action is unlikely in the absence of international cooperation. In other words, the precise nature of the domestic policy process – in interaction with negotiations at the regional and international levels – has important implications for the politics of climate change. Accordingly, this chapter draws elements from the interest group and bureaucratic models for analyzing environmental foreign policy as described in Barkdull and Harris (2002). Rarely, however, does policy simply reflect the aggregate of competing pressures from societal interests or the relative strength of bureaucratic actors. Rather, as some of the more recent institutionalist literature in comparative politics suggests (e.g., Hall 1986; Steinmo, Thelen and Longstreth 1992), societal and bureaucratic influences are mediated through such historically developed institutions as national styles of policymaking or the organization of the political system itself – electoral arrangements and the network of political parties that shape the electoral arena (see Barkdull and Harris 2002: 74–84).

With historical roots in corporatist arrangements typified by close collaboration between the state and functionally organized interest groups (Beyme 1985; Katzenstein 1985), policymaking in Germany is said to be characterized by its emphasis on consensus and consultation (Dyson 1992; Jänicke and Weidner 1997; Wurzel 2002; see also Chapter 14). When applied to GCC, it will be argued in this chapter that German policy has been shaped by a more collaborative approach that enabled a consensus to coalesce around the view that global warming was a serious threat that warranted concerted action. Out of this more consensual policy process emerged a commitment to dramatic reductions in GHG emissions. Domestic decisions over the extent of reduction targets and the means to achieve them, in turn, became intimately linked to negotiations at the regional and international levels on the shape of the FCCC and Kyoto Protocol.

THE GENESIS OF CLIMATE CHANGE POLICY IN GERMANY

The global warming debate in Germany had its origins in the controversy over nuclear power triggered by the 1986 Chernobyl nuclear accident.[2] With calls for either an immediate shutdown (from the Greens) or phase-out (from the Social Democratic Party [SPD]) of all nuclear plants, the construction of additional coal-fueled power plants was proposed to compensate for the lost capacity of nuclear facilities. The parties supporting nuclear power – most importantly, the governing Christian Democratic Union (CDU) and its Bavarian sister party (Christian Socialist Union [CSU]) – found in the issue of GCC what they hoped would be an effective counterbalance, arguing that nuclear power made good environmental sense when confronted with the ominous threats posed by global warming. Within the context of conflicting scientific claims and political polarization, it was decided to establish an *Enquete* (Inquiry) Commission – a parliamentary body occasionally created 'to deal with complex and often politically sensitive issues' (von Moltke 1991: 26).

The Enquete Commission on Preventive Measures to Protect the Atmosphere was created in December 1987. The first interim report, *Protecting the Earth's Atmosphere: An International Challenge* (Enquete Commission of the German Bundestag 1989 [Enquete Commission]), was based on hearings with scientists, politicians, representatives from industry, and federal ministries as well as discussions with environmental and consumer groups. It concluded that there was 'an extraordinary need for action' (Enquete Commission 1989: 43). A subsequent report followed in October 1990. It found that:

> 'There is such massive and unequivocal scientific evidence on ... the man-made greenhouse effect, the resulting climatic change and its repercussions ... that there can be no doubt that preventive action must be taken immediately, irrespective of any need for further research.' (Enquete Commission of the German Bundestag, 1990: 24)

Based on this conclusion, the Enquete Commission proposed a 30 per cent reduction of CO_2 emissions for Germany by the year 2005.

The Enquete Commission continued to hold hearings and produce reports; however, its work was most crucial in the formative years of German GCC policy. The Enquete Commission, in essence, stepped in to fill a political vacuum left by the political parties, interests groups and government agencies, none of which were prepared to deal with the global warming question, let alone provide leadership.

During the early 1980s, the SPD was not receptive to a message about the possible dangers associated with CO_2 emissions. In a party deeply divided over nuclear power, giving credence to the concerns about GCC threatened further division. Perhaps more importantly, strong ties to the powerful coal interests in Germany made for ready global warming skeptics. With Chernobyl, the party – out of federal office by this time – became more united in its rejection of nuclear power, but the global warming issue raised questions about the feasibility as well as desirability of a nuclear phase-out. It was to take several years of internal debate to work out a position on global warming that, to some degree, reconciled the various environmental/anti-nuclear/pro-coal factions present in the party.

The CDU hesitated to take up the issue in the early 1980s – despite its strong support for nuclear power – because of its close links to industry and the implications of global warming for fossil fuel use. After Chernobyl, the CDU (along with the nuclear lobby) touted nuclear power as a solution to the problem, precisely because it did not emit CO_2. If the real problem were CO_2 being emitted into the atmosphere, however, regulatory measures resulting in lower CO_2 emissions and/or higher prices for fossil fuels would seem to be the logical next step. The prospect of such actions initially found little support within important elements of the CDU/CSU, its Free Democratic Party (FDP) coalition partner, and industry.

The Greens and major environmental groups came relatively late to the global warming problem, in large part because of the nuclear power issue. They had their formative roots in the anti-nuclear movement of the 1970s and owed much of their strength to an uncompromizing rejection of nuclear power. They were less inclined to give much credence to the claims about global warming, especially since those claims were often associated with justifications for nuclear power.

The government, too, was ill prepared to address the global warming issue. In 1981, a study by the German Council of Experts on the Environment (an independent body created to provide the government with scientific assessments of environmental questions) gave practically no credence to anthropogenic influences on world climate. It was not until late 1987 that the Council finally acknowledged the possibility of GCC due to CO_2 emissions and other GHGs (Hatch 1995: 426). In other words, government officials did not receive much early warning on the development of this issue through channels established for this purpose. Perhaps more critically, however, the global warming question seemed to have fallen between the institutional cracks of government during much of the 1980s.

The Federal Ministry for the Environment (Bundesministerium für Umwelt, Naturschutz und Reaktorsicherheit, or BMU) had only come into existence in 1986 following Chernobyl and its (mis)handling by the Interior

Ministry, where responsibility for most environmental issues had resided. Climate change, however, was one of the areas outside the competence of the Interior Ministry. Through its control of meteorological questions, the Transport Ministry had been given responsibility for the climate change issue, where it was defined largely as a scientific question devoid of much political content. It was only after the ministry failed to provide for the effective participation of Germany in the initial deliberations of the Intergovernmental Panel on Climate Change (IPCC) – and the criticism that followed from the Enquete Commission – that the Chancellor's office transferred the climate issue to the BMU in late 1988.

In sum, it was within the framework of the Enquete Commission that the initial responses to GCC questions were formulated. Prominent figures from the scientific community and leading parliamentarians – chosen not just for their expertise but also for their ties to important social groups (von Moltke 1991: 27) – were brought together to deliberate. They were not simply agents of their political parties, interest groups or scientific bodies, however. Representatives from major industrial associations were consulted, studies were commissioned, politicians and ministry officials were heard, but Commission members set the tone and direction of deliberations. Out of this consultative process emerged a broad consensus for political action.

FORMULATING A CLIMATE CHANGE POLICY

Following the release of the Enquete Commission's first report, the locus of activity gradually shifted towards government as attention began to focus on appropriate policy responses. Even here, however, the views of the Commission influenced the debate, as they provided a baseline against which the positions of government bodies, political parties and organized interests could be measured. In 1989–90, the debate began to center on two questions: by what amount should Germany reduce its CO_2 emissions and what methods should be used to achieve the agreed target? In these areas, fundamental differences arising from the nature of the GCC issue itself had to be resolved: no single government ministry could control GCC policy, each ministry had different organizational responsibilities and constituencies, and, with coalition governments the norm in Germany, ministers frequently had different party affiliations.

The two major protagonists in governmental efforts to formulate a policy toward global warming were the BMU, which by this time had been given the GCC issue, and the Ministry for Economics (BMWi), where the responsibility for energy policy resided.

Reduction Targets

On the question of GHG reduction targets, the CDU Minister for Environment staked out a position calling for a 25 per cent cut in CO_2 emissions by the year 2005, much above what energy experts in the BMWi believed possible or desirable given their constituents in the energy sector and industry. The Economics Minister (FDP) opposed binding targets because of concerns about the loss of economic flexibility and dynamism in energy security and competitiveness. The two ministries came under increasing pressure to get an agreement, however, since Chancellor Kohl wanted to use the July 1990 Group-of-Seven economic summit to profile the German government's commitment to CO_2 reductions. As the summit approached, movement in positions became more noticeable.

While the BMU was holding to the general reduction target of 25 per cent, the BMWi had moved to a position that grudgingly accepted targets, but in so doing called for the early inclusion of eastern Germany in the calculations, since the former German Democratic Republic had been so inefficient in its energy use. The BMU, on the other hand, argued that German unification should not make it easier to achieve the 25 per cent target, especially in light of the message it might send to other nations about the seriousness of Germany's intention to combat global climate change. On 13 June 1990, the federal cabinet adopted the goal favored by the BMU: a 25 per cent reduction in CO_2 emissions by 2005 compared to 1987 levels. On the issue of East Germany, the question was fudged. In subsequent cabinet decisions, however, the reduction target was extended to 25–30 per cent by the year 2005 in light of German unification.

Measures to Meet Reduction Targets

At a June 1990 meeting, the cabinet also called for the creation of an inter-ministerial working group to formulate a set of proposals that would enable Germany to meet the emissions target. For the most part, debate focused on the application of a CO_2/energy tax and its linkage to a proposed EC-wide climate-protection tax. The BMU favored the adoption of a tax or levy, even in the absence of agreement at the EC level. The BMWi opposed a CO_2 levy, especially if it were undertaken unilaterally.

In October 1990, the EC adopted a target of stabilizing CO_2 emissions at 1990 levels by the year 2000 – a strategy designed, in part, to give the German and like-minded European governments greater influence in the international negotiations scheduled to begin in January 1991. The following September, the European Commission released a draft paper containing a set of proposals designed to achieve the agreed target. As part of this package, the Commission proposed a combined CO_2/energy tax that would be based 50 per cent on energy content and 50 per cent on carbon content. In response, European industrialists unleashed an intense lobbying effort in opposition (*The Economist*, 9 May 1992: 19, 85). It had an impact. The revised proposal of the Commission made any EC energy tax conditional on other OECD countries adopting similar measures.

Within the internal debate over GCC policy, the BMU had become the major proponent of Germany playing a leadership role (*Vorreiterrolle*) in the international fight against GCC. In the case of a CO_2 tax/levy, this meant Germany going it alone if necessary. For the BMWi, there should be no such role. If the German economy were to retain its economic competitiveness, there was no alternative to an EC-wide agreement. Up to mid-1991, the BMU had enjoyed the support of Chancellor Kohl in the various inter-ministerial struggles surrounding the global warming question. With economic growth slowing and the costs of German unification mounting, however, the BMU began to receive less backing from Kohl and others in the CDU. In December 1991, it was decided that the government would adopt a CO_2 tax only in combination with an EC-wide CO_2/energy tax.

Given that the George HW Bush administration in the United States (US) had been adamant in its rejection of such measures, however, it was clear that no CO_2/energy tax would be implemented by the EC any time soon. In essence, pressures from powerful industrial interests, combined with the divisions within government itself, made it much more attractive to shift responsibility to the EC for a decision that was bound to impose heavy political costs.

The loss of this CO_2/energy tax was the most visible setback to those wishing for quick action on global warming. Though the government had implemented other measures to reduce CO_2 emissions (see Hatch 1995: 431–2), taken together, they failed to represent the kind of commitment required if Germany were to meet its reduction goal of 25–30 per cent by 2005. Carbon dioxide emissions had declined by 14.7 per cent between 1987 and 1993 (BMU 1994b: 10), but this was due largely to the effects of unification: inefficient energy use in the former East Germany – combined with its reliance on dirty lignite (70–80 per cent of primary energy) – meant that the shift to other fuels and their more efficient use reduced CO_2 emissions substantially. The collapse of the economy in the East led to lower CO_2 emissions as well. According to studies commissioned by the government itself, however, the measures approved would fall well short of the 2005 target (BMU 1994a: 87). This became more problematic for a German government whose commitments to CO_2 reduction came under increasing domestic and international scrutiny in 1995.

DOMESTIC POLICY AND INTERNATIONAL COMMITMENTS: NEGOTIATING THE KYOTO PROTOCOL

The first session of the conference of the parties (COP1) to the FCCC was to be held in Berlin between 28 March and 7 April 1995. This high-profile event gave the GCC question renewed prominence on the domestic political agenda. As host, the German government felt compelled to make a credible demonstration of its ability to meet its reduction targets. As discussed earlier, efforts to put in place the set of measures felt necessary to achieve the announced target encountered several hurdles, not least of which was the German industry's resistance to a CO_2/energy tax, as well as regulatory measures that were viewed as adversely impacting on competitiveness. In an attempt to overcome this political impasse, a voluntary agreement between the federal government and industry to limit CO_2 emissions was announced in March 1995, just prior to COP1.

German Industry's Declaration on Preventing Global Warming

Negotiations between industry and representatives from the BMU and BMWi began in January 1995. Among the major points of contention were the explicitness of the commitments, how demanding they should be, and what concessions government would provide in return. Industry's opposition to a CO_2/energy tax remained. Of equal or greater concern, however, was a proposed Heat Utilization Ordinance for industrial companies. If adopted, it would have required companies to recover and utilize heat generated in their plants and make surplus heat available to others, an expensive process that – in the eyes of German industry – would severely compromise its competitiveness. Industry wanted these measures off the table. The government, for its part, wanted high absolute targets representing reductions that moved well beyond business as usual. Industry favored 'specific' rather than 'absolute' targets – reductions calculated on a per-unit-of-output basis ('specific') rather than in lower overall emissions ('absolute').

On 10 March 1995, the 'Declaration by German Industry and Trade on Global Warming Prevention' was issued. In this declaration, 15 industry associations agreed to use 'special efforts' to reduce their specific CO_2 emissions or specific energy consumption up to 20 per cent by the year 2005 (base year of 1987). The government agreed to hold in abeyance additional regulatory measures (such as the Heat Utilization Ordinance) and CO_2/energy tax. While welcomed by many concerned about the absence of action on GCC, the agreement was not without its critics (e.g., Fischedick et al. 1995; Kohlhaas et al. 1995).

In response to such criticism, further negotiations between government and industry resulted one year later in a revized agreement that pledged to reduce specific CO_2 emissions by 20 per cent, with a change in the base year from 1987 to 1990, which brought it into conformity with the base year employed in the FCCC negotiations while, at the same time, making it more ambitious, since many of the 'wall fall' benefits from German unification would be lost. Moreover, some of the associations switched their commitments from specific to absolute emissions reductions. Also, additional industrial associations joined, meaning that approximately 80 per cent of German industry's total energy consumption was now covered by the agreement. Finally, a monitoring system – to be administered by an independent third party (the Rhine-Westphalia Institute for Economic Research) – was established to provide greater transparency in evaluating compliance with the agreement.

At the heart of the review process was assessing the degree to which the commitment to undertake 'special efforts' to reduce CO_2 emissions was fulfilled. In the second monitoring report covering the period through 1997, it was found that several of the industry associations had almost achieved or exceeded their goals. Such results, the report concluded, were not 'unproblematic' for the persuasiveness of the voluntary agreements 'unless this were the motivation to define new and more ambitious goals' (Buttermann et al. 1999: 157). The clear implication of the monitoring report was that more must be done if the agreement was to move beyond 'business as usual'. This became increasingly salient as reductions in German GHG emissions became directly linked to binding commitments undertaken by the EU within the Kyoto Protocol.

Negotiating the Berlin Mandate and Kyoto Protocol

At the top of the agenda for COP1 was a review of the adequacy of the commitments contained in the climate change convention. Standing in the way of any movement on this issue, however, was the question of additional commitments for developing countries. Germany, along with other EU member states, helped remove this roadblock by pushing the position that future negotiations on a protocol should not include developing country commitments. This cleared the way for the adoption of the Berlin Mandate, which acknowledged that current commitments were inadequate and that the negotiation of further commitments, hopefully by COP3 in 1997, would not introduce new obligations for developing countries (Oberthür and Ott 1999: 46).

Among the issues to be resolved in the negotiations leading up to COP3 were the reduction targets, as well as the time frame for achieving them. To push the negotiating process on these questions, in March 1996 Germany proposed a reduction target of 10 per cent by 2005 and 15–20 per cent by 2010 (Oberthür and Ott 1999: 116). At the same time, efforts were made within the EU to formulate a common position that would allow it to assume a leading role in the negotiations. Initial discussions focused on an Irish proposal for a 5–10 per cent reduction in CO_2 emissions by 2005. Given the role of nuclear power in its energy mix, France argued for reductions based on per capita emissions. Germany, for its part, pushed for a 20 per cent reduction in CO_2 emissions, fearing that a low common target would weaken the EU's position in the international negotiations. Most other member states supported a less ambitious target for the EU. Greece, Ireland, Portugal, Spain and Sweden (because of a planned phase-out of nuclear power) asserted their right to increase domestic CO_2 emissions.

In the March 1997 meeting of the EU's Environmental Council, a common negotiating position was hammered out. It called for a 15 per cent reduction in emissions by 2010, though a precise burden-sharing arrangement was not agreed to at the time. Germany, however, committed to reductions that would cover approximately 80 per cent of the EU's overall target. Following the adoption at Kyoto of an 8 per cent reduction target for a basket of six GHGs by 2008–12, in June 1998 a modified burden-sharing agreement was adopted by EU environmental ministers. Germany's share translated into a 21 per cent reduction in the Kyoto Protocol's basket of GHGs. In other words, once the Kyoto Protocol would be ratified, Germany's emissions reductions would no longer simply be a national statement of intent, but part of the EU's legally binding commitment to reduce GHG emissions. Questions about the German government's ability to meet both its national and international obligations again arose, this time within the context of national elections scheduled for September 1998.

Climate Change and the Red-Green Coalition

While in opposition, the SPD and Green party had been critical of the Kohl government's over-reliance on voluntary measures. Among the policies they had been pushing were the introduction of ecological tax reform, adoption of further regulatory measures (e.g., the heat utilization ordinance), tightening existing measures (e.g., the insulation ordinance), and greater support for renewable energy sources. In the wake of the September 1998 elections, the new SPD/Green coalition government was now in a position to place its stamp on German GCC policy.

Ecological Tax Reform

From the mid-1990s, support for some type of ecological tax reform had been building in Germany as economic growth stagnated and unemployment rose. Proponents argued that an ecological reform of the tax system could provide a 'double dividend': environmental objectives could be achieved more efficiently through this market-based instrument while at the same time reducing the high cost of labor for companies, thereby encouraging economic expansion and job growth. The eco-tax reform was to be revenue-neutral in that increased taxes on energy would compensate for reductions in company social security contributions. Following several months of negotiations among coalition partners and various stakeholders, an eco-tax came into effect on 1 April 1999, gradually raising the price on gasoline, heating oil, natural gas and electricity in a series of steps.

Reflecting the influence of various interest groups and the need to get sufficiently broad support for this unilateral action, the government had to make several concessions: the consumption of coal was exempted and certain energy-intensive sectors were made eligible for reduced tax rates. In addition, electricity from renewables received only limited exemptions despite the desire to encourage the development of this type of energy source. These limitations on renewables, however, had more to do with EU requirements governing competition in a liberalizing European electricity market. The final increment in the eco-tax was introduced in early 2003. As a consequence of strong political resistance, the newly re-elected SPD/Green government decided to postpone any decision on additional increases in the eco-tax and an evaluation of its effects through its initial five stages (see Kohlhaas and Meyer 2005).

Renewable Energy

In an effort to encourage the expansion of renewable energy, the Renewable Energy Sources Act (EEG) was introduced in April 2000. In the wake of this legislation, Germany experienced an accelerated growth in the generation of power from renewables. The most impressive area of growth has been in wind power. Between the adoption of the 1991 Electricity Feed Act that required utilities to purchase electricity generated from renewables at a subsidized rate (introduced by the previous government) and the implementation of the EEG in April 2000, Germany had achieved an installed capacity of 4,500 MW. By the end of 2001, capacity had almost doubled to approximately 8,750 MW (BMU 2002: 14). At the beginning of 2003, over 12,000 MW of electricity was being generated by wind power, representing 3.5 per cent of all electricity consumption in Germany. In relation to other countries, this level of production translated into one-third of the world's wind power, making Germany the single largest producer, with the US and Spain following at 25 and 15 per cent respectively.

All told, the proportion of electricity from renewables increased from 5.2 per cent in 1998 to over 10 per cent in 2004. The announced goal of the government was to have this share rise to 12.5 per cent by 2010, which would represent a 70–80 metric tons (mt) reduction in CO_2 emissions. Over the longer term, the government set targets of 20 per cent of electricity from renewables by 2020 and one-half of all energy by 2050.

Voluntary Agreements under the Red-Green Coalition

Following the September 1998 elections, there were questions about the commitment of the new SPD/Green coalition government to voluntary agreements, since a number of representatives from both parties had been critical of the agreements while in opposition. As it turned out, the coalition agreement negotiated between the SPD and Greens following the elections included a statement supporting the use of voluntary agreements. Discussions on revisions to the voluntary agreements began in earnest during the first months of 2000.

Around the same time (March 2000), the Council of Environmental Advisors announced that the government would not be able to achieve its emissions-reduction goals unless additional efforts were undertaken, a fact subsequently acknowledged in statements by both the Ministers of Economics and the Environment. In October 2000, one month before negotiations on the Kyoto Protocol were to resume at COP6 in the Hague, the government announced a further iteration in Germany's Climate Protection Program (BMU 2000). In order to achieve its 25 per cent reduction target, the government estimated that an additional 50–70 million tons of CO_2 emissions had to be eliminated by 2005. Energy use and industry were expected to assume the largest share of the reductions (20–25 mt), followed by private households (18–25 mt) and the transportation sector (15–20 mt).

Within the context of the new national Climate Protection Program, in November 2000 a general agreement on further voluntary actions was signed between the government, the Federation of German Industry, and individual industrial associations. It committed industry to specific CO_2 reductions of 28 per cent by 2005 (the earlier agreement had set the target at 20 per cent) and a 35 per cent reduction in emissions of Kyoto gases (expressed in CO_2 equivalents) by 2012, compared to what they were in 1990. It was estimated that this would result in an additional 10 million ton reduction in CO_2 emissions by 2005 and a further 10 million tons CO_2 equivalent by 2012 (BMU 2000). Negotiations with individual industrial associations were to follow. With few exceptions, the subsequent sectoral agreements have either not been concluded or made public. The most notable exception was the agreement between the power sector and federal government. In accordance with the target established for the energy sector in the Climate Protection Program, the utilities committed to a 45 mt reduction in CO_2 emissions by 2010, 20 mt of which were to come from cogeneration.

FINALIZING THE KYOTO PROTOCOL

The protocol signed at Kyoto in December 1997 was a far-from-complete document. Left for later negotiations was the task of fleshing out the practical details required for its effective functioning. The magnitude of that task soon became apparent, as talks dragged on for another 4 years. Among the most contentious issues were the extent to which the 'flexible mechanisms' accepted at Kyoto (i.e., emissions trading, joint implementation (JI), the Clean Development Mechanism (CDM)) and carbon sinks could be used to meet the reduction targets adopted in the protocol.

Going into the negotiations, Germany and the EU had emphasized the importance of industrialized countries taking the lead in emissions reductions, meaning that those reductions should come primarily from domestic measures. The argument was couched both in ethical and practical terms: industrial countries had primary responsibility for the build-up of GHGs to current levels, and domestic action among those with historical responsibility was also a pre-condition for developing countries to assume obligations in the future (see Chapter 15). The US, along with such countries as Australia, Canada and Japan argued for maximum flexibility in the use of instruments, thereby lowering the costs of meeting reduction targets. The compromise at Kyoto resulted in acceptance of the flexible mechanisms, but their use was supposed to be 'supplemental' to domestic action. Also included were limited allowances for carbon sequestration/sinks (see Chapter 10). In subsequent negotiations, Germany sought to ensure that the flexible mechanisms were, in fact, supplemental to domestic reduction measures and that the use of project-specific instruments (JI and CDM) would result in real reductions. On the issue of sinks, Germany remained opposed to their use, arguing that a great deal of scientific and technical uncertainty in this area persisted. Domestic factors played a central role in shaping these priorities.

The position that Germany and other industrial countries must take the lead in combating climate change – and that this requires first and foremost domestic action – may have an ethical dimension that is informed by important elements in civil society as well as influential constituents within all the major political parties (see Chapter 15). There are, however, very tangible factors largely related to concerns about economic competitiveness that have shaped this approach as well. Though a significant share of Germany's ambitious reduction targets had come from 'wall fall' effects, the mix of policy instruments applied to the mitigation GCC has imposed substantial financial burdens on the domestic economy – reductions in GHG emissions from renewable sources, for example, are relatively expensive (Michaelowa 2003: 41). That the Kyoto Protocol might allow other industrialized countries to avoid domestic actions, thereby gaining competitive advantages in globalizing markets, has been central to the calculations of the German government in the negotiations.

In preparation for the negotiating session at COP6 in November 2000, Germany, in conjunction with several other EU member states and the European Commission, were successful in having the EU adopt a position calling for a 50 per cent ceiling on the use of flexible mechanisms as well as opposition to the inclusion of sinks during the first commitment period. With regard to the use of project-specific instruments (JI and CDM), Germany held the view that priority should be given to measures that would promote renewable energy and efficient energy use; sinks projects and nuclear power, on the other hand, should be excluded. In staking out this position, the German government made the argument that a central consideration in the choice of measures ought to be their sustainable effect on the climate. Added to this, however, was the calculation that these instruments might create a major market for the types of environmentally friendly technologies developed by German industry in response to the domestic regulatory environment. Regarding nuclear power, the red-green government did not want to encourage the expansion of a technology abroad that it considered dangerous and wanted to phase out domestically.

As it turned out, the differences over limits on the use of flexible mechanisms and sinks proved unbridgeable at the negotiations in the Hague. Following the withdrawal of the US from the Kyoto Protocol in March 2001, the talks assumed a new dynamic. Negotiations resumed in July 2001 in Bonn; the text was finalized at COP7 in Marrakech the following November. Since the conditions of ratification gave the countries previously aligned with the US considerable leverage, the EU made the major concessions required to get an agreement: no concrete ceilings were imposed on the use of flexible mechanisms, emissions reductions from carbon sinks were included, and afforestation and reforestation projects would be eligible for credits. The EU was able to have nuclear power projects excluded from JI and the CDM.

During negotiations on the Kyoto Protocol in 2001, debate at the European level began to intensify over a directive being drafted by the European

Commission for an emissions trading scheme for the EU. In general terms, German concerns about emissions trading in Europe mirrored to some degree those surrounding the Kyoto instruments. In a number of member states (e.g., Austria, Belgium, Denmark, Ireland, Italy, the Netherlands, Spain), emissions trends in recent years had raised serious questions about those countries' ability to meet their announced targets. From Germany's perspective, additional domestic action was required if questions of competitiveness were to be avoided (BMU 2000: 152). The concerns became more specific with the publication of a draft directive in October 2001.

Among the most contentious issues for Germany were the imposition of mandatory quotas on CO_2 emissions in selected industrial sectors and the level of permitted cuts. German industry was especially opposed to the draft, arguing that it imposed additional burdens that would further threaten its competitiveness. In December 2001, for example, nine large German firms from several industrial sectors sent a letter to the Chancellor threatening to terminate the voluntary agreement between industry and government if proposed EU guidelines for emissions trading were adopted. The BMWi, in turn, became the major advocate for industry's position within the government. While earlier expressing reservations about emissions trading, the Green Party and environmental non-governmental organizations (NGOs) now supported the proposal (Michaelowa 2003: 37). Among the specific changes sought by the government were recognition of actions undertaken by Germany since 1990 in the allocation of emissions permits; the free distribution of emissions permits on a permanent basis; compatibility with the flexible mechanisms of the Kyoto Protocol (i.e., credits earned through JI and CDM projects could be traded on the European market);[3] and a 'pooling' arrangement that would require sector-wide emissions quotas rather than the allocation of permits at the plant level (i.e., acceptance of a format compatible with the existing voluntary agreements).

The compromise agreed to at a meeting of the Environmental Council in December 2002 reflected the influence of Germany in the deliberations: acknowledging early efforts, 1990 could be used as the base year for the allocation of emissions permits; member states would be able to distribute those permits free of charge through 2012; emissions credits from JI and CDM projects could be sold on the European emissions market; permits could be allocated on the basis of earlier voluntary agreements between government and industry; certain sectors and companies could apply to opt out of emissions trading until 2008; and, in a modest concession made by the German government, companies would be permitted (rather than required) to pool their emissions rights (Press Statement 2002; *Süddeutsche Zeitung*, 10 December 2002).

FUTURE PROSPECTS

In recent years, attention has turned to the implementation of the Kyoto Protocol. However, recent trends in GHG emissions of many of the signatories to the protocol have brought into question their ability to meet the obligations agreed to in Kyoto, members of the EU included (UN: 2003). Germany is one of the few notable exceptions. Having achieved more than a 19 per cent reduction in emissions by 2004, it is well placed to meet its Kyoto target of 21 per cent. Nonetheless, German GCC policy has not remained immune from criticism. On the one hand, skepticism about achievements in reducing GHG emissions have been expressed, given the common impression that most reductions have been the unintended result of unification rather than government policy. At the same time, there was considerable doubt about the ability of the government to deliver on its commitment to reduce CO_2 emissions 25 per cent by 2005. Finally, some have pointed to inconsistencies in German climate policy resulting from a policy process overly influenced by strong lobbies. Michaelowa (2003: 34), for example, has described German climate policy as 'a hodgepodge of instruments that allows [Germany] to cater for [sic] the wishes of interest groups', implying that the set of measures selected to combat GCC is far from optimal.

Regarding the first two issues, a study commissioned by the BMU and Federal Environmental Agency (Eichheimer et al. 2001) came to the following conclusions: 40.3 per cent of CO_2 emission reductions between 1990–2000 were attributable to policy, the remaining 59.7 per cent to unification (Eichheimer et al. 2001: 39); for all Kyoto gases (including CO_2), policy was responsible for 53 per cent of the reductions compared to 47 per cent from unification (Eichheimer et al. 2001: 39); unification accounted for approximately 11 per cent of reductions in non-CO_2 GHGs (Eichheimer et al. 2001: 21); while Germany is likely to meet its Kyoto target, the prospects for reaching the more ambitious national target appeared 'rather slim' (Eichheimer et al. 2001: 38).[4]

Until relatively recently, the government – at least in its public statements – had maintained that the 25 per cent reduction target was still achievable (BMU-Presse 2002). This position has proven untenable, given that CO_2 emissions had failed to move beyond 16 per cent by the end of 2004. When dealing with an issue that will require a policy response extending over decades, however, whether this target was met by 2005 seems less critical than the general trajectory of German policy. In this regard, Germany appears to be among the few countries currently attempting to initiate a process that effectively addresses the challenges of GCC over the longer term. As the above figures suggest, significant reductions in GHG emissions have resulted from policies initiated by government. Moreover, emissions in the two most intractable sectors of the German economy (transport and housing) finally began to show declines after a decade of increases throughout the 1990s (BMU-Reden 2002).

The concerns expressed about the 'hodgepodge' of instruments resulting in the adoption of measures whose effectiveness is compromized by the influence of interest groups has some basis in reality: the concessions provided in the ecological tax reform for energy-intensive sectors and coal, the hardship clause that exempts the largest industrial users of electricity from the added costs of renewable energy, the modesty of the targets contained in industry's voluntary agreements, the rather generous allocation of allowances to industrial firms and electric utilities during the initial phase of the emissions trading program (estimated to achieve around 0.4 per cent CO_2 reductions from 2005 through 2007). On the other hand, compromise and second-best solutions are the inevitable outcome of a policy process based on consensus. At the same time, one could argue that this consensual policy style – nested in broader organizational arrangements alluded to in the earlier reference to historical institutionalism that structure societal pressures and bureaucratic influence – helps account for the success of Germany's climate policy to date.

The consensual approach embodied in the work of the Enquete Commission helped forge a common understanding of the science of global warming and its policy implications, thereby avoiding the often debilitating debates about scientific uncertainties and appropriate policy responses that have characterized the politics of GCC in such countries as the US. While the locus of policymaking subsequently shifted to the government, a consensual policy process sensitive to environmental concerns continued. This was due in no small part to the nature of the German electoral and party systems, where a single party rarely gains an absolute majority in national elections and smaller parties are potential coalition makers (and breakers), giving them influence beyond their numbers. At the same time, the 5-per cent-of-the-vote hurdle remains a constant in each party's electoral calculations. Under these circumstances, a relatively small number of voters for whom environmental issues are very important may be able to determine electoral outcomes. This can have the effect of enhancing the political salience of environmental issues like global climate change.

With the rise of the Greens, the CDU and SPD were compelled to respond to questions being raised about the environment. This, in turn, strengthened each party's environmental wing and made environmental issues important for both parties. The inclusion of the Greens in the governing coalition in 1998 had the added effect of further institutionalizing the presence of environmental interests in this consensus-oriented policy process. The result has been the adoption of ambitious reduction targets and a set of measures that reflect a search for consensus among various bureaucratic actors, powerful economic interests, and influential environmental NGOs.

Whether these measures can and will deliver the promised reductions in CO_2 emissions remains an open question. As indicated earlier, however, more salient may be how they operate over the longer term. Much depends on whether these measures represent static bargains or a dynamic process that allows for ongoing corrections and modifications. In this respect, developments over the past decade are instructive: from an initial emphasis on regulatory measures that encountered strong resistance from industry, voluntary agreements subsequently came to play a central role in the German approach to climate change; adamant opposition to market instruments in the form of a CO_2/energy tax has been supplanted by a grudging acceptance of energy taxes in the form of an ecological tax reform; while the voluntary agreements initially announced in 1995 were rather weak, the outcome of subsequent negotiations seems to indicate a process capable of movement towards more ambitious goals; efforts to encourage the expansion of renewable energy sources begun under a center-right CDU/CSU-FDP government and enhanced by the center-left SPD/Green coalition have made Germany a world leader in renewable energy and have resulted in the creation of an industrial sector able to exercise considerable influence in the policy process (see Michaelowa 2004).

Finally, the ambition of Germany's reduction targets – with an emphasis on domestic actions to meet them – has shaped its approach to the international GCC negotiations and has been a central factor in the EU's proactive role in those negotiations. In the absence of Germany's commitment to reduce its GHG emissions, the credibility of EU efforts to push for more ambitious targets within the Kyoto Protocol – and the importance of national measures in meeting them – would have been much diminished. At the same time, the inclusion of flexible mechanisms in the Kyoto Protocol and creation of an emissions trading regime for the EU, in conjunction with concerns about maintaining competitiveness in a world of globalizing economic relations, have compelled German stakeholders to give such market-based instruments greater salience in ongoing deliberations over climate protection policy.

CONCLUSION

Although elements of Germany's GCC policy have not gone unchallenged, the general approach has been and continues to be one based on consensus. No major interest group or political party appears inclined to challenge the view that the GCC threat is real, that political action is required to redress it, and that this action should be undertaken in conjunction with the obligations assumed by Germany and the EU within the framework of the Kyoto Protocol. Moreover, there has been a remarkable consensus on policies to address the threat: the center-left government continued and elaborated on the voluntary agreements and measures to encourage the development of renewable energy introduced by a center-right government; emissions trading, though initially opposed by environmental groups and the Greens, is now accepted by virtually all stakeholders; initial proposals for eco-taxes came to enjoy support within the major political parties by the mid-1990s, though the ecological tax reform adopted by the red-green coalition subsequently came under criticism from opposition parties. Such criticism notwithstanding, the eco-tax will likely be retained regardless of the parties in power, since the reductions in non-wage costs to employers (and hence, the positive impact on employment) would have to come from other sources. In addition, given the generous credits granted to industry in the emissions-trading system, the remaining reductions needed to meet the Kyoto target will, of necessity, have to come from the private household and transport sectors, which implies a continued reliance on some form of eco-tax. The September 2005 federal elections in Germany seemed to confirm this continuity in German climate change policy.

The federal elections resulted in the formation of a Grand Coalition composed of CDU/CSU-SPD, and headed by a CDU chancellor. Despite calls for certain 'course corrections' in energy policy (primarily some reduction in subsidies for renewables) by the CDU/CSU during the electoral campaign, the central elements of Germany's climate change policy remained in place. According to the government's coalition agreement, the production goals established for renewable energy by the previous government are to be retained, along with the basic structure of the Renewable Energy Sources Act. The eco-tax continues to operate, emissions trading is to be further refined in preparation for the second allocation period, and increased emphasis will be given to energy efficiency.

In sum, Germany has been at the forefront of efforts to meet the challenges of climate change. Domestically, its program to reduce GHG emissions has been among the most ambitious in the world. This program has been driven, for the most part, by forces within Germany: societal and state actors, notably institutions that have evolved historically (e.g., a consensual style of national policymaking, electoral arrangements and the network of party politics). Thus the German experience shows how approaches to environmental foreign policy – and international relations generally – that focus on the role of domestic politics direct our attention to the forces and actors that can be central to shaping a state's GCC policy (see Barkdull and

Harris 2002: 74–84). It also demonstrates how domestically derived foreign policy can influence regional and global policy. Indeed, Germany has been a central actor in shaping Europe's approach to climate change and a key to the successful implementation of the EU's international commitments. As indicated by a German proposal to reduce GHG emissions by 40 per cent for a second commitment period of the Kyoto Protocol ending in 2020 – provided the EU commits to a 30 per cent reduction target (BMU-Reden 2003) – Germany is clearly determined to remain at the center of regional and international efforts to combat the threat of global climate change.

NOTES

1. For further discussion of the German case, see especially Chapter 14.
2. For a discussion of early developments in the climate change issue, see Hatch (1995: 420–21) and Cavander and Jäger (1993). For detailed accounts of the controversy surrounding nuclear power in Germany, see Hatch (1986, 1991, 1996).
3. In line with its efforts to limit JI and CDM projects to those emphasizing renewable energy and energy efficiency, the German government felt that German firms were especially well placed to earn credits through JI and CDM projects.
4. For some evidence that German emissions cuts may not be so heavily the result of unification, see Chapter 15.

REFERENCES

Barkdull, John and Paul G Harris (2002), 'Environmental change and foreign policy: a survey of theory', *Global Environmental Politics* **22**: 63–91.

Beyme, Klaus von (1985), 'Policy-making in the Federal Republic of Germany: a systematic introduction', in Klaus von Beyme and Manfred G Schmidt (eds), *Policy and Politics in the Federal Republic of Germany*, New York: St Martin's Press.

BMU-Presse (2002), 'Klimaschutziel bleibt erreichbar' ['Climate protection goal remains attainable'], Nr.222/02, (8 September), available at www.bmu.de/presse/2002/pm222. php.

BMU-Reden: Rainer Baake (2002), 'Eröffnung des 13th Internationalen Sonnenforums' ['Opening of the 13th International Solar Forum' (12 September), available at www.bmu.de/reden/rede_baake020912.php.

BMU-Reden: Jürgen Trittin (2003), 'Klimaschutz braucht technische innovation und gesetzliche reformen' ['Climate protection needs technological innovation and legal reforms'] (10 März), available at www.bmu.de/reden/rede_trittin030310.php.

Bundesministerium für Umwelt, Naturschutz und Reaktorsicherheit (BMU) (1994a), *Environmental Policy: Climate Protection in Germany, First Report of the Government of the Federal Republic of Germany Pursuant to the United Nations Framework Convention on Climate Change* September, Bonn: BMU.

Bundesministerium für Umwelt, Naturschutz und Reaktorsicherheit (BMU) (1994b), *Environmental Policy: The Federal Government's Decision of 29 September 1994 on Reducing Emissions of CO_2, and Emissions of other Greenhouse Gases in the Federal Republic of Germany*. November, Bonn: BMU.

Bundesministerium für Umwelt, Naturschutz und Reaktorsicherheit (BMU) (2000), *Act on Granting Priority to Renewable Energy Sources*, Berlin: BMU.

Bundesministerium für Umwelt, Naturschutz und Reaktorsicherheit (BMU) (2000), *Nationales Klimaschutzprogramm: Beschluss der Bundesregierung vom 18 Oktober 2000* [*National Climate Protection Program: Decision of the Federal Government of 18 October 2000*]. Fünfter Bericht der Interministeriellen Arbeitsgruppe 'CO_2-Reduktion', [Fifth Report of the Interministerial Working Group 'CO_2-Reduction'], Berlin: BMU.

Bundesministerium für Umwelt, Naturschutz und Reaktorsicherheit (BMU) (2002), *Bericht über den Stand der Markteinführung und der Kostenentwickelung von Anlagen zur Erzeugung von Strom aus erneuerbaren Energien* [*Report on the Status of the Market Introduction and Cost Development of Installations for the Production of Electricity from Renewable Energy Sources*]. Berlin: BMU.

Buttermann, Hans Georg, Bernhard Hillebrand and Ulrike Lehr (1999), '*CO_2-Emissionen und wirtschaftliche Entwicklung: Monitoring-Bericht 1998*' ['*CO_2 Emissions and Economic Development: Monitoring Report 1998*']. *Untersuchungen des Rheinisch-Westfälischen Instituts für Wirtschaftsforschung* (28) [*Studies of the Rhineland-Westphalia Institute for Economic Research* (28)], Essen: RWI.

Cavander, Jeannine and Jill Jäger (1993), 'The history of Germany's response to climate change', *International Environmental Affairs* **5**(1): 3–18.

Eichheimer, Wolfgang, Ulla Boede, Frank Gagelmann, Eberhard Jochem, Barbara Schlomann, John Cheshire and Hans Joachim Ziesing (2001), 'Greenhouse gas reductions in Germany and the UK – Coincidence or policy-induced? An analysis for international climate policy', *Study on behalf of the German Federal Ministry of the Environment and German Federal Environmental Agency*: Karlsruhe, Brighton, Berlin, (June).

Enquete Commission of the German Bundestag (1989), *Protecting the Earth's Atmosphere: An International Challenge*, Bonn: Deutscher Bundestag.

Enquete Commission of the German Bundestag (1990), *Protecting the Earth: A Status Report with Recommendations for a New Energy Policy/Summary and Recommendations of the Third Report*, Bonn: Deutscher Bundestag.

Enquete Commission of the German Bundestag (1992), *Klimaänderung Gefährdet Globale Entwicklung: Zukunft Sichern, Jetzt Handeln* ['*Climate Change Endangers Global Development: Ensure the Future – Act Now*'], Karlsruhe: CF Müller.

Fischedick, Manfred, Kora Kristof, Stephan Ramesohl, Stefan Thomas (1995), '*Erklärung der Deutschen Wirtschaft*': *Königsweg oder Mogelpackung?*' ['*Declaration of German Industry: Royal Path or Deceptive Packaging?*'], Wuppertal Institut für Klima, Umwelt, Energie Nr 39, July.

Hall, Peter (1986), *Governing the Economy: The Politics of State Intervention in Britain and France*, New York: Oxford University Press.

Hatch, Michael T (1986), *Politics and Nuclear Power: Energy Policy in Western Europe*. Lexington KY: University Press of Kentucky.

Hatch, Michael T (1991), 'Corporatism, Pluralism and Post-Industrial Politics in West Germany'. *West European Politics* **14**. 1: 72–97.

Hatch, Michael T (1995), 'The politics of global warming in Germany', *Environmental Politics* **4**(3): 415–40.

Hatch, Michael T (1996), 'Nuclear power and postindustrial politics in the west', in John Byrne and Steven M Hoffman (eds), *Governing the Atom: The Politics of Risk*, New Brunswick, NJ: Transaction Publishers.

Jänicke, Martin and Helmut Weidner (1997), 'Germany', in Martin Jänicke and Helmut Weidner (eds), *National Environmental Policies: A Comparative Study of Capacity-Building*, Berlin: Springer.

Katzenstein, Peter J (1985), *Small States in World Markets: Industrial Policy in Europe*, Ithaca: Cornell University Press.

Kohlhaas, Michael and Bettina Meyer (2005), 'Ecological tax reform in Germany: economic and political analysis of an evolving policy', in Michael T. Hatch (ed.), *Environmental Policymaking: Assessing the Use of Alternative Policy Instruments*, Albany, NY: State University of New York Press.

Kohlhaas, Michael, Barbara Praetorius and Hans-Joachim Ziesing (1995), 'German industry's voluntary commitment to reduce CO_2 emissions – no substitute for an active policy against climate change', *Economic Bulletin* **32**(5).

Michaelowa, Axel (2003), 'Germany: a pioneer on earthen feet?', *Climate Policy* **3**: 31–43.

Michaelowa, Axel (2004), 'The German wind energy lobby: how to successfully promote costly technology', HWWA Discussion Paper 296. Hamburg: Hamburg Institute of International Economics.

Oberthür, Sebastian and Hermann E Ott (1999), *The Kyoto Protocol: International Policy for the 21st Century*, Berlin: Springer.

Press Statement (2002), 'Federal Government reaches breakthrough in emissions trading', 291/02, Berlin (4 December), available at www.bmu.de/english/news/pressrelease021204.php.

Steinmo, Sven, Kathleen Thelen and Frank Longstreth (eds) (1992), *Structuring Politics: Historical Institutionalism in Comparative Analysis*. New York: Cambridge University Press.

United Nations (UN) (2003), *Compilation and Synthesis of Third National Communications*, FCCC/SBI/2003/7, Bonn: FCCC, (16 May).

von Moltke, Konrad (1991), 'Three reports on German environmental policy', *Environment* **33**(1).

Wurzel, Rüdiger KW (2002), 'The Europeanisation of German environmental policy: from environmental leader to member state under pressure?' FFU-report 09–2002: Forschungstelle für Umweltpolitik, Freie Universität Berlin.

3. The indispensable awkward partner: the United Kingdom in European climate policy

Loren R Cass

INTRODUCTION

The United Kingdom's (UK) role in the development of a common European response to global climate change (GCC) presents a fascinating case study of the ability of individual member states to pursue their national preferences within the context of the broader European environmental foreign policy process. During the 1970s and 1980s, European states and nongovernmental organizations (NGOs) denounced the UK as the 'Dirty Man' of Europe. British governments consistently dragged their feet as their European partners sought to press ahead with international environmental negotiations and the formulation of a common European environmental policy. Though it balked at international environmental commitments, the UK emerged as an overachiever in fulfilling many of those same commitments. It also significantly altered its strategic approach to environmental policy in the 1990s and emerged as a leader in shaping European and international environmental law.

This chapter is structured around an analysis of the UK's role in shaping European foreign policy in the GCC negotiations. Global climate change provides an issue that spans a period of reform in British environmental policy and provides insight into the shifting nature of British environmental foreign policy (EFP) as well as the shifts in its relations with the European Union (EU) on environmental issues. The British, along with the Germans (see Chapter 2), provided the vast majority of CO_2 emissions reductions for the EU as a whole. Without the British reductions, it would have been much more difficult for the EU to pursue an aggressive position on international CO_2 reduction commitments. Successive British governments also strategically positioned themselves between the United States (US) and EU, acting as a bridge between the two sides. As a result, the UK played a critical role in shaping both the European negotiating position as well as functioning as an intermediary between the US and the EU in the negotiations for the Framework Convention on Climate Change (FCCC), the Kyoto Protocol and the Hague/Bonn agreements to implement the Kyoto Protocol.

This chapter is divided into two major sections. The first section outlines the evolution of the British approach to environmental regulation since the 1980s and compares it with the regulatory approach that emerged at the EU level. The conflict between the British and the German/northern European approaches initially compelled the UK to more aggressively obstruct attempts to expand European competence over environmental affairs. The British government chose to use other international institutions to balance the influence of the EU and its member states. Recently, the British government has had more success in shaping European environmental policy and has thus been more willing to work through the EU, rather than around it.

The second section analyzes the evolution of British policy in the case of climate change. This section focuses on three periods of GCC policy. The first, and most important, period (1988–92) traces the evolution of British and European positions in the climate negotiations leading up to the FCCC. This period was characterized by hostility between the UK and the European Commission (Commission) over environmental policy. The Margaret Thatcher and John Major governments attempted to restrict the role of the Commission in the climate negotiations and protect the government's ability to pursue British preferences. The second period (1995–97) entails the intense negotiations leading up to the signing of the Kyoto Protocol. During the Kyoto negotiations, British and European positions aligned more closely, and the UK chose to cooperate more fully with its European partners on international GCC policy. However, it continued to balk at common European policy initiatives to reduce greenhouse gas (GHG) emissions, which it viewed as threatening British economic interests. Finally, the third period focuses on the failure of the Hague negotiations to implement the Kyoto Protocol. During the negotiations, the British position was closely aligned with the EU on most issues. However, the British government clashed with its EU partners over issues of American participation in the climate agreement. The Blair government desperately tried to bridge the gap between the EU and US, but the effort ultimately failed. The British were forced to move forward with the Kyoto Protocol without the US, which reduced the UK's ability to shape the Bonn/Marrakech negotiations to implement the protocol as well as the subsequent negotiations surrounding the post-2012 commitment period.

This case study offers interesting insights into the forces shaping EFP for countries operating within the constraints of a common European environmental foreign policy. Within the typology of EFP theories presented by Barkdull and Harris (2002), this case study provides strong evidence for the role of ideas in national preference formation and the associated links between domestic ideational dialogue and international normative debates (the 'Systemic-Cognition/Ideas' and 'Societal-Cognition/Ideas' boxes in the Barkdull and Harris matrix; see Chapter 1: Figure 1.1). British environmental policy was heavily influenced by domestic political norms that have often been at odds with continental norms. These norms are viewed as legitimate by domestic political actors and limit the range of international policy outcomes acceptable domestically. They thus constrain the international negotiating position of the government and create pressure for the government to internationalize domestic policy norms to limit adjustment costs. The normative debates surrounding the political response to GCC thus occurred at three levels – the domestic, the EU and the international – with important links among the three levels.

This case study also offers evidence of the important role played by the state in shaping the British response to GCC (the 'State-centric/Interests' box in the Barkdull and Harris matrix; see Chapter 1: Figure 1.1). Within the normative parameters surrounding GCC policy, successive British governments strategically maneuvered in what could be described as a 'three-level game' to pursue what were perceived to be British national interests related to GCC policy. The government linked GCC policy to other domestic and European political initiatives and manipulated the negotiations to achieve important political objectives. The analysis of the interaction between British and European positions within the climate negotiations suggests that EU member states may possess significant room to maneuver within the common European position. Successive British governments supported European initiatives when they served their interests, but they were also willing to use the US and other international institutions to constrain European initiatives that ran counter to Britain's broader interests. The UK played an indispensable role within European GCC policy by providing the sizable emissions reductions that allowed the EU to seize an international leadership position, but British governments also used this leverage and their relationship with the US to constrain European foreign policy. Overall, this case study suggests an important role for the state in shaping EFP, but it also suggests a significant role for ideational variables in establishing the parameters within which domestic and environmental foreign policy must be pursued.

BRITISH ENVIRONMENTAL POLICY AND CONFLICTS WITH EUROPE

Each state possesses a unique approach to state sponsored environmental action. This approach is a product of the institutional framework of the state, regulatory style, cultural differences and historically contingent choices made in both the domestic and international policy processes (Boehmer-Christiansen and Skea 1991; Knill 2001; Münch et al. 2001; Spretnak et al. 1984; Vogel 1986). The state's approach to environmental action is not static; over time, it will change incrementally, and in times of crisis it may change very rapidly. Once regulations are enacted, the various domestic actors respond to the incentives and strategic opportunities created by those regulations. Any international proposal that threatens the domestic status quo must overcome the inertia of established domestic policies or must be made consistent with existing domestic policies. The effect is that previous policy decisions constrain the range of acceptable international policy initiatives.

The UK has a long history of environmental regulation. The British people have in general experienced improvement in environmental quality over time. The UK is blessed with fast flowing rivers and strong winds that rapidly move pollutants away from the British islands. The British approach to environmental regulation has thus been much more relaxed than either German or American environmental policy. Richard Macrory (1989: 31) captures the essence of British environmental policy: 'Discretion and practicability are the key attributes that over a lengthy period of time have characterized the design and application of much of British environmental law and policy, with a degree of parochialism and administrative complacency running a close second'. Before 1995, Parliament largely confined its efforts to qualitative measures and avoided the designation of environmental goals and national emissions or quality standards. It delegated authority to the bureaucracy to set goals and implement appropriate policies. Bureaucrats worked closely on a case-by-case basis with polluters to find solutions to specific environmental problems. The focus of environmental policy has been on the effects of pollutants on human health and the environment. The mere release of a pollutant is not necessarily perceived to be a problem; rather, the focus of regulation is minimizing the negative externalities of pollutants. The Royal Commission on Environmental Pollution attributes the success of British environmental policy to '[a] policy of persuasion and cooperation with industry … an aggressive policy of confrontation, involving prosecution for every lapse, would destroy the basis of cooperation' (Macrory 1989: 40).

British environmental policymakers are very conscious of the costs of regulation to industry. The government has thus been loath to establish mandatory technology standards for industry. It has preferred to look at the most appropriate environmental controls in each circumstance. The result is that British policy has tended to have very few technology requirements or emissions standards. The 1990 Environmental Protection Act acknowledged two primary principles for pollution control. First, firms were to use the 'Best Available Technology Not Entailing Excessive Costs'. Business and regulators negotiate mutually acceptable pollution control procedures by balancing the demands of competitiveness and pollution control. Secondly, the discharge of pollutants – whether solid, gaseous or liquid – should be controlled by the 'Best Practicable Environmental Option' (BPEO). The BPEO required regulators and regulated companies to compare potential approaches to pollution control on a case-by-case basis to find the best fit between environmental protection and the capabilities of the particular industry and firm.

The 1995 Environment Act and the 1997 National Air Quality Strategy created a system of ambient air quality standards intended to protect human health throughout the country. This represents a departure from earlier environmental control efforts. There has been greater centralization and standardization, but British environmental policy continues to be characterized by substantial flexibility in its implementation. Cooperation between the government and industry is the essence of British environmental regulation. It has also been the foundation of British EFP. The British government has traditionally refused to sign international environmental treaties until domestic implementation is determined to be feasible. This often involves a long process of domestic consultation with domestic polluters and other affected interests (Boehmer-Christiansen 2000: 288). The British negotiating style often produces delays and ambiguity until the domestic position is established. Once the consultation process is completed, the British government tends to be a tenacious defender of domestic interests.

The British approach to environmental regulation differs radically from the regulatory approaches of most of its European partners and particularly the German approach. German environmental policy was negligible from the end of World War II through to the 1960s. The public largely accepted that a certain level of pollution was necessary to regain pre-World War II standards of living. Serious attempts at environmental regulation did not begin in earnest until scientists publicized threats to German forests from acid rain. Modern German environmental regulation thus developed under a threat to the natural environment rather than primarily because of public health concerns (Pehle 1997: 165; Boehmer-Christiansen and Skea 1991, 61).

This fact significantly shaped modern German environmental regulation, and the German approach to environmental policy significantly influenced early European initiatives on the environment.

Two primary principles provide the foundation for German environmental policy: precaution and cooperation. The commitment to prevent pollution from becoming a health or environmental hazard exemplifies the precautionary principle. The focus is not on the effects of pollutants as in the British case, but rather the mere release of pollutants is something to be avoided. It is thus not necessary to demonstrate an environmental harm to trigger regulation.

The German government views environmental protection as a process of developing the technologies to prevent the release of pollutants. This technological approach focuses on automatically improving environmental quality through the adoption of ever-improving technology standards (Stark 2001: 114). Furthermore, the principle of cooperation is vital to the functioning of the regulatory system. Government, industry and labor play the primary roles in negotiating environmental and technology standards. However, unlike the British case, the negotiations take place to establish common environmental and technology standards that apply to all economic actors. There is no negotiation on a case-by-case basis as in the British case, and Germany requires very strict technology standards that are to be applied uniformly.

Thus, the German focus on uniform emission and technology standards conflicts with the British emphasis on environmental quality and flexibility in regulation. This conflict in philosophical approaches to environmental policy extends to the European level. Germany undertook dramatic reforms of domestic environmental policies in the late 1970s and 1980s. The German government sought to extend this reform process to the European level by seeking to export its policies to minimize the effects on competitiveness. The UK has been reluctant to accept binding emission standards and mandatory environmental control technologies. This approach would undermine the flexibility and cooperative nature of British environmental policy. The German government, on the other hand, has been reluctant to accept uniform quality standards as the foundation for environmental policy because the density of industrial development and Germany's location in the heart of Europe would make it extremely costly to achieve such standards throughout Germany. Successive British governments were willing to use the threat of a veto to block European initiatives or to force revisions to make the directives consistent with the British regulatory style (Golub 1996: 706–10).

During the 1980s, as the European Commission and environmentally progressive member states sought to expand European competence over environmental regulation, the UK's domestic regulatory culture and approach to environmental policy conflicted with the dominant EU approach that was largely driven by Germany (Golub 1996; Rose 1990). The conflict over both the content and philosophy of environmental regulation isolated the British on many issues. It also produced expensive domestic adjustments – both politically and economically. The government created numerous political conflicts through its threatened vetoes of proposed environmental directives such as the regulation of the disposal of waste at sea and pollution from paper pulp mills. In cases such as the Large Combustion Plant Directive (LCPD), the British government held out – against concerted political pressure – to achieve derogations from aspects of the LCPD that violated its domestic regulatory norms (Boehmer-Christiansen 2000: 287–88). However, following the 1990 Environmental Protection Act, the Conservative government altered its strategy and aggressively worked through the Commission to reshape the European approach to environmental policy (Héritier et al 1996: 229–32).

The British government proved to be a tenacious defender of the interests of British industry, and it has been reluctant to pursue environmental policy changes without securing the support of industry. The British EFP positions reflect the prominent role of consultation with affected interests in the domestic policy process. To defend British interests, successive governments have pursued a number of strategies that have profoundly affected the content of the European position as well as the ability of the European Commission to negotiate on behalf of the member states. British governments have strategically pursued international environmental negotiations through multilateral institutions to preserve British sovereignty and create a counter balance to internal European efforts to regulate environmental problems. As part of this strategy, the UK has utilized its 'special relationship' with the US to both provide it greater influence within European deliberations and also to counter European policy initiatives. The result has been that the UK has had a greater influence over European EFP than its economic and political position within the EU might suggest.

THE UNITED KINGDOM AND GLOBAL CLIMATE CHANGE

The UK's experience with international environmental policy significantly shaped its position on GCC. Internationally, the UK had faced difficult pressures throughout the 1980s regarding its environmental policies. It was pilloried over its role in the acid rain problems of Europe. The European Community (EC) pressured the UK to improve its environmental practices and imposed significant domestic adjustment costs through the environmental directives flowing out of Brussels. Climate change appeared to be the next major environmental problem to reach the diplomatic agenda. By assuming an early leadership role, Prime Minister Margaret Thatcher's government could significantly influence the development of the international response and divert climate change away from the EC and into a broader international forum.

The Thatcher government sought to minimize the costs of GCC policy by pursuing multilateral negotiations outside of the EC. It positioned itself as a broker between the US and Europe, which strengthened its influence in the negotiations. The reluctance of the Ronald Reagan and George HW Bush administrations to address GCC would also act as a check on European GCC policy. Climate change had the added advantage of being linked to other Conservative Party policy goals, including the Thatcher's promotion of nuclear power and her pressure on the coal-miners unions. The government also sought to create the foundation for reducing GHG emissions in the future.

In short, the Thatcher government became a strong advocate for international negotiations on GCC, but it aggressively sought to control the issue's evolution. Climate change was a new and 'sexy' issue. It provided the opportunity for the government to assume a high profile leadership position that could help to improve its green credentials both at home and internationally. It also had very low short-term costs. The GCC negotiations were likely to be a long process. Short-term changes in government policy would be unnecessary.

Early Climate Initiatives

The Thatcher government faced its first strategic challenge to its GCC policy when France, the Netherlands and Norway called a summit in March 1989 to prepare for a convention on the atmosphere. The Hague Summit on the Atmosphere was intended to create a new world environment organization to be backed by the International Court of Justice. The US and the Soviet Union (USSR) were not invited to the meeting to minimize opposition to the proposal. The Thatcher government rejected the initiative and forced the EC to withdraw from the meeting. The prime minister refused to attend – even though 17 heads of government participated (Bodansky 1993: 466). The government emphasized that it would not accept a body with supranational powers that could supersede national sovereignty on environmental issues (Boehmer-Christiansen 1995: 12). The Thatcher government repeatedly rejected proposals for an 'international ecology authority' and instead was able to guide the conference toward an affirmation of the existing UN structures.

In April 1989, the British Foreign Office began actively promoting the negotiation of a climate convention. At a May 1989 meeting of the UN General Assembly, the British permanent representative to the UN, Sir Crispin Tickell, called for the negotiation of a framework convention on climate change laying out principles for 'good climatic behavior' (IER 1989a: 281). He argued that the convention should be completed as soon as possible with protocols to be negotiated later 'as scientific evidence requires and permits' (ibid.). He then called for a strengthening of the UN Environment Program (UNEP) and suggested that it be made a specialized agency with improved funding. He contended that climate policy had to be handled at a high level of diplomatic involvement. The British proposal put the UK at the forefront of GCC policy and established its preference for negotiations to be conducted in an intergovernmental forum. In addition, because it was proposing only a framework convention, no immediate policy changes were necessary.

The UK government followed the UN initiative with a similar proposal before the governing council of UNEP in May 1989, which called for the development of a framework convention on climate change to be negotiated within UNEP. Protocols would be developed after completion of the convention. While the UK sought to advance GCC policy, it also managed the negotiating process to mollify its American partner. At the Nairobi meeting, a group of countries, including West Germany, called for an immediate start to negotiations. The British delegation played an important role in drafting a compromize on the issue that delayed the start of negotiations until after the IPCC produced its report (IER 1989c: 279). The UK action illustrated the limits to its support for action on climate change. Encouraging early negotiations would have damaged relations with the US and would concede the initiative to more aggressive supporters of international GCC policy.

The one forum in which the UK was not actively pressing for action on GCC was the EC. The Thatcher government was determined to keep climate change within an international forum and under domestic policy control. In June 1989, the EC Environment Council committed member states to individual and collective action to combat climate change. Germany and the Netherlands sponsored the initiative (Dickson 1989: 24). However, the EC ministers were unable to agree on any specific policies.

The spring of 1989 was a period of intense environmental interest in the UK. The prospect of the June 1989 European elections and the strength of the British Green Party in contesting those elections created further pressure for the Conservative Party to improve its environmental image. The Green Party did not win any seats, but it succeeded in polling nearly 15 per cent of the vote. The Conservative Party took a lesson from the Green Party's success and moved environmental policy much more to the forefront of priorities. Chris Patten became Environment Minister in 1989 and promised a vigorous British role in immediately pushing forward a GCC convention with 'binding agreements on subjects like energy efficiency and forestry, as the science and political will come into balance' (IER 1989b).

Growing international support for action to address climate change created a difficult dilemma for the Thatcher government. At the November 1989 Noordwijk ministerial conference on climate change, several European countries proposed binding commitments to reduce emissions of CO_2 and to limit the burning of fossil fuels. The draft declaration called for signatory countries to commit themselves to the stabilization of CO_2 emissions by the year 2000. It also called for research into the possibility of reducing emissions by 20 per cent by the year 2005 (Neal and Wastell 1989: 4). The UK was reluctant to go along with specific targets – both because it was wary of the costs to its economy and because of opposition from the US, but it also did not wish to be perceived as undermining GCC policy. The junior environment minister, David Trippier, argued that any specific agreement would be 'premature':

> 'All countries should commit themselves to the work of the IPCC and not allow that work to be diverted at this stage by arbitrary and unsubstantiated targets. In any case, we have no basis for setting a target of 20 percent reductions when so many other figures have been cited elsewhere'. (Neal and Wastell 1989: 4)

The British delegation instead called for all nations to support a framework convention with specific goals to be negotiated after the Intergovernmental Panel on Climate Change (IPCC) provided a scientific basis for action.

The Noordwijk conference put the Thatcher government in a very awkward position. Throughout the summer, it had hailed its position as a leader in promoting GCC policy, yet when pressed to commit to a precise emissions reduction target, it faced several difficult tradeoffs. The US was unwilling to commit to any specific policies, but the UK did not want to see the US excluded from the process. The British government also did not have any specific policies prepared to meet such a commitment. In fact, accepting an emissions reduction commitment could undermine the government's attempts to privatize the electricity industry. On the other hand, the government could not be seen to be obstructing progress and thus undermine its domestic and international reputation on GCC policy. British negotiators offered a compromize. They proposed that the parties commit to stabilizing CO_2 emissions by the year 2000 with the levels of stabilization left undefined (Raun 1989: 2). However, even this compromize was unacceptable to the US, USSR and Japan, and the final statement only asserted that 'in the view of many industrialized nations ... stabilization of [CO_2] emissions should be achieved as a first step at the latest by the year 2000' (Abramson 1989: 1). The other states in attendance committed themselves to a non-binding goal of stabilizing CO_2 emissions by the year 2000.

The UK continued to balk at European GCC initiatives. At an EC Environment Council meeting in March 1990, the UK agreed that the Commission could co-ordinate the EC negotiating position, but the content of the EC position was more controversial. The Commission proposed that the EC countries agree to an emissions freeze by 2000 and significant cuts by 2010. The Netherlands, Denmark and Germany criticized the Commission proposal as too weak (*Europe Environment* 1990: 7). British Environment Minister Patten rejected the emissions targets and called for stricter emissions limits on trucks and deregulation of the transportation and energy sectors – reforms that the British were already pursuing at the EC level. He argued that action on these issues would go a long way toward reducing carbon emissions and would reduce inefficiencies in the EU marketplace. He asserted that the British proposals could cut CO_2 emissions by 100 million tons a year and would be more effective than 'simplistic political formulas' (Hunt 1990a: 4). The Thatcher government attempted to link GCC policy to its broader domestic and EU policy agenda, which effectively blocked undesirable EC commitments.

The Conservative government consistently argued that it would act when the IPCC had finished its work. The government had little choice but to accept the group's findings as the basis for future action. In anticipation of the release of the IPCC report and its conclusion that GCC was indeed occurring and was likely linked to anthropogenic emissions of GHGs, the prime minister announced that the UK would conditionally agree to stabilize its CO_2 emissions at 1990 levels by the year 2005: 'Provided others are ready to take their full share, Britain is prepared to set itself the demanding target of a reduction of up to 30 per cent in presently projected levels of carbon dioxide emissions by the year 2005. This would mean returning emissions to their 1990 levels by that date' (Brown and Williams 1990: 1). This was 5 years later than the other European countries had agreed and according to the prime minister, would only occur if other countries (presumably the US) acted as well.

The advantage of 2005 for the British government was that most of the actions necessary to bring the UK in line with the EC's LCPD would take place between 1990 and 2003, with the majority occurring after 1998. Though directed at sulfur dioxide and nitrous oxide (NO_X) emissions, the LCPD created incentives for British electricity generators to switch from coal to natural gas to avoid expensive retrofitting of existing plants with flue-gas desulfurization. Natural gas plants were also cheaper to build and run.

As older coal-burning plants were closed, and newer, cleaner natural gas-burning plants came on line, British emissions of CO_2 would naturally fall. An earlier target would have required a renegotiation of the Environmental Protection Act of 1990 and would likely have created greater pressure for reductions in transport emissions, which the Thatcher government hoped to avoid before the next election cycle.

The UK's new commitment to stabilize CO_2 emissions by the year 2005 clashed directly with EC commitments. In September 1990, the EC Environment Council debated a declaration of intent to stabilize emissions by the year 2000. Prior to the meeting, Junior Environment Minister Trippier called attempts to act by the year 2000 'ludicrous' and said that Britain would not budge from its target (Hunt 1990b: 13). At the time, it appeared the UK was willing to block the entire package. In October 1990, the European Environment and Energy Councils, meeting jointly, were able to agree that the EC as a whole would stabilize CO_2 emissions at 1990 levels by the year 2000. However, the declaration also stated that not all member countries were willing to meet that target. British negotiators succeeded in maintaining the UK's 2005 target date at the cost of weakening the common European negotiating position.

Following Margaret Thatcher's 1990 resignation, the new John Major government attempted to capitalize on GCC to further its policy agenda. It tried to link the promotion of nuclear energy to its GHG reduction commitments. The government also desperately needed to increase government revenue to reduce the growing budget deficit. Climate change provided a justification for substantially increasing gasoline and home heating fuel taxes. International leadership to address GCC would also enhance the government's green credentials. However, there were significant limits to the policies that it was willing to accept to achieve GHG emissions reductions. It wanted to protect domestic industry from European efforts to cut carbon emissions. It also continued to object to the expansion of European control over taxation through the proposed European carbon tax. The government faced a growing tension between its desire to address GCC and the potential adverse effects of reducing CO_2 emissions on domestic economic interests. It wanted to maintain control of GCC policy at an intergovernmental level and to deny the expansion of European policy competence. To achieve these objectives it was imperative that the US be a party to any GCC treaty. Achieving these goals was complicated by a more activist EC policy, a consensus among its major European partners on the need for concrete international goals and timetables, and a hardened American position against these very same targets and timetables. The result was a vast chasm between the American and European positions.

The FCCC Negotiations

British negotiators aggressively worked to shape the negotiating process from the very beginning of the FCCC negotiations in 1991. At the first meeting of the Intergovernmental Negotiating Committee on Climate Change (INC), the UK (as co-coordinator of the IPCC's Legal Measures Group) presented a possible format for the climate convention. The proposal was based on the Vienna Convention on ozone depletion and envisioned no concrete commitments or policy actions. It anticipated a framework convention with protocols to be negotiated later (see Chapter 1). The convention was acceptable to the US and provided limited justification for common action at the EC level. However, it was unacceptable to its European partners, who demanded the inclusion of concrete emissions reduction targets and timetables either in the treaty or in a protocol to be adopted simultaneously. It was apparent that finding a compromize acceptable to the US and the EC would be a daunting task.

On 8 June 1991, Environment Secretary Michael Heseltine traveled to the US to propose a compromize plan for establishing firm targets and timetables for cutting GHG emissions. The government proposed a 'phased, comprehensive' program for reducing GHG emissions. Under the proposal, countries would be allowed to make 'national commitments' to cut a range of GHGs, including methane and chlorofluorocarbons, as the US had proposed. The plan had the potential for securing American participation in a climate convention. It could also force the EC to alter its negotiating position and obviate the need for common European polices to reduce GHG emissions (Hunt 1991: 18). The proposal went a long way toward meeting the American demands. However, the Bush White House was unwilling to budge from its rejection of concrete limits on emissions – including commitments that could be easily met under the British plan.

The British government would subsequently try again to appease the Americans with a proposal to create a system of 'pledge and review' (IER 1991: 380). Under the system, states would make individual commitments to cut GHG emissions. An independent body would evaluate the pledges to assure compliance. The convention itself would be a framework with no specific targets or timetables. The proposal was a further concession to the US, and it had the added advantage of allowing the UK to maintain its 2005 date for stabilization. Environmental NGOs derided the proposal as 'hedge and retreat' (IER 1991: 380). The EC and its member states also rejected it, and the British government was forced to disavow it.

Conflicts within the John Major cabinet over the British emissions reduction target and energy efficiency proposals, as well as the distraction of the April 1992 elections, undermined the government's negotiating position.

However, a post-election cabinet shake-up facilitated a shift in the British negotiating stance. Following a 30 April Cabinet meeting, Prime Minister Major announced that the UK would move its target date for stabilization of carbon emissions from 2005 to 2000, in line with the rest of the EC. He reiterated that the target remained contingent upon other countries (i.e., the US) accepting a similar target.

The revized commitment played well to the public, but new emissions data suggested that the UK would likely meet the 2000 target without significant additional policy actions (Schoon 1991: 8). British rhetoric also belied the simultaneous efforts of British diplomats to bridge the gap between the US and the EC. In those discussions, the UK made it clear that it was willing to accept an agreement without binding targets or timetables. During late April 1992, Environment Secretary Michael Howard went to the US and participated in negotiations with State Department officials that led to the language that the INC chairman eventually offered at the final INC meeting in New York. The British helped prepare an agreement without targets or timetables – despite the recent acceptance of the 2000 target date. The Major government thus played a critical role in negotiating the final content of the FCCC. The government's efforts secured American acceptance of the FCCC, but, in doing so, it abandoned the binding targets and timetables demanded by its European partners. The British and Americans presented the Europeans with a fait accompli that fell far short of the European position on GCC.

The British government effectively shaped the negotiation of the FCCC to fit its interests. It capitalized on an activist GCC policy to deflect criticism of its environmental policies. This stance was useful for domestic political purposes but became increasingly untenable as international momentum grew for concrete action. Rather than maintain its leadership role, the UK stepped back and aligned itself more closely with the US. In finally adopting a specific target for CO_2 emissions reduction, it embraced a commitment that would likely be met as a result of policies already in place to address other problems. The British government effectively stymied EC attempts to capture GCC policy and successfully maintained the climate negotiations in a multilateral setting where the UK would not be isolated and where the US would thwart costly initiatives.

The Kyoto Negotiations

As the climate negotiations intensified leading up to the 1997 Kyoto conference of the parties to the FCCC, the UK was well placed to meet its international commitments. British CO_2 emissions were falling more rapidly than predicted as a result of an accelerated shift from coal to natural gas for energy production. The British government had ample room to meet the 2000 stabilization target as well as accept a more aggressive emissions reduction target for the post-2000 period. The fact that additional domestic measures would not be necessary to meet international commitments minimized domestic opposition to GCC policy. Industrial interests supported the government's policy and undertook voluntary commitments to cut CO_2 emissions. The UK's ability to fulfill its commitment with existing domestic policies also strengthened its position vis-à-vis its European partners. The government effectively argued that it did not need to participate in common European policies to reduce GHG emissions because it would already fulfill its commitments. It was thus more willing to support a common European position in the negotiations without worrying about costly European policy measures.

The Major government actively worked to secure a leadership position in the negotiations, but it also wanted to secure American participation in a new protocol. To accommodate American concerns with binding emissions reduction commitments, British negotiators proposed a system of 'hard commitments to achieve soft targets'. The proposal would allow countries to commit to a set of domestic policies to reduce GHG emissions. As long as the country fulfilled its domestic policy commitments, it would be considered in compliance with its international commitments – even if it missed its emissions reduction target because of unexpected levels of economic growth or other shocks (ENB 1995: 5). The proposal had the potential to be acceptable to the US, because it would not bind it to specific emissions reductions; rather, the US would be bound to implement its domestic policy plan. It had the additional advantage of insulating the British government from potential shocks that could force it to miss its target as well as obviating the need for common European emissions reduction policies. The EU rejected the British proposal and demanded concrete emissions reduction commitments.

At the EU level, GCC continued to be a divisive issue. Among the member states, there was broad support for an aggressive international negotiating position, but when discussions turned to EU policies, the consensus broke down. The member states split into two primary groups. The most environmentally proactive members grew increasingly frustrated with the slow pace of action. Sweden, Finland, the Netherlands, Germany, Austria and Denmark sought to press forward more aggressively and emphasized the need for a carbon/energy tax among other common policies and measures. The UK, with frequent support from Spain, France and Italy, was skeptical of EU regulation of GHG emissions. The European Commission, on the other hand, continued its aggressive approach to GCC policy. It maintained pressure for a hard-line negotiating position internationally and put forward a number of policies to reduce European GHG emissions.

The EU debate centered on two primary sets of issues. The most important was how to define the EU's GHG reduction commitment. The commitment posed a number of difficult questions. How large should it be? Which GHGs should be included? What was the appropriate timeframe for the reductions? Secondly, once the member states agreed on an EU commitment, what common policies would be required to achieve it. The member states accepted that every state could not adopt the same emissions reduction target. Variations in levels of development and other domestic circumstances made such an approach impossible, thus making 'burden sharing' critical to the European negotiating positions. Though there was general acceptance of the principle of burden sharing, there was little consensus on what it would mean. The problem for the member states was that only Austria, Germany and the UK had achieved CO_2 emissions reductions by 1996. The rest of the member states were unlikely to meet the 2000 stabilization target. The British reductions were critical to the member states' ability to meet their common target. The UK was in a strong position to influence the European negotiating position because the other member states would need to rely on the British contributions to the burden sharing agreement.

As the Kyoto negotiations approached, the Council agreed to a 15 per cent GHG emissions reduction target by 2010. This target was unrealistic. First, the European burden sharing agreement would only provide for a 10 per cent joint emissions reduction (IER 1997a: 205). Additionally, the US and Japan were unwilling to accept such a target. The US was reluctant to commit to anything beyond a stabilization of emissions at 1990 levels. The environment ministers also made it clear that the EU reductions were contingent upon similar action by the other industrial states. The EU position reflected a desire to put pressure on the US and Japan in the upcoming Ad Hoc Group on the Berlin Mandate meeting. The member states did not have to seriously consider narrowing the 5 per cent gap between member state's commitments and the burden sharing commitment. It was unlikely that the EU would even need to meet the 10 per cent reduction commitment that the member states were offering. The UK went along with the European position and sought to put pressure on the US to accept meaningful emissions reduction commitments.

The Commission anticipated that the new burden sharing accord would create momentum toward new EU policies to reduce GHG emissions. The agreement contained an list of potential common and co-ordinated policies. The list included renewable energy initiatives, combined heat and power, fuel switching, improved energy efficiency standards, improvements in transportation emissions and taxation mechanisms (IER 1997b: 207). The Commission hoped to utilize the agreement as the foundation for additional EU measures, but it failed to enact any meaningful policies to reduce GHG emissions (with the exception of a long-delayed refrigerator efficiency directive) and faced constant obstruction of its policy proposals by a number of member states – often led by the UK. Yet, despite the internal struggles, the EU maintained a forceful position in the GCC negotiations. As individual states, only a minority of governments would have been willing to propose aggressive international GHG reduction targets, but jointly the member states provided a counterweight to the US and its allies in the negotiations. The UK, along with the EU, established a strong position of moral authority in the negotiations – even though there was little substance to the internal European policy. Beyond the moral high ground, the EU had little leverage over the US beyond the ability to veto an agreement, but the threat was of limited use because it would play into the hands of opponents of the climate treaty.

As the final round of the Kyoto negotiations began, the EU and US remained deadlocked and far apart. The US refused to accept significant GHG emissions reduction commitments. It also demanded a number of provisions that the Europeans perceived to be loopholes that would fundamentally undermine the Kyoto Protocol. For example, American negotiators demanded the inclusion of 'flexibility mechanisms' such as emissions trading (ET) and joint implementation (JI). They also called for credits for carbon sequestration through carbon sinks. The US refused to budge from its position, and as one American delegate put it, the EU was 'having more fun being green than in being practical' (ENB 1997: 15).

It was only in the final 2 days that both sides initiated significant compromizes. The parties took all of the issues to the wire and then resolved them in a single package. The final deal that emerged could not have been predicted based on the prior negotiations. The US committed to a much larger emissions reduction target than it had originally offered. It agreed to reduce its emissions by 7 per cent between 2008 and 2012, and the EU member states agreed to reduce their emissions by 8 per cent jointly. This represented a major concession on the part of the US, but, in return, it received significant concessions from its negotiating partners. First, the basket of GHGs was expanded from the EU's proposed three gases (CO_2, methane and NOx) to include three others favored by the US (hydrofluorocarbons, perfluorocarbons and sulfur hexafluoride). The inclusion of all six substances provided additional flexibility in reducing emissions. The second major concession was that the targets were defined as net emissions reductions. The expansion of CO_2 sinks could thus offset CO_2 emission increases. According to the US's 1997 emissions submission to the FCCC secretariat, the inclusion of sinks reduced American emissions by 8.2 per cent (428 million metric tons) in 1995 (Bureau of Oceans and International Environmental Scientific Affairs 1997: 56). This represented substantial savings that could be applied toward the American 7 per cent GHG reduction commitment. The US also received a number of compromizes in terms of the flexibility mechanisms. During the final debate China, India and Indonesia blocked American proposals for ET. In a last minute compromize brokered by the UK, the delegates accepted Article 17, which acknowledged the principle of emissions trading but deferred decisions on its operation to future negotiations (ENB 1997: 12–13). The UK played a much more constructive role within the EU than it had in earlier negotiations. The result was that the UK had a less significant influence on the outcome at Kyoto, but the rapid decline in British GHG emissions meant that any agreement acceptable to the US and the other EU states would be achievable without further domestic policy changes.

The Hague Negotiations

The Kyoto Protocol created a framework for future emissions reductions. However, it left the rules that would govern emissions reductions to future negotiations. What sink enhancement measures would qualify for credits? How much of a state's emissions reduction commitment could be met through the use of flexibility mechanisms? What would happen if a state failed to meet its emissions reduction commitment? These were some of the critical issues that negotiators had deferred to future negotiations. The agreement was unlikely to be ratified by the developed states until all of these issues were resolved. It was initially believed that these unresolved issues could be resolved by the 1998 fourth conference of the parties (COP4) in Buenos Aries. However, it rapidly became apparent that the final agreement would require a package deal on all outstanding issues. The FCCC secretariat sponsored frequent workshops and other meetings to promote progress on the various technical issues, but governments were not eager to compromize on the critical issues until a final deal could be achieved. At COP4 the parties could only agree to complete a final agreement by the 2000 COP6 meeting at The Hague.

The new British government, led by Prime Minister Tony Blair, placed the UK in a renewed leadership position in the post-Kyoto negotiations. The Blair government's commitment to reduce British emissions by 20 per cent was among the most ambitious of the developed states. As COP6 approached, the climate negotiations continued to founder on the same set of priority issues left over from Kyoto: caps on the use of flexibility mechanisms and rules for their implementation as well as accounting rules for sinks. National negotiating strategies undermined the resolution of the critical obstacles to an agreement prior to COP6. All of the delegations took provisions important to other states 'hostage' to use as bargaining chips in the final negotiations (ENB 1998: 11–12).

As The Hague talks moved toward a conclusion, it was apparent that the negotiations were deadlocked. In an attempt to bridge the divide, COP6 President Jan Pronk put forward the outline of a final agreement. The proposal was negatively received by most delegations, and Pronk was roundly criticized for offering the proposal too late in the process and with insufficient detail to permit meaningful negotiations. As the negotiations drew to a close, the British government made one final attempt to bridge the EU–US divide. Following a lengthy telephone discussion between President Clinton and Prime Minister Blair, British Deputy Prime Minister John Prescott put together a draft agreement developed with input from the US and members of the EU (Drozdiak 2000: 1). The agreement would set quantitative limits on the credits that could be claimed from sinks. The limits were lower than the US desired but higher than what the EU had been willing to accept. The two groups were essentially arguing over a few million tons of carbon, a relatively insignificant amount. While this small group of delegates (reportedly composed of officials from France, Germany, Sweden, the UK and the US) was able to reach agreement, the EU delegation as a whole later rejected it after further internal deliberations (Fidler and Houlder 2000: 17).

The rejection led to acrimonious accusations among American, British, French and Commission officials. In a rare public exchange of insults, Prescott betrayed the deep divisions that the final hours of negotiations had created. He blamed French environment minister and head of the EU delegation, Dominque Voynet, for the failure: 'She got cold feet, felt she could not explain it, said she was exhausted and tired and could not understand the detail and then refused to accept it. That is how the deal fell' (AFP 2000). Commenting on the agreement, American negotiator, Frank Loy, asserted that after the negotiations with the EU representatives, 'we physically shook hands. ... I asked, 'Are we now in full agreement; is this a deal?' 'I was pointing to a piece of paper. They said "yes"' (Revkin 2000: 1).

Following the EU's rejection of the agreement, Loy presciently noted, 'I think it is fair to say that that was a pretty important opportunity that was not cashed in on' (Revkin 2000: 1). Commenting on the proposed compromise, Voynet argued that 'this compromise is completely one-sided; it transfers the Protocol on reducing greenhouse gas emissions into a Protocol that, on the contrary, would authorize an increase in emissions by 2010. If we accept it, it will open loopholes that would then be very difficult to close' (*European Report* 2000: 2547).

Participants in the negotiations provided very interesting explanations for the failure of the negotiations. Voynet asserted that the problem was cultural: 'It's extremely difficult to negotiate between groups where political cultures are so different' (Revkin 2000). Voynet asserted that the American insistence on free market principles in environmental control was perceived in France as the 'law of the jungle'. Europeans tended to expect regulatory and fiscal methods to achieve environmental goals. According to Voynet, 'These differences account for why our positions were so radically antagonistic' (Drozdiak 2000: 1). Both the German and French delegations were led by Green Party ministers, which created political obstacles to achieving an agreement with the US. Loy blamed the failure on the domestic political interests of important European actors: 'We showed real willingness to compromise. But, too many of our negotiating partners held fast to positions shaped more by political purity than by practicality, more by dogmatism than pragmatism' (Drozdiak 2000: 1).

The participants' emphasis on cultural divides highlights a critical role that the British government has played in mediating US–EU relations. Successive British governments have placed great emphasis on maintaining strong relations with the US to both protect British interests and enhance British influence. In the Hague negotiations, British and American interests were not closely aligned. However, the Blair government continued to place great emphasis on securing American participation in the final agreement. In similar fashion to the FCCC negotiations, the British and Americans hoped to salvage an agreement and present the EU with another fait accompli. However, the other member states refused to allow the UK to bypass them again. The British government failed to secure US participation in Kyoto, and this would appear to be a significant defeat for British interests. However, American participation became less important as British emissions continued to fall. The Blair government was willing to accept larger emissions reduction commitments than most of its fellow EU member states. It thus had little to fear from the Kyoto agreement and the American exclusion.

CONCLUSION

The case of GCC policy illustrates some of the profound effects that the UK and other individual member states can have on the EU's environmental foreign policy. British environmental policy was heavily conditioned by domestic political norms that have often been at odds with continental environmental policy norms. The conflicts in approaches to environmental policy have often slowed the process of forming a common European position.

The UK tends to emphasize the importance of substantial scientific evidence of negative externalities of pollutants before taking action. The British tradition of assuring that international commitments are achievable before agreeing to them further delays acceptance of international environmental agreements. The British approach often conflicts with the greater emphasis that Germany and other EU members place on the precautionary principle. These states and the European Commission tend to be more willing to take on new commitments in the near-term.

The British government is thus often placed in a position of obstructing common European negotiating positions on environmental issues. This pattern of behavior can be seen in the acid rain, ozone and climate negotiations. Strategically, the British government has been predisposed to promote multilateral forums for environmental negotiations as a means of curbing the influence of the European Commission and the more environmentally proactive member states. In the case of GCC, the UK also used its special relationship with the US to increase its influence in the negotiations, and it used the US as a counter-weight to the EU. The reluctance of the UK to immediately assume new environmental commitments does not reflect an unwillingness to act. In fact, once the British government takes on a new environmental commitment, it has tended to over-achieve in meeting its commitments. Many of the UK's fellow European partners have been much more willing to agree to international commitments but more reluctant to fulfill them.

Conflicts in regulatory style are important variables shaping EFP in Europe. It is imperative to understand Britain's role to explain European EFP on GCC. Important variables in the UK's EFP fall within the 'Societal-cognition/Ideas' and 'Systemic-cognition/Ideas' boxes in the Barkdull and Harris (2002: 69) matrix (see Chapter 1: Fig 1.1), and their importance suggests that scholars need to pay more attention to how variation in national approaches to environmental policy affect foreign policy development within the EU. The member states attempt to shape the EU's EFP to make it consistent with their own national approaches and interests. The internal EU debates typically occur concurrently with international normative debates related to how the international community should respond to the emergent global environmental problem. It is the simultaneity of these processes that creates opportunities for member states to exploit forces at the EU and/or international level to pursue their preferred approach to addressing the environmental problem.

Members of the EU face a complicated mixture of domestic, European and international normative constraints on foreign policy. However, the creation of the normative context is a dynamic process, and there appears to be some latitude for states to shape that context and to strategically utilize European and international negotiations to secure domestic foreign policy interests.

This case study also offers evidence for the important role of the state in shaping the British and EU responses to climate change and EFP more broadly. These forces fall within the 'State-centric/Interests' box in the Barkdull and Harris (2002: 69) matrix (see Chapter 1: Fig 1.1). Successive British governments strategically maneuvered in what could be described as a 'three-level game' to pursue British preferences related to GCC policy. The government linked climate policy to domestic and European political initiatives and manipulated the negotiations to achieve important political objectives. The government variously utilized the EU and the US to pursue its preferences depending on which actor was most closely aligned with British interests.

This analysis of the interaction between British and European positions within the climate negotiations suggests that EU member states may possess significant room to maneuver within the common European position. Overall, this case study demonstrates an important role for the state in shaping EFP, but it also suggests a significant role for ideational variables in establishing the parameters within which domestic and EFP must be pursued. It is this combination of variables that would appear to provide important opportunities for future scholarship on environmental foreign policy.

REFERENCES

Abramson, Rudy (1989), 'US and Japan Block Firm Stand on Global Pollutants', *Los Angeles Times* (8 November): 1.

Barkdull, John and Paul G Harris (2002), Environmental Change and Foreign Policy: A survey of theory, *Global Environmental Politics*, **2**, 2 (May): 63–91.

Bodansky, Daniel M (1993), 'The United Nations Framework Convention on Climate Change: a commentary', *Yale Journal of International Law* **18**: 451–558.

Boehmer-Christiansen, Sonja (1995), 'Britain and the International Panel on Climate Change: the impacts of scientific advice on global warming. Part 1: Integrated Policy Analysis and the Global Dimension', *Environmental Politics* **4**(1) (Spring): 1–18.

Boehmer-Christiansen, Sonja (2000), 'The British case: overcompliance by luck or policy', in Arild Underdal and Kenneth Hanf (eds), *International Environmental Agreements and Domestic Politics: The Case of Acid Rain*, Burlington, VT: Ashgate.

Boehmer-Christiansen, Sonja and Jim Skea (1991), *Acid Politics: Environmental and Energy Policies in Britain and Germany*, New York: Belhaven.

Brown, Paul and Nigel Williams (1990), 'UK: Thatcher's global warming plan criticised by opposition MPs', *Guardian* (26 May): 1.

Bureau of Oceans and International Environmental Scientific Affairs (1997), 'Climate Action Report: 1997 Submission of the United States of America

Under the United Nations Framework Convention on Climate Change', July, Department of State Publication 10496.
Dickson, Tim (1989), 'EC "will take environment action"', *Financial Times* (9 June): 24.
Drozdiak, William (2000), 'Global warming talks collapse; US, EU fail to resolve dispute over curbing emissions', *Washington Post* (26 November): 1.
Earth Negotiations Bulletin (ENB) (1995), 'Report of the Second Session of the Ad Hoc Group on the Berlin Mandate: 30 October – 3 November 1995', *Earth Negotiations Bulletin* **12**(24) (7 November).
Earth Negotiations Bulletin (ENB) (1997), 'Report of the Third Conference of the Parties to the United Nations Framework Convention on Climate Change: 1–11 December 1997', *Earth Negotiations Bulletin* **12**(76) (13 December).
Earth Negotiations Bulletin (ENB) (1998), 'Report of the Meetings of the UNFCCC Subsidiary Bodies: 2–12 June 1998', *Earth Negotiations Bulletin* **12**(86)(15 June).
Europe Environment (1990), 'Environment council: CO_2, car emissions, and CFC debates; nitrates and water proposals blocked', *Europe Environment* **1574** (24 March): 7.
European Report (2000) 'EU/UN: unholy compromise to drag climate talks out of the mire', *European Report* **2547** (25 November), available at www.lexisnexis.com/universe.
Fidler, Stephen and Vanessa Houlder (2000), 'Europeans defend rejection of climate deal', *Financial Times* (28 November): 17.
Golub, Jonathan (1996), 'British sovereignty and the development of EC environmental policy', *Environmental Politics* **5**(4) (Winter): 700–28.
Héritier, Adrienne, Christoph Knill and Susanne Mingers (1996), *Ringing the Changes in Europe: Regulatory Competition and the Transformation of the State. Britain, France, Germany*, New York: Walter de Gruyter.
Hunt, John (1990a), 'Patten proposes EC limitations on emissions of carbon dioxide', *Financial Times* (24 March): 4.
Hunt, John (1990b), 'Opposition to EC on environment', *Financial Times* (21 September): 13.
Hunt, John (1991), 'UK to offer deal over greenhouse gases to US', *Financial Times* (3 June): 18.
International Environment Reporter (IER) (1989a), 'UK representative to United Nations proposes global convention to council', *International Environment Reporter* **12** (14 June): 281.
International Environment Reporter (IER) (1989b), 'Patten promises "white paper" on agenda for remainder of century', *International Environment Reporter* **12** (8 November): 538.
International Environment Reporter (IER) (1989c), 'Delegates to UNEP meeting say treaty on global change should be top priority', *International Environment Reporter* **12** (14 June): 279.
International Environment Reporter (IER) (1991), 'Negotiators of global climate accord identify major issues to be tackled', *International Environment Reporter* **14** (3 July): 360.
International Environment Reporter (IER) (1997a), 'Environment Council reaches mandate for negotiations at climate change meeting', *International Environment Reporter* **20** (5 March): 205.

International Environment Reporter (IER) (1997b), 'Commission hopeful climate change accord will help boost effort for CO_2/energy taxes', *International Environment Reporter* **20** (5 March): 207.

Knill, Christoph (2001), *The Europeanisation of National Administrations: Patterns of Institutional Change and Persistence*, Cambridge: Cambridge University Press.

Macrory, Richard (1989), 'The United Kingdom', in Turner T Smith (ed.), *Understanding US and European Environmental Law: A Practitioner's Guide*, Boston: Graham & Trotman.

Münch, Richard, Christian Lahusen, Markus Kurth, Comelia Borgards, Carsten Stark and Claudia Jauß (eds) (2001), *Democracy at Work: A Comparative Sociology of Environmental Regulation in the United Kingdom, France, Germany and the United States*, New York: Praeger.

Neale, Greg and David Wastell (1989), 'Britain fights Green demands from EEC', *Sunday Telegraph* (5 November): 4.

Pehle, Heinrich (1997), 'Germany: domestic obstacles to an international forerunner', in *European Environmental Policy: The Pioneers*, Mikael, Skou Andersen and Duncan Lifferink (eds), New York: Manchester University Press.

Raun, Laura (1989), 'States to back global warming convention', *Financial Times* (11 November): 2.

Revkin, Andrew C (2000), 'Treaty talks fail to find consensus in global warming', *New York Times* (26 November): 1.

Rose, Chris (1990), *The Dirty Man of Europe: The Great British Pollution Scandal*, London: Simon and Schuster.

Schoon, Nicholas (1991), 'Cabinet row brews over "greenhouse" gas cuts', *The Independent* (7 December): 8.

Spretnak, Charlene, Fritjof Capra and Wulf-Rüdiger Lutz (1984), *Green Politics*, London: Paladin.

Stark, Carsten (2001), 'Germany: rule by virtue of knowledge', in Richard Munch (ed.), *Democracy at Work: A Comparative Sociology of Environmental Regulation in the United Kingdom, France, Germany, and the United States*, London: Praeger.

Vogel, David (1986), *National Styles of Regulation: Environmental Policy in Great Britain and the United States*, Ithaca, NY: Cornell University Press.

4. Middle power leadership in the climate change negotiations: foreign policy of the Netherlands*

Norichika Kanie

INTRODUCTION

This chapter examines the foreign policy of the Netherlands in the context of negotiations on the Kyoto Protocol's Quantified Emission Limitation and Reduction Objectives (QELROs). The Netherlands, despite being only a 'middle power', exerted significant leadership in the process of formulating the European Union's (EU's) proposals for the 1997 Kyoto conference of the parties to the Framework Convention on Climate Change (FCCC). The EU showed leadership in the negotiations on global climate change (GCC), setting an ambitious emissions reduction target for the Kyoto conference. Careful investigation shows that the Netherlands' leadership during this process was a result of its domestic policies, national institutions and a set of ideas that had developed through important research projects. The framework of the EU, as a coalition for the negotiations and as an economic block, worked in the process as a power device between a European middle power and the international GCC power game, making the influence of the Netherlands' foreign policy larger than one would normally expect from a state of its size.

The kind of leadership that was exerted by the Netherlands can be identified as 'leadership through unilateral action' or 'directional leadership', consistent with the way leadership in multilateral negotiations is construed by such scholars as Arild Underdal (1994), Oran Young (1983, 1989), and Joyeeta Gupta and Michael Grubb (2000). Case studies on socio-economic and global environmental regime-building processes have shown that while this type of unilateral leadership occurs less frequently than expected in actual multilateral negotiations, leadership through exemplary policy experimentation and social persuasion is a role to which smaller power countries may occasionally aspire.[1] Equally important for this case was 'instrumental leadership', which builds on the supply of high-quality diplomatic and intellectual resources. An enabling domestic institutional environment that supports participation in external policymaking is also crucial to instrumental leadership.

Domestic policies, institutions and ideas matter in determining foreign policy on global climate change. The case presented in this chapter shows that interest-based and cognition/ideas-based approaches to environmental foreign policy (EFP) (Barkdull and Harris 2002) help explain the process through which GCC policy emerges and is formulated. Once foreign policy is implemented, however, power-based accounts of foreign policy also provide explanations for related international dynamics. Although policies, institutions and ideas still matter in explaining EFP and behavior of a country, the power-based approach is important for explaining the influence of a particular state in the process of building international environmental regimes.

BACKGROUND

A Space for Middle Power Foreign Policy

'Niche diplomacy' is the term coined for understanding the foreign policy and diplomatic behavior of middle powers (Cooper 1997). Although middle powers are not always in a position to control international order (such as the hegemony of big powers), their influence has been more than negligible. Case studies have shown that middle powers' foreign policy and diplomacy can be influential in world politics, and middle powers may aspire to leadership on occasion. Such a 'niche' can be found in issues that require international cooperation, including the global environment.

As Keohane put it, cooperation can emerge and regimes can be maintained even after (or without) hegemony 'not only because the conditions for maintaining existing international regimes are less demanding than those required for creating them', but also because shared interests can lead to the creation of regimes without hegemony (Keohane 1984: 50). International regimes established in the absence of hegemony may be called *negotiated* regimes, in contrast to *imposed* regimes (Young 1983; 1989).[2] *Negotiated* regimes are said to be 'relatively common at the international level', especially after the 1960s (Young 1983).

International regimes of this kind may be recognized as 'intervening variables' that stand between fundamental characteristics of world politics, such as the international distribution of power, on the one hand, and the behavior of state and non-state actors, on the other hand. Thus the relative power relations between states are not necessarily fully reflected in the behavior of membership of a particular international regime. Consequently, some of the non-hegemonic countries, including so-called 'middle power' countries like the Netherlands, have been able to exert disproportionate influence. Studies of small countries, middle powers and 'middlepowermanship' have also identified the bridge-building role of small or middle-power countries such as Norway, Sweden, Canada and the Netherlands in situations of conflict (Cooper et al. 1993; Cooper 1997). Cox has argued that this tendency has been encouraged by the transition from a post-war reconstruction regime to the current form of multilateralism, which increasingly depends upon the principle of power dispersion and social equity (Cox 1989).

There are at least four ways of defining a middle power: (1) a state's position in the international hierarchy – states are ranked from big to small according to criteria such as strength of economy and military capability; (2) location – a middle power derives its status from being a relative power in a region, or from its geographical location between big powers; (3) a normative view – middle powers are potentially wiser or more virtuous than states positioned either 'above' them or 'below' them; (4) diplomatic behavior, which can be termed 'middlepowermanship' – the bridge-building role between two conflicting parties, the breaking role in a negotiation stalemate, the leadership role in international cooperation, and so on. The definition adopted here combines (1) and (4).[3]

Modes of Middle Power Leadership in Terms of Foreign Policy

Within such a maneuvering space, middle powers can even aspire to leadership. Existing theories of leadership use different terms when describing modes of leadership. The terms are not so distinct from one another. Gupta and Grubb (2000) have summarized three attempts at elaborating typologies of multilateral leadership. The first typology came from Young (1991), followed by Underdal (1994) and Malnes (1995). Each of them outlines three modes of leadership: structural, entrepreneurial and intellectual leadership for Young; coercive, instrumental and unilateral leadership for Underdal; and 'stick and carrot', problem-solving and directional leadership for Malnes (see Table 4.1). In summarizing these three attempts, and in order to provide a leadership typology designed particularly for analyzing the issue of climate change, Gupta and Grubb proposed a fourth typology characterized by three modes of leadership: structural, instrumental and directional.

Among modes of leadership, leadership based on unilateral policy action ought to be increasingly important in tackling global issues, given the importance of bottom-up multilateralism in addressing global issues in the current form of multilateralism. Directional leadership, or leadership through unilateral action, operates as a form of social persuasion so that the state demonstrates 'that a certain cure is indeed feasible or does work', or 'set[s] a good example for others to follow' (Underdal 1994: 185). Underdal argues that 'the mechanism of setting an example is advocated by some groups of environmentalists who claim that by unilaterally imposing on one's own society strict standards of pollution control a government may help strengthen public demand in other countries for equally strict measures', and that 'by imposing or threatening to impose unilateral environmental protection measures, a government can strengthen demand within its own society for international regulations' (Underdal 1994: 185).

Table 4.1 Typologies of leadership modes

Young	Underdal	Malnes	Gupta & Grubb	Description
Structural	Coercive	Carrot and Stick	Structural	Use of incentives based on political and economic power
Intellectual	Unilateral	Directional	Directional	Use of ideas and domestic implementation to influence the perception of other countries as to what is desirable and possible
Entrepreneurial	Instrumental	Problem-solving	Instrumental	Crafting of structures and applying diplomatic skill

Source: Gupta and Grubb (2000).

According to Underdal, leadership through unilateral action combined with social persuasion and instrumental leadership can be exerted not only by big powers, but also by small or middle-power countries like the Netherlands. In other words, the size of the power does not matter in exerting domestic resource-based multilateral leadership. In terms of approaches to foreign policy study, this means that interests and cognition/ideas account more for middle power leadership than does a power-based approach to foreign policy, as categorized in Barkdull and Harris (2002; see Chapter 1).

In practice, however, different modes of leadership are often found in some sort of combination, and obviously some modes of leadership are closely related to the (material) power of a country. For example, coercive (or structural) leadership often entails 'stick and carrot' tactics by which the leading country opens the negotiations on the basis of its economic and/or military power. Therefore, it is important to continue to distinguish the modes and the sources of leadership when examining a particular case in practice, particularly in the GCC case examined here, which requires rigid clarification of the sources of leadership. This is because the Netherlands participated in the GCC negotiations leading up to Kyoto as part of the EU coalition. This means that attention should also be paid to the role of the EU as a unified coalition that stands between multilateral negotiation and the Netherlands' foreign and domestic policy.

In analyzing the case of the Netherlands' GCC leadership, I concentrate on the structure of state-society interface in which major actors such as environmental nongovernmental organizations (ENGOs), epistemic communities and business or industry organizations share both domestic policy innovations and the international diffusion of policy. This may reveal additional insights into interests in society and states in understanding EFP (as described in Barkdull and Harris 2002; see Chapter 1). Now I turn to a review of the development of the Netherlands' GCC policy from its inception until 1997. In particular, I focus on changes in the Netherlands' greenhouse gas (GHG) emissions reduction targets. Then I turn to an empirical study of the multilateral negotiation process leading up to the third conference of the FCCC parties (COP3) and analyze the mechanism through which developments in the domestic policies affected the Netherlands' ability to take a leading role in the negotiations.

CLIMATE CHANGE POLICY IN THE NETHERLANDS: EARLY DEVELOPMENTS

Due to mounting concern with the state of the environment, the Netherlands has struggled for years with GCC development and related institution building. Therefore, by the time of the Kyoto Protocol negotiations there was a certain level of consensus among the major stakeholders involved in domestic policy making that it was important to tackle issues related to GCC. It was this domestic consensus, enhanced and made more sophisticated by domestic policy and institutions, which pressed for further development in the processes of reaching international agreement on climate change. It is important to briefly examine Dutch GCC policy, which developed within the wider context of the Netherlands' environmental policy.

Since the 1970s, Dutch environmental policy has moved towards a more comprehensive approach to environmental problems. Discussion of the relationship between the environment and economic growth began with the publication of the Club of Rome report, *Limits to Growth*, in 1972. This report had considerable impact on the population of the Netherlands, which is considered to be relatively environmentally conscious in comparison to people in other countries, due in part to the country's geographic location at the end of Rhine and the fact that approximately one third of its land is below sea level. The Club of Rome report became a best seller in the Netherlands, with roughly one-half of all copies worldwide being sold there (Liefferink 1997: 213). The report argued that if the current pace of industrial development were to continue, the growth of human society would reach the limits of environmental capacity within the next hundred years, and inequity between the rich and the poor would grow. The legitimacy of industrial society and its practices was already a common topic of discussion in the early 1970s, which led, in part, to the government considering development assistance policy to be a high priority. Such was the widespread reaction to the report, in the spring of 1973 Queen Juliana even invited a large number of specialists from the political, industrial and academic scenes to the Royal Palace in Amsterdam to discuss the repercussions of *Limits to Growth* (Hajer 1995: 176).

As a consequence of the rising concern for the environment, the Netherlands made much progress in environmental policy and its institutionalization. The Ministry of Public Health and Environmental Hygiene (*Volksgezondheid en Milieuhygiene*) was established in October 1971. Also, following the Villach Conference in Austria, during which climate scientists called on political movements to prevent climate change and which is widely recognized as the first step toward international politicization of the GCC issue, the term 'climate change' appeared for the first time in Dutch environmental policy in the 1985 'Indicative Multi-year Programmes' report. However, a serious economic recession in the 1980s prevented the country from introducing concrete and comprehensive GCC policy at the time. The Netherlands would have to wait until 1989 for such a policy, when the first National Environmental Policy Plan (NEPP) was introduced.

Beginning in the mid-1980s political concern for the state of the environment grew remarkably within the international community; the Netherlands was no exception. Although environmental politics had become something of a 'non-issue' during the late 1970s and early 1980s, 'a sudden change' in the situation arose in the mid-1980s (Van der Straaten 1992: 49). Several factors contributed to a growing public interest in environmental issues, including major disasters – such as the 1985 Chernobyl explosion, accidents involving oil tankers, salt discharges and the dumping of toxic waste – and the publication of scientific research connecting depletion of the stratospheric ozone layer, deforestation of tropical rain forests, loss of biological diversity and global warming or the 'greenhouse effect' (Kaufmann and Schrijver 1990: 23). Moreover, in 1987 the United Nations'

(UN) World Commission on Environment and Development (WCED) published its report, *Our Common Future*, which had a considerable impact on the Dutch public. As the people of the Netherlands were recovering from an economic recession they were able to turn their attention to issues other than the economy. The Dutch Minister of the Environment, like his counterparts in other countries, found that the WCED report bolstered his position. After its publication, he immediately argued that 'the principle of sustainable development, as defined in that report, should be the starting point of every environmental policy in the Netherlands' (Van der Straaten 1992: 50). Dutch environmental policy attracted international attention because of the speed with which the government, backed by growing public concern, implemented environmental measures.

In December 1988, after only a year's preparation, the National Institute of Public Health and Environmental Protection (*Rijkesinstituut Voor Volksgezondheid en Milieuhygiene* (RIVM)) published the first integrated and comprehensive environmental survey in the Netherlands, *Concern for Tomorrow: A National Environmental Survey 1985–2010,* which was to be a background document for the NEPP. The survey discussed a range of environmental problems on five different levels within the context of sustainable development.[4] The survey considered climate change, setting out four scenarios about future carbon dioxide (CO_2) emissions. *Concern for Tomorrow* showed that the Dutch government took the message sent by *Our Common Future* seriously. The Minister of Housing, Spatial Planning and the Environment (VROM) argued: 'It is my opinion, on the basis of the present report (*Concern for Tomorrow*), that a far-reaching adjustment of our management of the environment may not be delayed, even if we are still faced with numerous uncertainties' (RIVM 1987: VII).

When the results of the scientific survey were turned into policy, however, serious conflict occurred. That is to say, politicization of environmental policy caused conflict in the Netherlands' coalition government. On 2 May 1989, disagreement over the financial resources to be devoted to the NEPP became a factor in the government's collapse. These domestic controversies delayed adoption of the NEPP, but at the same time drew public attention to environmental policy (Wintle 1994). In fact, before the general election, 45 per cent of the Dutch surveyed said that environmental issues were their highest priority (van Holstein and Irwin 1992: 68).

A general election was held on 6 September 1989, and the political parties that advocated stronger environmental policies (D66 and GroenLinks) were the only ones to gain seats compared to the previous election. On 7 November 1989, a center-left coalition emerged. After taking office, the new coalition government announced in the NEPP-Plus on 27 November that environmental policy was to be the third main component of government policy and that certain policy areas required additional action, including the reduction of CO_2 emissions and the saving of energy (NEPP-Plus 1990: 5). The first NEPP to include GCC policy (with the subtitle 'To Choose or To Lose') was presented under the auspices of four ministers, namely the Minister of VROM, the Minister of Economic Affairs, the Minister of Agriculture and Fisheries and the Minister of Transportation and Public Works. Thus, climate change as a global issue was accommodated in Dutch environmental policy for the first time in the NEPP. The NEPP, which initially was controversial and eventually led to a general election and caused the Dutch public to pay attention to the improved environmental policy, provided the necessary momentum to advance policy in this environmental arena.

Recognizing the 'global character of the CO_2 problem' and the 'special responsibilities for the industrialized nations' simultaneously, the NEPP included an action plan for international action along with domestic policy strategies. It explicitly declared its intention to 'undertake international initiatives' to reach an international consensus regarding CO_2 emissions. It then set the target that 'CO_2 emissions in the industrialized nations must be stabilised at their 1989/1990 level in around the year 2000' (Second Chamber, session 1988–89, 21 137, nos1–2: 131). Accordingly, it set a target for the Netherlands: 'for the moment, stabilisation of CO_2 emission in 2000 at their average 1989–1990 level, which is estimated at 183 million tons per year', which may be 'supplemented and modified depending on the results of international consultations' (Second Chamber, session 1988–89, 21 137, nos1–2: 130). Energy conservation was identified as the primary means of achieving the unilaterally set CO_2 reduction target. The Netherlands was only the second country, after Sweden, to establish such a target (IEA 1992).

Domestic consensus on environmental policy, including GCC policy developed due to adoption of the NEPP. This, in turn, led to the Dutch initiative to help set the international GCC agenda. The most important initiative was the Ministerial Conference on Atmospheric Pollution and Climatic Change held in the Netherlands at Noordwijk on 6–7 November 1989, shortly after the general election. The Dutch government hosted this first political conference on GCC at the ministerial level, which reputedly was one of the most important GCC-related at that time (Bodansky 1993: 467; Paterson 1996: 38). In fact, this conference was one of the most important events for agenda-setting on the GCC issue before the UN-initiated negotiations.

Once GCC policy obtained strong public support, it developed alongside the Netherlands' political culture. In the literature this kind of political culture is described as the 'green polder model', where social groups and governmental organizations interact so closely that they can negotiate, find solutions together and compromise to a certain extent on many environmental issues (van Putten 1982; van den Bos 1991; Visser and Hemerijck 1997; Hanf and van de Gronden 1998; de Jongh 1999). A good example of this kind of arrangement may be a negotiated agreement between a government and industry on energy efficiency (see below).

Nevertheless, the policy set out in the NEPP was criticized by environmental groups as too weak. An additional action plan was produced by the 'NEPP-plus' in 1990, 'with a view to the international consultations on this issue', including the views of Intergovernmental Panel on Climate Change (IPCC) (NEPP-plus, 1990: 20) (see Table 4.2). The goal of the CO_2 emissions reduction policy was modified and set 'to stabilize emissions by 1995, and if possible as early as 1994, at the 1989–90 level of 182 million tons' annually (NEPP-plus 1990: 20). Moreover, an absolute reduction target in the year 2000 was set at a level of 173–177 million tons of CO_2 per annum (an absolute reduction of between 3–5 per cent). The targets were calculated against a projected increase in CO_2 emissions in the Netherlands of approximately 2 per cent per year if no measures were introduced. Under that scenario, approximately 195 million tons of emissions in 1994 and 220 million tons in 2000 would have been released. These figures were linked to a forecast of 2.5 per cent annual economic growth. These policy goals, which were based in *Our Common Future* and *Concern for Tomorrow*, were 'internalized' domestically in the areas of fuel use, energy saving, traffic and transport, and waste prevention.

Table 4.2 Changes in the Dutch CO_2 emissions reduction target (1989–97)

	Target (CO_2)	**Year**
NEPP1 (1989)	0%	2000
NEPP+ (1990)	0%	1995
	3–5% reduction	2000
NEPP2 (1993)	3% reduction	2000
	5% reduction (depending on international developments)	
1995	3% reduction	2000
1996	3% reduction (depending on international developments)	2000

Although GCC policy was initiated within the first NEPP, unexpectedly high economic growth made the projected reductions excessively difficult to achieve. Based on the NEPP's action plan, the first Policy Memorandum on Climate Change was sent to the Second Chamber of the Parliament on 4 September 1991. The memorandum set out policy measures to achieve the target. In addition, it summarized research activities on the further reduction of GHGs in order to implement the Toronto and the Noordwijk Declarations on combating climate change (Second Chamber session 1990–91).[5] The target became slightly stricter in NEPP2 in 1993, requiring a 3 per cent reduction in 2000 relative to the 1990 level (182 million tons), leaving open the possibility of a 5 per cent reduction depending on attainment of an international agreement. However, the possibility of increasing the target to 5 per cent disappeared in 1995. A policy document from 15 September 1995, which became known as 'the CO_2 letter', upheld the 3 per cent reduction target with no increase to 5 per cent, 'in view of the laborious progress made in international negotiations, the agreement within the EU to realize stabilization without joint implementation and the difficulty encountered with keeping the 3 per cent reduction target within reach' (VROM 1995). Moreover, the 1996 Second Memorandum on Climate Change stated that domestic stabilization of CO_2 emissions after the 2000 should be at least –3 per cent relative to the 1990 level, provided international circumstances allowed. It also expressed a willingness to make further reductions, depending on international agreement and equitable burden sharing between countries.

Additional reductions of CO_2 emissions would require, for example, economic incentives such as the introduction of a CO_2 tax on industry. However, the unilateral introduction of this type of tax might well have undermined the international competitiveness of Dutch industry. Conditioning further domestic changes upon international agreement represented a sustainable compromize between GCC policy and domestic industry, which was naturally keen to remain internationally competitive.[6] At the same time, the 'international approach' of Dutch GCC policy was strengthened as additional stringent domestic GCC policies were made contingent upon the conclusion of an international agreement. Unilateral policy action was reaching its limit; at this point it became directly connected to the multilateral FCCC protocol-making process, which culminated at the Kyoto conference in 1997.

PREPARATIONS FOR THE KYOTO CONFERENCE OF THE PARTIES

Because the European Community (EC) is a party to the FCCC, most of the negotiations by EU states were undertaken by the so-called troika, comprizing the former, current and next presidents of the Council of Ministers. Although the EU took certain initiatives throughout the FCCC process, the negotiation process itself did not make substantive progress until COP2. Some people doubted whether the protocol would be adopted by COP3 at Kyoto. The EU perceived the protocol negotiation as a great opportunity to take a leadership role in the FCCC process (Huber 1997). In this context, a middle-power country such as the Netherlands, negotiating as part of the EU, potentially obtains greater negotiating power in global multilateral diplomacy than it would alone. Comparing the situation to the 'like-minded coalition' with Sweden, Norway and some other European countries (created in the 1970s in the context of UNCTAD negotiations), the Dutch chief negotiator for the Kyoto COP said: 'The "like-minded" coalition was a more loose arrangement in the 1970s and 1980s than the EU, and certain member states of the EU preferred to work with non-EU member states rather than the members of EU. But, that has changed dramatically. If the Netherlands speaks with a few other countries, people may listen politely. But, if the EU speaks, people listen to it seriously. It is more powerful'.[7]

BEFORE THE DUTCH PRESIDENCY OF THE EUROPEAN COUNCIL

Since the EC became involved in the multilateral GCC negotiations in the early 1990s, the EU member states and the Commission have realized that emissions allocation or burden sharing within the EU is necessary. It was not until the first half of 1997, however, that the EU succeeded in reaching its internal burden-sharing agreement for GHG reductions. The burden-sharing arrangements, particularly in preparation for the COP3 protocol negotiations, were tabled by the Italian presidency during the first half of 1996. The discussion intensified during the second half of 1996 under the Irish presidency. An EU GCC policy workshop called 'Towards a European Consensus' was held on September 2–3 1996, in Dublin. Negotiators from member states participating in the EU Ad Hoc Group on Climate, officials from the Commission's Directorate General XI (DGXI, the EU's environmental directorate), Dutch researchers on GCC policy and professional facilitators attended the workshop (Ringius 1997).

At the workshop DGXI proposed an overall CO_2 reduction target of 10 per cent by 2005 and an internal differentiation scheme among member states. Under the proposal, nine member states would reduce their emissions, two would be allowed to increase them, and the rest would stabilize their emissions. However, most of the member states objected to the proposal and argued that an emissions reduction target of around 5 per cent by 2005 would be more realistic. Although these proposals were rather similar to those agreed in March 1997, the Commission's proposal was thought to lack substantive content. Describing the atmosphere at the Dublin workshop, Ringius stated 'the DGXI proposal set a too ambitious common target, appeared to be ill prepared, and was not persuasively prepared in Dublin. For these reasons, especially the lack of substantive content, the proposal failed to build consensus among the member states. Explained an astute Irish negotiator: "The proposal did not have the necessary intellectual rigor"' (Ringius 1997: 19).

At the European Environment Council meeting, held in Luxembourg on 15 October 1996, most of the time was spent on agendas of the Irish presidency: the fifth action plan for sustainable development, a fuel program for vehicles, and a strategy for water management. Agreement on EU burden sharing had not been reached. The Presidency called on delegations to work 'actively' on the burden-sharing arrangements 'in order to reach conclusions at the Environment Council in December' (*Europe Daily Bulletin* 1996a). The July 1996 Ministerial Declaration of COP2 had already called for proposals from the parties to be submitted by the fifth meeting of the Ad Hoc Group on the Berlin Mandate (AGBM5) in December in order to make substantive progress in the negotiation process.[8] For the EU this represented another great opportunity to take a leadership role in the global setting.

However, the results of the Environment Council, held in Brussels on December 9–10 1996, were disappointing. Climate change was on the agenda. However, the Irish presidency's proposal for burden sharing, which included a 5–10 per cent reduction of GHGs by 2005 and a 10–15 per cent reduction by 2010 (relative to 1990), failed because of disagreements on the method for calculating CO_2 emissions and incongruity in the burden-sharing pattern among member states (*Europe Daily Bulletin* 1996b).[9] Therefore, the EU was unable to present a proposal to AGBM5, which ran until 13 December and the issue was deferred for further discussion under the Dutch presidency. Meanwhile, the political pressure to find a solution to the burden-sharing issue was mounting. This enabled the EU – and, as it happens, the Netherlands – to take a lead in the Kyoto Protocol negotiation process.

Domestic Political Agreement for the Dutch Presidency

By the time the Netherlands assumed the European Council's presidency in the first half of 1997, it was clear that despite its ambitious target and policy measures on climate change, domestic CO_2 emissions were increasing, mainly due to low energy prices and unexpected economic growth. Indeed, emissions in the Netherlands increased by 7 per cent between 1990 and 1996 (*Change* 1998). This increase led to popular scepticism regarding the prospects for GCC policy. Yet scepticism itself did not make the Netherlands reluctant to tackle the burden-sharing issue. There had been a substantive domestic debate concerning an ambitious CO_2 reduction target. For example, when the Temporary Second Chamber Commission on Climate Change concluded in September 1996 that 'the CO_2 emissions reduction in the Netherlands in 2020 must be around 30–40 per cent', the employers' organization, VNO-NCW, strongly objected, arguing that this conclusion as 'absolutely unfeasible' (*Het Financieele Dagblad*, 12 September 1996). However, the Director of Environmental Affairs at VNO-NCW, who had been lobbying 'at every level' of the policy-making process, still felt that conditions were favorable for the environmentalists and said, 'the Parliament and the society were very much in favour of those environmental targets. We felt that pressure'.[10]

Although the 'environmental boom' since 1989 – when the environment was one of the main issues of the election – disappeared, the general political atmosphere was still in favor of stricter environmental measures. One example of the continued Dutch interest in environmental issues was that almost 11 per cent of the population were members of the three major ENGOs – WWF Netherlands, Greenpeace Netherlands and Milieudefensie (Friends of the Earth Netherlands). The domestic atmosphere, together with the inherent 'international approach' of Dutch GCC policy, made it politically possible for the VROM minister, Margaretha De Boer, to propose an ambitious target for the forthcoming EU negotiations. At the highest level of the decision-making process, many members of the cabinet agreed that it was not possible to realize an ambitious GHG-reduction target. At the same time, however, no one strongly objected to presenting such a presidency proposal to the Council of Environmental Ministers. As De Boer said, 'nobody believed it could be realized. Everybody thought, "no, impossible", but said, "if you want to go for it, go"'.[11]

In the end, the Dutch EU president gave GCC policy the highest priority among the five important agendas related to the environment. An EU presidency memorandum noted as follows:

> 'Climate policy plays an important role nationally in the Netherlands as has been regularly emphasized by the Lower House of Parliament. It is clear that we will have to have European and internationally coordinated instruments for a further reduction in CO_2 emissions at the national level. Internationally as well we are being confronted with a major turning point. During the third Conference of Parties to the World Climate Convention (December 1997, Kyoto, Japan) agreements will have to be made on post-2000 emissions reduction. The preparations for this at the European level will have to take place under the Dutch Presidency, which will entail a major effort on the part of the EU president. … It is clear that the role of the European Union and the influence of the Union on international environmental policy will depend on whether the EU succeeds in deciding on a common position in March 1997' (VROM DGM/IMZ/ 96074893).

Dutch Contributions to the EU Burden Sharing Agreement

It was clear from the failure of the Irish presidency that the key to an arrangement on burden sharing was a convincing approach backed by rigorous science. The EU Ad Hoc Group on Climate organized a workshop at Zeist, the Netherlands, on 16–17 January 1997. The Dutch presidency presented its strategy: a combination of an EU-wide abatement target and a proposal for an internal burden sharing mechanism called the *Triptych Approach*. Bert Metz, then the Dutch chief negotiator on climate change, had an original idea that a sectoral approach would work. However, he needed a good scientific brain to realize it. A research group at Utrecht University, led by Kornelis Blok, made the approach materialize. The research group was well respected throughout the EU for its work in the field of energy research, having produced a report on EU-wide policies and measures in collaboration with WWF Netherlands for submission to the EU Ad Hoc Group on Climate. Professor Blok was also an energy policy consultant to the Dutch government. Therefore, the research group had undertaken several research projects under the Dutch Governmental Research Program: Global Air Pollution and Climatic Change. The group's knowledge base related to GCC policy had been enhanced through program-funded projects. The group knew where to go in order to obtain necessary data by the time Metz gave the idea.

The main characteristic of the *Triptych Approach* was its sectoral methodology: each member state was divided into three sectors: the light domestic sector, the energy-intensive export sector and the electricity-generation sector. An emissions reduction target in each country was set by adding the potential for emissions reductions in every sector. The EU-wide target was set by adding these figures. This approach was based on lessons learned from the earlier failure to agree on burden sharing, where an agreement had been reached on the total figure for member states but allocating reduction amounts between them proved to be impossible. By using the sectoral method, *Triptych Approach* successfully penetrated the member states' 'black boxes' (Gupta, Jepma and Blok 1998; Phylipsen, Bode, Blok, Herik and Metz 1997).

The *Triptych Approach* gave rise to considerable discussion (see Table 4.3). Some member states called for changes, and some poorer member states were not satisfied with their quota. Yet the overall response to the mechanical formula of the approach was basically positive; Danish government officials in particular fully supported it. Officials from member states said that the 'approach presented a new, promising angle on the whole question of burden sharing' and 'accelerated the process of differentiation' (Ringius 1997: 25–6). On 27 January 1997, the Dutch presidency sent a proposal based on the *Triptych Approach* to the member states. The burden-sharing formula was modified and revised based on the discussions held at the Zeist workshop. Consequently, the emissions increase in the poorer member states was countered by additional reductions in Finland, Italy and the Netherlands. In an accompanying letter Minister de Boer pointed out the importance of the member states' contributions to establishing leadership at the forthcoming global negotiations and asked for further compromize: 'A challenging period is ahead of us and I firmly believe the world is looking to the European Union to play an ambitious leading role in the international negotiating process. That role can never be more than the sum total of the commitments individual member states are willing to make.'[12]

Table 4.3 Greenhouse gas reduction targets for 2010 (relative to 1990)

	Triptych Approach	**Presidency Proposal**	**February Informal Pledges**	**Council Decision**
Belgium	–15	–15	–10	–10
Denmark	–15	–25	–15	–25
Germany	–21	–30	–15	–25
Greece	2	5	10?	30
Spain	9	15	15	17
France	–4	–5	–5	0
Ireland	–5	5	10?	15
Italy	–7	–10	–5	–7
Luxembourg	–20	–40	–30	–30
Netherlands	–9	–10	–10	–10
Austria	–5	–25	–15	–25
Portugal	20	25	25?	40
Finland	–8	–10	–5?	0
Sweden	20	5	5	5
UK	–20	–20	–10	–10
EU Total	–12	–15	–11	–15
Gases	CO_2	CO_2,CH_4,N_2O		CO_2, CH_4, N_2O

Note: Question marks (?) indicate unclear pledges (Ringius 1997).

Sources: Gupta et al. (1998) and Ringius (1997).

Member states at the EU Ad Hoc Group meeting on 17–18 February discussed the proposal, but no conclusion could be reached. The discussion made clear that it would be very difficult to achieve an EU-wide 15 per cent reduction and that a target of 10 per cent would be more realistic. The

German Minister of Environment and the European Commission were concerned that setting an EU target below 10 per cent would mean that the EU's position in the global negotiations would not be politically credible (Ringius 1997). International ENGOs such as WWF and Friends of the Earth also expressed their anxiety about the situation, but thought that the 15 per cent target was in fact impossible. However, because the member states' reduction targets had been lowered in the course of discussion, it was clear that the total reduction for GHG emissions would be approximately 10 per cent relative to 1990, and not 15 per cent as suggested in the proposal. Subsequently, the Dutch proposal at the EU Council of Environmental Ministers on March 3–4 was revised to an EU-wide reduction target of 12 per cent. The Council seemed unable to agree on the more ambitious target until the Danish Environment Minister suggested keeping the EU-wide target at 15 per cent but accepting a 10 per cent internal burden sharing for the time being. It was suggested that if the Kyoto agreement were to exceed a reduction of 10 per cent, then the remaining 5 per cent would have to be renegotiated after the Kyoto conference. Thus, the member states 'agreed that 10 per cent would be enough until Kyoto', and so avoided acrimonious discussion about the division of the remaining 5 per cent (*Europe* 1997).

The EU proposal of 15 per cent reduction in emissions of three GHGs by 2010 relative to 1990 had a strong impact on the sixth meeting of the Ad Hoc Group on the Berlin Mandate (AGBM6) negotiations held in Bonn.[13] The proposal facilitated concrete discussion, and thus led the negotiations in the direction that the EU presidency had intended. By proposing the first QELROs for the Annex I parties (developed countries), the EU, led by the Dutch presidency, contributed to facilitating the global negotiating process (*Earth Negotiations Bulletin* 1997).

THE DUTCH RESPONSE AND THE INSTITUTIONAL BACKGROUND

The Council's decision of March 1997 required the Netherlands to achieve a 10 per cent reduction by 2010. This was a very tough target for the Netherlands, a country which had introduced GCC policy as early as 1990 and yet faced difficulties in reducing GHGs. Policy debate in the Netherlands on the implementation of the Council decision and the strategy for the Kyoto Protocol negotiations started in the summer of 1997. As in the case of the NEPP, the debate was held among the four main ministries involved – the ministries of Housing, Spatial Planning and the Environment, Economic Affairs (EZ), Agriculture, Nature Management and Fisheries (LNV) and Transport, Public Works and Water Management (V&W). (The ministries of Development Cooperation and Finance were also invited.) The meeting was chaired by a representative from the climate change department of the Ministry of VROM. In addition, as task force called the Task Force Protocol Negotiations was formed between the ministries of VROM and Foreign Affairs (BuZa). This body met weekly for in-depth discussions in preparation for interdepartmental meetings. The final decision with respect to the Netherlands' ultimate position at the international negotiations was made by the Co-ordination Committee on International Environmental Issues, chaired by BuZa. There was a sub-committee on climate change that met every two or three weeks and was attended by all the ministries involved. In essence, the issues were discussed and common ground was found at the committee chaired by the Ministry of VROM, then the committee chaired by the BuZa finalized the Dutch position in preparation for the international GCC (Kyoto) negotiations.[14]

Domestic co-ordination among target groups was mainly undertaken by the Ministry of VROM, as is generally the case with environmental policy under NEPP. Other ministries accountable for specific issues were also charged with negotiations among target groups in their own areas of responsibility. As for the ENGOs, which did not fit into the framework of target groups, formal talks and exchanges of information were institutionalized as 'DOMILO' (*Directeuren Overleg Milieuorganisaties* [Environmental Organization Director Consultation]), which met every two months with a director-level civil servant at the Ministry of VROM. There was also 'DOMILO-plus', which met four times a year with the Minister of VROM. These institutionalized meetings were, however, designed for general lobbying on environmental policy; more substantive and effective exchanges of information and views regarding GCC were undertaken through daily informal contacts. The main channel for information discussion was the international policy department of the Directorate-General of Climate Change in the Ministry of VROM. With regard to contacts and exchanges of information, everything went smoothly and satisfactorily for both the ministry and ENGO sides, as Metz saw ENGOs as 'allies' rather than enemies.[15] In addition, those who were interested in the issue, including civil servants, non-governmental organization (NGO) representatives and researchers, gathered at a policy study group meeting (*beziningsgroep*) to exchange views and information. These participants attended the meeting not as representatives of their organizations but in their personal capacity.[16]

The government sent a letter concerning the Netherlands' negotiating target in Kyoto, generally called the 'Kyoto letter', to Parliament on 17 October 1997. The letter described projections made in the Fourth National Environmental Outlook and presented by RIVM that showed that the national target of –3 per cent for CO_2 reduction in 2000 would not be met under current policies and that additional policies would be necessary in order to meet the target. The Kyoto letter also concluded that it would be difficult to meet the objective of 10 per cent reduction for CO_2, CH_4 (methane) and N_2O (nitrous oxide) in 2010. It also gave two policy options as examples, both of which emphasized the need for an internationally co-ordinated policy at the EU level (Second Chamber, session 1997–98).

In general, the industrial sector opposed a strict GCC policy because it was afraid of the negative impact on international competitiveness. This opposition prevented many countries from tackling climate change. The Netherlands was not exceptional in this sense, but the 'internalization' approach of the NEPP helped prevent the industrial sector from undermining an ambitious stance in multilateral negotiations. The industrial sector introduced Long-Term Agreements (LTAs) on energy efficiency in 1992 and sought to improve energy efficiency by signing LTAs with the government. This approach gave the sector a free hand in improving its own energy efficiency in the most cost-effective way, while the government promised not to introduce additional policy measures as long as industry improved energy efficiency in line with the LTA.

In addition, in 1997 the introduction of a new covenant, called the Benchmarking Covenant, was under discussion for the energy-intensive industries. This covenant bound the government not to introduce additional policies on the condition that the Dutch industrial sector reached the 'top level in the world' in energy efficiency within at least 3 years relative to the same industry sectors abroad. The government can introduce additional policies only if industry fails to reach the target (Ministry of Housing et al. 1997). On the one hand, this strategy permitted the industrial sector to maintain international competitiveness and prevented the unilateral introduction of additional policy measures suggested in the Kyoto letter. On the other hand, these agreements firmly established the policy measures to be applied to industry and prevented it from being too fearful of future regulations. Chris Dutilh, the co-ordinator of environmental policy at Unilever, a multinational chemical company, and a member of the environmental board of VNO-NCW, said:

> 'regarding the Kyoto target, the industry was allowed to have it because we thought it was a political debate. The government sorts out its business, we sort out something with the government so that it knows what to expect from us. If the government wants to go further, it is their own responsibility. It is reasonable to think that if the Dutch society is politically motived to do more, then the people should pay the cost. The Benchmarking Covenant will lay the bottom line for the industry. This is what the industry contributes to the government's plan. If society plans more, then society has to pay.'[17]

Interdepartmental deliberations regarding the Dutch strategy were held twice at BuZa before the Kyoto conference. Some points were left unresolved, and the final decision was made at a cabinet meeting attended by the prime minister in the last week of November 1997.

THE KYOTO CONFERENCE

After the seventh meeting of the Ad Hoc Group on the Berlin Mandate (AGBM7) in July–August and AGBM8 in October, it became apparent that the EU proposal contained the most ambitious target for the reduction of GHGs. The Group of 77 (developing countries) expressed support for the EU proposal at AGBM8. A group of influential ENGOs under the umbrella of the Climate Action Network also expressed support by the time the Kyoto Conference was convened (ECO 1997). The EU troika negotiated on behalf of the EU at Kyoto. The troika consisted of Luxembourg, the Netherlands and the United Kingdom (UK). In practice, however, the Netherlands and the UK led the negotiations because of the small size of the Luxembourg delegation.[18] The Netherlands' chief negotiator until the high-level meetings was Bert Metz, the main architect of the *Triptych Approach* and the main Dutch negotiator since COP1, where he was nominated by the French presidency to be an EU negotiator together with British and German delegates.[19] He was a well-known negotiator, with a reputation of being one of the best in the EU. Although the Dutch delegation to COP3 did not include an NGO representative, as had been the case at the Rio Earth Summit in 1992, contacts with NGOs occurred often throughout COP3, in addition to the formal gatherings held every morning.[20]

The EU had the most ambitious target among Annex I (developed countries) at Kyoto, and it won the support of many developing countries and ENGOs. The Dutch network of governmental and non-governmental officials worked together at the conference to advance the negotiations towards a target that was more ambitious than many parties wanted. For example, a Dutch negotiator deliberately informed a Dutch NGO member of one government's proposal at a closed-door meeting to weaken the targets. Using the global network of ENGOs, the NGO member brought the proposal to the public's attention in order to generate public pressure.[21] Likewise, a representative of WWF Netherlands supported the EU proposal and tried to sway American and Japanese delegates by drawing their attention to the intellectual background of the proposal, and to the policies and measures that had already been taken to attain the 15 per cent target included in the EU proposal. Moreover, the sole role of one Friends of the Earth Netherlands (*Milieudefensie*) representative at this stage was to convince Norwegian and Australian representatives to support the European proposal.[22] In fact, after the DOMILO-plus meeting before the Kyoto Conference, Teo Wams, the leader of *Milieudefensie*, took a 'big step' and made up his mind that his organization would basically support the EU position at Kyoto in order to make the conference successful.[23]

It is difficult to measure the impact of these patterns of behavior and persuasion on the outcome of the negotiations. However, it is clear that both governmental and non-governmental organizations generally supported the EU proposal at this stage. This provided the Dutch government with another channel to influence the inter-governmental negotiations using the network of ENGOs that had been actively involved in Dutch GCC policymaking at the domestic level. In the end, at the Kyoto conference 38 industrialized countries agreed to reduce six types of GHGs – carbon dioxide, methane, nitrous oxide, HFCs, PFCs and SF6 – by 5.2 per cent relative to 1990 levels between 2008 and 2012 (see Chapter 1). The shares were 6 per cent for Japan, 7 per cent for the United States (US) and 8 per cent for the EU. As Sjostedt argued, it is possible that 'EU diplomacy has contributed to producing the inadequate but still encouraging results obtained in Kyoto' (Sjostedt 1998: 227).

CONCLUSION

There is a mutally supportive relationship between the leading role of the Netherlands in the Kyoto Protocol negotiating process and its relatively early and ambitious domestic policy development. What are the mechanisms that link these two arenas? What implications can we draw from this case for the study of European EFP? Two definitive, and mutually supportive, mechanisms by which domestic policies, and institutional and structural factors facilitate the leadership-taking potential of a country may be learned from this case. One is the importance and necessity of building 'domestic capacity' (leadership potential) to exert leadership in multilateral negotiations. Theoretically, this points to Barkdull and Harris's (2002) category of ideas in society shaping EFP and contributing to leadership: ideas in foreign policymaking, such as the Triptych Approach, can lead to leadership. This is particularly true with respect to environmental issues, the recognition of which requires scientific knowledge (as the history shows us). However, ideas do not necessarily mean the attainment of a leadership role. A second factor is at work as well.

Although the extent of a country's power is not taken into account in the directional mode of leadership, power bases are still necessary to achieve leadership in practice. In this case, the framework provided by a regional organization (the EU) functioned as a power device that enabled a middle power to materialize its leadership-taking potential. Hence, ideas do matter in the making of EFP, but when it comes to leadership and the international influence of those ideas, a power-based approach still needs to be considered to get a fuller account of EFP.

With respect to building leadership potential, the most important factor that increased the Netherlands' leadership-taking ability was that it succeeded in building a certain level of domestic consensus regarding GCC policy and external pressure to further advance international policy development. More importantly, this consensus was maintained, facilitated and enhanced by domestic policies and institutions. For example, by signing an environmental covenant with the domestic industrial sector, the Netherlands successfully eliminated a potential negative influence on its ambitious negotiating posture. Since industry, which is generally conceived as the biggest 'opposition group' in tackling GCC-related issues (at least it was in 1997), had negotiated an agreement with the government in advance – which stated explicitly the burden industry would have to bear – it did not have a great incentive to object to the development of a relatively ambitious international environmental agreement (i.e. the Kyoto Protocol as agreed). Also, as GCC policy developed, the Netherlands gradually built a policy network that helped maintain general consensus in this issue area. Moreover, ENGOs such as WWF, Greenpeace and Friends of the Earth, which have branch organizations in the Netherlands are part of a global network. The ENGO network actually functioned to help link 'the local and the global' (Princen and Finger 1994). This created a new type of global interest network that fed into the global environmental foreign policymaking process. In this regard, this chapter reaffirms that the interests in society described in Barkdull and Harris's typology (see Chapter 1, Figure 1.1) do matter in the process of making EFP.

Policy measures functioned to transform domestic consensus into external pressure. After implementing unilateral measures in domestic GCC policy, additional domestic developments in the Netherlands became conditioned upon reaching international agreement. In this way, the failure of Dutch domestic environmental policy to achieve further domestic CO_2 reductions can be attributed to the failure of the international GCC negotiations. Thus, political pressure on the former could well be transformed to political pressure on the latter. At the same time, the difficulty of achieving the GHG reductions target facilitated GCC-related research, both within and between governmental organizations and NGOs. Thus, intellectual capacity expanded by the time the Kyoto Protocol negotiations began. The research GCC program also facilitated dialogues and networking between government and NGOs.

'Domestic capacity' (leadership potential) is something that non-hegemonic powers (e.g. Denmark, Norway, Canada, Australia) may have in some issue areas, notably that of the environment (as described above in relation to studies of middle powers). This, again, means that there are instances where ideas do matter in the foreign policy process. However, transforming potential into actual leadership occurs less frequently than one would expect in actual multilateral negotiations. In many cases, middle powers may have potential, but potential does not necessarily lead to successful leadership. Middle-power countries have been perceived as being able to exert, at most, 'directional leadership' or 'leadership through unilateral action' stemming from setting a good example and indirectly leading multilateral negotiation in the long run, but not directly leading negotiations in a visible, short-term manner.

The case of the Netherlands shows how a middle power could exert substantial short-term leadership in the international negotiations. The reason why this was possible points to another major aspect of the Netherlands' leadership in this case. The Netherlands' leadership was only possible because it could negotiate on behalf of the EU within a unified coalition. The timing and structure of power diffusion within the EU institutional setting on the one hand, and the EU's status as one of the three biggest world economic powers on the other, were necessary factors that allowed the Netherlands to actually exert influence during the Kyoto negotiations. On this point, this case study may be compared with the COP6 negotiations held in The Hague in November 2000. Of course, we would need a deeper investigation for a rigorous comparison between the two negotiation processes, but there appear to be very interesting comparisons. Although the Netherlands hosted and chaired COP6, it was neither the president of the EU nor a part of the EU troika. Therefore, the Netherlands could not chair the COP on behalf of the EU coalition. This may be one reason why the EU and the US could not reach an agreement in the final stage of COP6. The Netherlands could bring some inspiration to the COP negotiations, such as inviting indigenous peoples' groups to the meeting, because of certain ideas embedded in its foreign policy. However, it could not lead the EU internal negotiations and thereby make the EU 'one voice' as happened (more or less) at Kyoto. The upshot is that domestic policies and institutions can facilitate leadership-taking potential of a country in a multilateral negotiation. The main lesson to be learned from this chapter is that transforming potential leadership into actual leadership through unilateral action requires bases of power.

The EU as an institution has of course changed since 1997, with the number of member states expanding to 25. Nevertheless, ideas in EFP will continue to matter, particularly due to the complexity of global climate change. The story of the Netherlands in the early GCC negotiations can therefore be helpful in planning for the future. In diplomatic negotiations on GHG reductions targets beyond the Kyoto Protocol's commitment period (ending in 2012), which officially took off in the late-2005 Montreal COP11/MOP1, idea-rich countries with capable domestic policy institutions have greater potential to lead, even if they are not – and perhaps *because* they are not – great powers. This may become even more important in forthcoming rounds of negotiation than it was during the diplomatic process leading up to the Kyoto Protocol, because implementation of the protocol and new GCC-related initiatives (mainly those led by the US) will occur simultaneously. The success or failure of the Kyoto Protocol's implementation, as well as the ideas to improve related international institutions, will matter in deciding the direction of future activities, whether they be an extension of the Kyoto Protocol or totally new arrangment, as the US has been demanding. Power will always count, of course, but the case of the Netherlands in the early GCC negotiations shows how greatly ideas, in this case ushered by a so-called middle power, can matter in the making of environmental foreign policy.

NOTES

* Portions of this chapter were orginally published in Kanie (2003a, 2003b).

1. For example, unilateral action can show that environmental protection measures are feasible and work in real life (Underdal 1994: 185).
2. Young clarified the difference between negotiation, imposition and spontaneous processes. See Young (1983 and 1989: 202).
3. However, many have admitted that it is difficult, or even meaningless, to define this concept rigidly, even though it is clear that a number of academics and practitioners use the concept of 'middle power' when they refer to a certain group of states or to the behavior of these states. (See Cooper et al. 1993; Pratt 1989, 1990; Stokke 1989; Helleiner 1990; Holmes 1965.)
4. Those five levels are global, continental, fluvial, regional and local.
5. The Toronto Declaration, the first international declaration with a numerical CO_2 emissions reduction target, set out a target if -20 per cent by 2005 compared to 1987. The Noordwijk Declaration set the target as stabilizing CO_2 emissions by 2000.
6. In fact, the Long-term Agreement on Energy Efficiency between the Dutch government and each industry points out that the government would not levy additional taxes to improve energy efficiency at the expense of industry.
7. Interview with Bert Metz, 20 March 1998. For discussion of the 'like-minded' countries, see Dolman (1984: 1–13).
8. Kyoto Protocol negotiations on GHG reduction targets were held in the Ad Hoc Group on the Berlin Mandate (AGBM) meetings.
9. France, Sweden, Spain, Greece, Italy and Portugal recommended a reduction per person. The other countries preferred a reduction in absolute value.
10. Interview with Wiel Klerken, Director of Environmental Affairs, VNO-NCW, The Hague, 1 December 1998.
11. Interview with Margaretha De Boer, The Hague, 18 November 1998.

12. Letter of 17 January 1997; quote from Ringius (1997: 27).
13. The three conditions for the EU proposal were the same level of reduction by other developed countries, internationally co-ordinated policies and measures to be decided at Kyoto, and joint implementation between Annex 1 countries (European Council Decision of March 4, 1997).
14. Interviews with Henk Merkus, Ministry of VROM, 13 May 1998, and Andreas te Boekhorst, Ministry of BuZa, February 24, 1998.
15. Interview with Bert Metz, 20 March 1998.
16. This kind of cooperation between government and NGOs is often referred to in connection with Dutch political culture. See, for example, Visser and Hemerijck (1997); Kickert, Klijn and Koppenjan (1997); and Kalders (1998).
17. Interview with Chris Dutilh, Rotterdam, 9 November 1998.
18. Interview with Margaretha de Boer, The Hague, 18 November 1998. For a detailed discussion of the UK's role, see Chapter 3.
19. The French presidency did not participate in the troika at COP1.
20. According to interviews, NGO delegations were satisfied with their contact and exchanges of information with government delegations during COP3.
21. Interview with a Dutch delegate, 12 May 1998.
22. Interview with Sible Schone, WWF Netherlands, 24 November 1998, and with Wim Kersten, Milieudefensie (Friends of the Earth Netherlands), 12 March 1998.
23. Interview with Teo Wams, Milieudifensie, 26 October 1998.

REFERENCES

Bodansky, Daniel (1993), 'The United Nations Framework Convention on Climate Change: a commentary', *Yale Journal of International Law* **18**(2): 451–558.

Barkdull, John and Paul G Harris (2002), 'Environmental Change and Foreign Policy: A Survey of Thought', *Global Environmental Politics*, **2**, 2 (May): 63–91.

Change (1998), *Change: a newsletter of the Dutch NRP* (July-August), available at www. nop.nl/.

Cooper, Andrew F (ed.) (1997), *Niche Diplomacy: Middle Powers after the Cold War*, London: Macmillan.

Cooper, Andrew F, Richard A Higott, and Kim R Nossal (1993), *Relocating Middle Powers: Australia and Canada in a Changing World Order*, Victoria: Melbourne University Press.

Cox, Robert W (1989), 'Middlepowermanship, Japan, and future world order', *International Journal* **44** (Autumn): 823–62.

De Jongh, Paul E and Sean Captain (1999), *Our Common Journey: A Pioneering Approach to Cooperative Environmental Management*, London: Zed Books.

Dolman, Andrew J (1984), 'Disarmament, development, dual-purpose technologies and the like-minded countries', *Cooperation and Conflict* **XIX**: 1–13.

Earth Negotiations Bulletin **12**(45) (10 March 1997).

ECO (1997) 'COP-2', *ECO* 2 **XCVIII** (2 December 1997), available at www. climatenetwork.org/eco/eco2_cop3.html.

Europe (1997), *Europe* **6927** (5 March).

Europe Daily Bulletin (1996a), *Europe Daily Bulletin* 6835 (18 October).

Europe Daily Bulletin (1996b), *Europe Daily Bulletin* 6871 (11 December).

Gupta, Joyeeta and Michael Grubb (eds) (2000), *Climate Change and European Leadership: A Sustainable Role for Europe?*, Dordrecht: Kluwer Academic Publishers.

Gupta, Joyeeta, Catrinus J Jepma, and K Blok (1998), 'International climate change policy: coping with differentiation', *Milieu* **3**(5): 264–74.

Hajer, Maarten A (1995), *The Politics of Environmental Discourse: Ecological Modernization and the Policy Process*, New York: Oxford University Press.

Hanf, Kenneth and Egbert van de Gronden (1998), 'The Netherlands: joint regulation and sustainable development', in Kenneth Hanf and Alf-Inge Jansen (eds), *Governance and Environment in Western Europe: Politics, Policy and Administration*, New York: Addison Wesley Longman: 152–80.

Helleiner, Gerald K (ed.) (1990), *The Other Side of International Development Policy: The Non-Aid Economic Relations with Developing Countries of Canada, Denmark, the Netherlands, Norway, and Sweden*, Toronto: University of Toronto Press.

Holmes, John W (1965), 'Is there a future for middlepowermanship?', Paper given at the Third Annual Banff Conference on World Development: *Canada's Role as a Middle Power*, Canadian Institute of International Affairs (August 1965).

Huber, Michael (1997), 'Leadership in the European Climate Policy: Innovative Policy Making in policy networks', in Duncan Liefferink and Mikael Skou Andersen (eds), *The Innovation of EU Environmental Policy*, Copenhagen: Scandinavian University Press.

International Energy Agency (IEA) (1992), *Climate Change Policy Initiatives*, Paris: OECD/IEA.

Kanie, Norichika (2001), *Chikyu Kankyou Gaikou to Kokunai Seisaku* (*Global Environmental Diplomacy and Domestic Policy*), Tokyo: Keio University Press.

Kanie, Norichika (2003a), 'Domestic capacity, regional institutions and global negotiations: lessons from the Netherlands-EU Kyoto protocol negotiation', in Michael Faure, Joyeeta Gupta and Andries Nentjes (eds), *Climate Change and the Kyoto Protocol: The Role of Institutions and Instruments to Control Global Change*, Cheltenham, UK: Edward Elgar.

Kanie, Norichika (2003b), 'Leadership in multilateral negotiation and domestic policy: the Netherlands at the Kyoto protocol negotiation.' *International Negotiation* **8**(2): 339–65.

Kaufmann, Johan and Nico Schrijver (1990), *Changing global needs: expanding roles for the United Nations system*, New Haven, CT: Academic Council on the United Nations System.

Keohane Robert (1984), *After Hegemony: Cooperation and Discord in the World Political Economy*, Princeton: Princeton University Press.

Kickert, Walter JM, Erik-Hans Klijn and Joop FM Koppenjan (eds) (1997), *Managing Complex Networks: Strategies for the Public Sector*, Thousand Oaks, CA: Sage.

Liefferink, D (1997), 'The Netherlands: a net explorer of environmental policy concepts', in Mikael Skou Anderson and Duncan Liefferink (eds), *European Environmental Policy: the Pioneers*, Manchester University Press.

Malnes, Raino (1995), '"Leader" and "Entrepreneur" in International Negotiations: A Conceptual Analysis', *European Journal of International Relations*, **1**,1: 87–112.

Ministry of Housing, Spatial Planning and the Environment; Ministry of Economic Affairs; Ministry of Agriculture, Nature Management and Fisheries; and Ministry of Transport, Public Works and Water Management (1997), *Policy Document on Environment and Economy: Towards a Sustainable Economy* (June).

Paterson, Matthew (1996), *Global Warming and Global Politics*, New York: Routledge.

Phylipsen GJM, JW Bode, K Blok, Merkus Henk and Bert Metz (1997), 'A triptych sectoral approach to burden sharing: GHG emissions in the European Bubble', Unpublished paper, Ministry of Housing, Spatial Planning and the Environment.

Pratt, Cranford (ed.) (1989), *Internationalism Under Strain: The North-South Policies of Canada, the Netherlands, Norway, and Sweden*, Toronto: University of Toronto.

Pratt, Cranford (ed.) (1990), *Middle Power Internationalism: The North-South Dimension*, Montreal: McGill-Queen's University Press.

Princen, Thomas and Matthias Finger (1994), *Environmental NGOs in World Politics: Linking the Local and the Global*, London: Routledge.

Richardson, Jeremy (ed.) (1982), *Policy Styles in Western Europe*, Boston: Allen and Unwin.

Ringius, Lasse (1997), *Differentiation, Leaders and Fairness: Negotiating Climate Commitments in the European Community*, Oslo: CICERO.

Rijkesinstituut Voor Volksgezondheid en Milieuhygiene (RIVM) (1987), *Concern for Tomorrow: A National Environmental Survey 1985–2010*.

Second Chamber (the Netherlands), session 1988–89, **21** 137, nos 1–2.

Second Chamber (the Netherlands), session 1990–91, **22** 232, nos 1–2.

Second Chamber (the Netherlands), session 1997–98, 24785, no. 4.

Sjostedt, Gunnar (1998), 'The EU negotiates climate change: external performance and internal structural change', *Cooperation and Conflict* **33**(3): 227.

Stokke, Orav (ed.) (1989), *Western Middle Powers and Global Poverty: The Determinants of the Aid Policies of Canada, Denmark, the Netherlands, Norway and Sweden*, Uppsala: Scandinavian Institute of African Studies.

Underdal, Arild (1994), 'Leadership theory: rediscovering the arts of management', in William Zartman (ed.), *International Multilateral Negotiation: Approaches to the Management of Complexity*, San Francisco: Jossey-Bass.

van den Bos, JMM (1991), *Dutch EC Policy Making: A Model-Guided Approach to Coordination and Negotiation*, Amsterdam: Thesis Publishers.

Van der Straaten, J (1992), 'The Dutch national environmental policy plan: to choose or to lose', *Environmental Politics* **1**(1) (Spring): 45–71.

Van Holstein, JJM and GA Irwin (eds) (1992), *De Nederlandse kiezer 1989*, Amsterdam.

van Puttern, J (1982), 'Policy styles in the Netherlands: negotiation and conflict', in J. Richardson (ed.), *Policy Styles in Western Europe*, Boston: Allen and Unwin.

Visser, Jelle and Hemerijck, Anton (1997), *A Dutch Miracle: Job Growth, Welfare Reform and Corporatism in the Netherlands*, Amsterdam: Amsterdam University Press.

VROM (Ministry of Housing, Spatial Planning and the Environment) (1995), 'CO2 Letter', sent by the Government to Parliament, September 1995. Government Publication: VROM.

Wintle, Michael (ed.) (1994), *Rhetoric and Reality in Environmental Policy*, Aldershot, UK: Avebury.

Young, Oran R (1983), 'Regime dynamics: the rise and fall of international regimes', in Stephen Krasner (ed.), *International Regimes*, Ithaca: Cornell University Press.

Young, Oran R (1989), *International Cooperation: Building Regimes for Natural Resources and the Environment*, Ithaca: Cornell University Press.

5. Climatic issues in Polish foreign policy

Anita Bokwa

INTRODUCTION

This chapter explores the environmental foreign policy (EFP) of Poland, a former communist state that became a member of the European Union (EU) on May 1, 2004.[1] In order to understand present Polish foreign policy on global climate change (GCC) it is necessary to recall the development of its environmental policy after the Second World War and to explore factors contributing to its present shape. Poland's evolution from a socialistic to a democratic system caused huge changes in the main actors and ideas that have been decisive for its GCC policy. A weak national economy and a still-developing public awareness of environmental issues are factors inherited from the pre-1989 period that strongly influence the present situation. In this chapter, the historical, political and economic causes of environmental degradation in Poland are first presented. They serve as a background to the country's environmental policies from 1945 to 1989. Next, more recent developments are discussed, including domestic climate policy, the main actors involved in shaping it, new geopolitical conditions and the resulting Polish foreign policy concerning GCC. Poland's policies on climate change are examined with reference to the theoretical typology described by Barkdull and Harris (2002), which juxtaposes systemic, societal and state-centric theories of EFP with considerations of power, interests and ideas (see Chapter 1).

POLISH ENVIRONMENTAL POLICY, 1945–89

After the Second World War, Poland was incorporated into a group of states whose foreign and domestic policies were dominated by the Soviet Union (USSR). Industrialization became the utmost goal of the state economic and of all political activity. However, environmental issues, including climatic ones, were not neglected. Unfortunately, they were the subject of spectacular political gestures rather than real action.

Historical and Political Background

Stalinist doctrine, socialist industrialization and environmental damage

As shown by Sowa (1990), there are two main types of authorities: those serving the social reality and those creating the social reality. Stalin's creationism was an example of the latter. In Poland after 1945, Marxism interpreted by Stalin determined the authorities' decisions. The decisions were independent from real expectations and social needs, as well as experts' opinions. The authorities were creating the social reality; goals and forms of social life were known in advance, since they resulted from doctrine and not from social experience. Political power was not based on law; it replaced law. The society was not an objective social reality but only raw material to create that reality. The authorities did not care about facts and means of action accompanying them; they cared only about goals defining directions and means of action. There was only one constant and unchanged goal: to keep power and to maintain it as a creative factor and the main basis of the system. The authorities were not limited in any way; they were even not bound by their own earlier decisions.

At the same time, Poland was one of the countries most heavily damaged during the Second World War. The destruction in industry amounted to 40 per cent of fixed assets (Leszczycki 1965). Therefore, in the years 1946–54 the main effort of the whole nation was directed to the reconstruction of industry, rebuilding towns and cities, and similar activities. Additionally, important changes in the organization and management of the economy took place, especially after January 3, 1946, when the act of nationalization was announced. All larger factories and companies became state owned (Bywalec 1990). Further development of the Polish economy was strongly influenced by the fact that Poland was incorporated into the Soviet block after the Second World War. Already in the 1930s, Stalin was pushing the image of the Soviet Union as a besieged fortress, endangered continuously by capitalist enemies. That approach was also adopted after 1945 in other countries of so-called 'real socialism', such as Poland and the German Democratic Republic (GDR) (Delorme 1990). The cold war and the threat of global conflict constituted the grounds on which the USSR and other socialist countries built their fundamental economic goals. The main task was to achieve, in a short time, 'defensive ability' (Bywalec 1990). Weapons production demanded the development of heavy industry and such other branches of industry that were needed to support it. Stalin was forcing the idea of greater importance of the development of production methods than the production of consumption goods. In the 1950s, that concept became the economic principle of socialism. It caused the emergence of huge industrial centers and rapid expansion in the workers' social class, which led to strengthening of the state's (ie authorities') power (Delorme 1990).

In Poland, the spatial structure of industry did not change significantly until 1950, compared to the period before the Second World War. Production developed and was concentrated in those industrial regions that had suffered less destruction during the war, notably Upper and Lower Silesia, as well as the Lodz area (Leszczycki 1965). However, Stalin's doctrine of massive industrialization meant that industry was supposed to be developed both by constructing huge factories and by creating new industrial centers in regions not already industrialized. The national economy was centrally controlled and therefore all aspects of industrialization were realized in the same way all over the country. The main features of the industrial development were as follows:

1. It was an extensive process entailing diversified natural resources and based on enormous investments and large increases in the employment rate. The latter was first a positive tendency, but it turned out that other sectors of economy (e.g., different services, craft) were insufficiently developed and sometimes suffered from the strongly encouraged industrial development.
2. Industrialization was a highly unbalanced process, as only a few selected branches of an industry were developed while others were neglected.
3. Profits of the so-called large-scale effect were overestimated. The size and production abilities of the constructed factories were determined following political and not economic assumptions.
4. Urbanization developed at a lower rate than industry, causing serious shortages in basic infrastructure and negative impacts on the quality of people's lives, and resulting in the need for transportation systems to bring workers from rural areas (Kortus 1990).

The centrally planned economy was realized in long-term economic plans (1947–49, 1950–55, 1956–60, 1961–65, 1966–70, 1971–75, 1976–80, 1983–85, 1986–90). Each of them was supposed to be completed with an increase in industrial potential and employment (Leszczycki 1965). Within the framework of Stalin's system, however, every effect was achieved at enormous costs. The system could function only because there were huge natural and human resources in the USSR that seemed endless and unlimited. Other socialist countries followed those assumptions directly. However, the environmental outcomes were much more disastrous than in the USSR, as those countries occupied much smaller areas than the Soviet Union. Heavy industry and raw-materials production were heavily developed, but both of them are the most harmful to the natural environment. Political and economic priorities were realized irrespective of the costs and of people's needs, and while ignoring environmental conditions.

In the 1960s and 1970s, the technical and scientific progress also influenced the military sector. New weapons were based mainly on modern technologies and the role of heavy industry and raw materials production was significantly diminished. Therefore, Stalin's model economy lost its role as the basis for weapon production and national security. However, the whole infrastructure remained and turned into a self-driven system. In Poland, mainly steelworks and power plants used coal. Steelworks production was sold mainly to coalmines and power plants, whereas the latter produced energy mainly for coalmines, steelworks and their transportation systems. Additionally, thousands of people were hired to work within that system, both in production processes and various services accompanying them. The whole structure was becoming more and more inert, expensive and influenced by political and economic bureaucracy, which depended on Stalin's model of economy. It had to lead to an economic crisis (Delorme 1990).

Consequently, in 1989 Poland started a new period of its history with not only a ruined economy but also with a very polluted natural environment in many areas, and with many sources of air pollution affecting other European countries. (Paradoxically, Poland is a country with the last truly wild forest complexes in Europe, untouched by human activity (e.g., Bialowieza National Park in eastern Poland.) Before 1989, environmental issues, including climatic ones, were insignificant to the state in comparison to seeking national power, an example of the power-systemic approach in the Barkdull and Harris typology (2002; see Chapter 1: Figure 1.1). Poland's policy-makers were usually designated by the socialistic party and did not worry about the next elections; there was simply no competition. So, unlike in the West, their lack of interest in environmental issues was not due to a short-term view of matters and preparation for next election campaign; it was rather the matter of adopting Stalin's model of economy and considering natural resources as endless and unlimited. Building on the cognition-societal approach in the Barkdull and Harris (2002) typology, it can be seen that the policymakers' ideas could be sharply detached from the beliefs and opinions widely held in society, while being in accordance with Marxism as interpreted by Stalin.

State information policy and public awareness of climatic issues

After the Second World War, public understanding and perception of environmental problems evolved differently in Poland than it did in Western Europe. At the beginning of the 1970s, the Western world realized that economic progress and the increasing exploitation of nature no longer correlated with the growing welfare of societies. Changes caused by humans in the natural environment turned out to be dangerous for the health and quality of life for millions of people. The issue became the subject of public discussions and numerous publications (Kocik 2000: 204–6). In Poland, however, the situation was quite different. From the 1950s onwards, industrialization of the country was considered to be the main means of achieving social and economic goals.

Polish public opinion began to register the dramatic state of Poland's natural environment when United Nations Secretary General U Thant's report on the environment was published in 1969. Still, most papers describing these were not normally published in Poland because they implicitly criticized the economic system that was responsible for environmental degradation (Delorme 1990). A good example of what was published is the information policy concerning the Lenin Steelwork in Cracow (today the Sendzimir Steelwork), one of major achievements of the socialistic industrialization in Poland. Environmental issues were mentioned by the experts from the beginning of planning and construction of the steelworks, which was completed in 1954. All documents stated that environmental protection and people's living conditions were taken into consideration, and that appropriate actions were undertaken to provide for the highest standards. That was the official propaganda, but the truth was quite different. Even though from the 1960s onwards the impact of the steelworks was becoming continuously more and more negative, the factory was enlarged and environmental problems were neglected (Raczynski 1990).

In the 1970s, uncontrolled and rapid industrial development led to environmental degradation in many regions of Poland. At that time, only some groups of intellectual elites were aware of the actual environmental situation (Kozlowski 1983: 9–10; Raczynski 1990). Until 1979, air pollution in Poland was constantly increasing, and at the beginning of the 1980s 27 areas with permanently excessive pollution concentrations were distinguished as ecologically endangered. They covered 11.3 per cent of the country's area, and were populated by 35.5 per cent of Polish citizens (Kozlowski 1997: 9). In 1979, an independent report was prepared by a group of Polish scientists and submitted to the government. It stated that official propaganda was creating a false image of the environmental situation in Poland, and it postulated a new approach to the natural resources that could be taken by the state.

During the time of the Solidarity movement, in 1980–81, further reports and environmental strategies were prepared, and the regions of ecological disaster were officially recognized (e.g., in Cracow and at the seaside in Gdansk). Additionally, as a result of the first free and democratic citizens' environmental initiative, in 1980 the Polish Ecological Club was established in Cracow. On December 13, 1981, the democratic changes in Poland were stopped as General Wojciech Jaruzelski announced marshal law. Critical publications on environmental issues were prohibited until 1989. All changes in the structure or operation of the Polish economy were stopped as well (Kozlowski 1983: 9–11). The conservative forces tried to preserve Stalin's economic system in Poland and continued investments in heavy industry for another 8 years ('Katowice' Steelwork in Dabrowa Gornicza was constructed in the Silesia region) (Delorme 1990).

As suggested by Barkdull and Harris (2002), ideas operating at the global level can affect foreign policies of states (cognition-systemic approach), and environmental beliefs widely held in society can shape policies of states (a cognition-societal approach). In Poland, before 1989, it was quite the opposite. The country was separated from global environmental discussion by the Iron Curtain, and opinions on environmental issues were excluded from public debate and ignored by the state. However, they survived within some elite groups and emerged as an element of public awareness immediately after 1989. However, even today, environmental and climatic issues are still not widely known in Polish society, partially because they appeared in the public debate so late and partially because present economic problems affecting most of society are more salient in people's minds.

Polish Ecological Policy: Its Origins and Main Actors

Polish ecological policy originated in the early 1970s. During that period, the protection of the natural environment was included for the first time in governmental decisions concerning state development. It was the result of two independent events that occurred concurrently. First, U Thant's report was published in 1969 and thus people all over the world became more interested in environmental issues. Secondly, starting in December 1970, a new political elite of the Polish United Workers' Party ruled Poland. The leadership of Wladyslaw Gomulka was put aside due to massive protests, and a new group headed by Edward Gierek, tried to be perceived as much more modern, open to the world and better educated. Therefore, they declared their interest in ecological problems and promoted various activities in both domestic and foreign policies (Alberski 1996: 88–9).

Another important factor was the international ecological debate initiated during the 1972 Stockholm Conference on the Human Environment (Pietras 1996: 231). In 1975, the Polish government approved a 15-year program of environmental protection, which became the official approach to national environmental policy. The program, however, did not present the actual state of the natural environment in Poland; on the contrary, it was rather optimistic and diminished the role of environmental issues in spatial planning (Kozlowski 1983: 9–11). Therefore, although environmental protection was mentioned in many significant documents, the planned activities were of a rather symbolic character. It was the result of the accepted social and economic policy. The main goal was to improve living conditions and the main means to achieve it was intensified industrialization. It inevitably led to increased pressure on the natural environment. In short, environmental protection was understood mainly as diminishing some negative effects of industrialization, but in such a way as to avoid slowing economic growth. The well-known saying was that industry would repair the losses it caused; industry was perceived as having no negative aspects or features.

During this period, public political actors, namely institutions of the central and local structures of the state government, had the greatest impact on environmental policy in Poland. All other actors (e.g., local authorities, political parties, directors of huge factories) tried to act through public ones, because that guaranteed the highest chances for success of planned actions. The policy of the 1970s brought profits and prestige to the groups engaged in the development of mining, metallurgy, energy production and the chemical industry. They formed a lobby, became powerful political actors and efficiently blocked any decisions that could weaken their positions. Moreover, ecological issues and goals were the responsibility of many other public actors, including seven ministries and several governmental councils. Generally, in the 1970s, the public actors influencing environmental policy were characterized by high dynamics (i.e., constant changes in structure and organization), poor cooperation, weak national political structures and the inability to establish and realize efficient ecological policy (Alberski 1996: 103–15).

The origin of Polish ecological policy points to several approaches in the theoretical typology of environmental foreign policy described by Barkdull and Harris (2002). What happened in Poland in the 1970s concerning environmental policy was the outcome of the interaction between various forces operating at international and domestic levels. First, ecological issues became systemic ideas that strongly influenced the policy of many countries, even the socialist ones like Poland (the systemic and cognitive approach). Poland's new political elite at that time was the only social group holding power and they decided to follow the newly emerged systemic ideas about the environment (the societal and power approach). Then, gradually, some bureaucratic institutions became the most powerful in the state (the state-centric and power approach), and their interests ultimately controlled Poland's EFP (the state-centric and interest approach).

Early Environmental Foreign Policy

In the early 1970s, Poland actively participated in many international environmental programs, such as the Scientific Committee on Problems of the Environment, Man and Biosphere; the United Nations Environment

Program; the Helsinki Convention etc (Alberski 1996: 89). However, that participation was limited to those programs, and only such agreements that did not contradict interests of the main political actors in Poland. Poland refused to sign the Sulfur Protocol of 1985, which called for the reduction of sulfur dioxide (SO_2) emissions by 30 per cent. The mining and metallurgy sectors dominated in Poland's economy at that time, and they blocked joining protocol. Joining the protocol's so-called '30-per cent club' would mean weakening their position and enormous costs involved in modernizing old technologies. To show the scale of the problem, annual SO_2 emissions in Poland increased from 2.5 million tons (mt) in 1978 to over 4 mt in 1988, making Poland the third largest European polluter after the USSR and the GDR. In 1988, Poland emitted more SO_2 than the United Kingdom (UK) (3.5 mt), West Germany (3.4 mt) or France (1.8 mt). At the same time, the quantity and quality of Poland's industrial production were significantly lower than in those countries.

According to the typology of Barkdull and Harris (2002), Poland was a 'dragger' (the systemic and interest approach to EFP). This was the outcome of high costs of environmental-damage abatement and the power possessed by lobbies within bureaucratic state institutions (the combination of societal-power and societal-interest approach factors). However, despite the Sulfur Protocol and similar events showing that Poland was not willing to participate in some international environmental actions, mainly due to economic limitations, it nevertheless tried to be active in some bilateral cooperation. Ecological agreements were signed with Finland (in 1986), Sweden (1986) and the Netherlands (1987) (Pietras 1996: 117, 229–32).

SOCIALIST HERITAGE AND NEW TENDENCIES IN POLISH CLIMATE POLICY AFTER 1989

Kocik (2000: 208) argues that the majority of Polish society is still at the stage of gradually realizing the threats to the natural environment caused by human activity, although people still do not understand many ecological connections. Public awareness of ecological problems is therefore still being created, established and shaped. However, some political and scientific elite groups have already achieved a high level of environmental awareness, and they are more and more determined to modify Poland's policy towards more ecological protection. In the 1990s, an increase of ecological awareness in Poland's society could be observed. People became more sensitive to environmental pollution, and the quality of the natural environment turned out to be an important element of people's quality of life. However, the differences in ecological awareness between the Western and Eastern

European countries are still one of the largest barriers to environmental protection. It results mainly from insufficient ecological education in the former socialist countries.

A serious disadvantage in raising public awareness on environmental issues in Poland is a relatively low standard of living of the majority of the population. In 2001, the gross national product (GNP) reached US$4,600 per capita, which is much lower than the average for the world and for Organization for Economic Cooperation and Development (OECD) countries. In 2002, the unemployment rate increased to 18 per cent (in some regions it exceeds 25 per cent) and possibilities for the Polish population to be involved in general consumption are very modest (Napiorkowski et al. 2002: 33–4, 61). Another problem is the increasing gap between the richest and poorest social groups. Most families spend 37–56 per cent of their income on food and regular monthly payments (e.g., electricity, apartment rental, etc) (Olecka et al. 2001: 16). High unemployment is of a structural character, and therefore its reduction could take decades. Increasing unemployment caused pauperization of many families in the 1990s, and that in turn has led to a decrease of public acceptance for the ecological activities to be given high importance. In this situation, where people's basic needs cannot be satisfied, there is hardly any chance to convince people to make sacrifices for the natural environment (Kocik 2000: 209). Indeed, many people are ready to sacrifice the natural environment if doing so can improve their living conditions (Kassenberg 2002: 42).

Another important issue is that during the socialist period, Poland was isolated from Western-style goods. The general lifestyle was completely different than that in the West. Because all factories were state-owned, there was no competition, no advertising and no publicity. As mentioned earlier, the development of the means of production (meaning mainly the heavy industry) was much more important than the production of consumer goods, which were of rather poor quality anyway. Therefore, the changes that have been occurring since 1989 have created very high consumption expectations in Polish society. However, it has turned out that most people could not afford even moderate consumption, which has only strengthened those desires. Furthermore, a rapid increase in the number of cars can be observed, together with a growing amount of waste and the degradation of the landscape by huge shopping centers (Kassenberg 2002: 85–8).

Sustainable development and the protection of the atmosphere are usually viewed as tasks demanding a certain reduction in consumption in the rich Western World. Indeed, people living in those countries use enormous amounts of energy (especially in comparison with some developing countries) in order to make their lives even more comfortable. Therefore, reducing consumption would mean for them only a few less fancy gadgets, without any significant impact on their real quality of life. In the case of Poland, demanding that people lower their levels of the consumption would be regarded as absurd by most of Polish society, as most of the families can hardly satisfy their basic needs. Moreover, the growing consumption of a wide range of goods from all over the world is one of the symbols of Poland's regained freedom and independence; it is a sort of compensation. Poland is a relatively large country, which means that it is also an attractive market for large international companies. The danger that Poland faces is that it will blindly follow the Western way, instead of learning the lesson of how consumption affects the environment. Poland runs the risk of increasing consumption, following false needs created by advertizements, instead of fulfilling real needs. However, even if this happens, Poland should be judged differently than the West, owing to its history.

Reflecting on the cognition-societal approach summarized by Barkdull and Harris (2002), we can say that GCC issues are not yet widely recognized in Polish society, although there are some elite groups that strongly promote them (as described further below). This results from excluding environmental issues from public debate before 1989, and from economic problems experienced by many families in the present free-market system. However, Polish foreign GCC policy is first of all an outcome of external, geopolitical conditions, and only to a small degree a result of public awareness of those issues.

DOMESTIC CLIMATE POLICY AND ITS MAIN ACTORS

Even today, there is no strong ecological lobby in the Polish parliament. In 1991, three green parties participated in elections, and together they obtained 2.26 per cent of the vote (Pietras 1996: 237). None of them, however, had a well-prepared program. The candidates were not reliable, often connected with the former socialist party. There were no young people truly engaged in ecological issues. Therefore, all the green parties failed to place even one deputy in the parliament. In a way, the ecological initiatives of the 1980s were wasted and lost their significance in public life. However, since 1992, two tendencies may be observed. One of them is the rise of public awareness expressed, for example, by a growing number of publications on ecological issues and larger involvement of local authorities in environmental protection. An alternative tendency is the slowing down of ecological economic development by decision makers in the government and parliament, proved, for example, by accepting the strategies of economic development that ignore ecological issues, and the weakening of the role of the State Inspection of Environmental Protection (Poland's so-called 'ecological police'). These tendencies result from the fact that Poland's democracy is still being formed and thus each political party winning elections begins to manage environmental protection itself, without a clearly defined program.

Every election means an almost complete replacement of the personnel working in political bodies responsible for environmental protection and ecological economic development (the situation is the same in all other political bodies). The result is politics – not competence – that matters when new personnel are recruited. That has a disastrous effect on Poland's environmental policy (Kozlowski 2000: 19, 77–8, 292–3). However, even though environmental protection is still at the political margin (since issues like unemployment take much more public attention, as argued previously), the process of widening the social context of environmental protection and diversification of the actors participating in it is happening continuously (Alberski et al. 2002: 16–27). To be sure, none of the Polish political parties promotes ecological issues as important ones; those issues are usually omitted or treated instrumentally. However, the number of Polish ecological non-governmental organizations (NGOs) has reached 800. Even though they have gradually become more professional and created many young leaders, they have not been able to establish a 'green lobby' that would be able to influence state policy.

On the other hand, there are strong economic and political lobbies inherited from the communist past that act against sustainable development and against solving GCC-related problems. Those lobbies include coalmines and huge power plants, as well as companies producing cars and fuels. They tend to increase their production, disregarding greenhouse gas (GHG) emissions. The coal-energy production lobby profits from strong connections between owners, managers and trade unions (Kassenberg 2002: 76). Most coalmines and power plants are still state owned; the share of foreign investors in electricity production is about 10 percent, and in the case of heat production it reaches 35 per cent (Domanski 2001: 367). The lobby blocks the development of renewable sources of energy or tries to buy cheap electricity in the countries of the former USSR in order to export it to the EU. Such cheap electricity is often produced with old, environmentally harmful technologies. Furthermore, the transportation lobby favors the increasing role of cars instead of the role of public transportation (Kassenberg 2002: 77).

In 1991, the Polish parliament adopted the First National Environmental Policy. It was the first strategic environmental action program in Poland (and in Central and Eastern Europe), and one of the first documents in the world that treated environmental protection in an integrated manner. It covered all of the most important components of the environment, along with the most significant environmental pressures, indicating their sources in the most prominent areas of daily life and economy. It also adhered closely to the principles of sustainable development, referred to at that time in Poland as 'eco-development'. The Second National Environmental Policy of 2000 is completely different to the first one, and it takes new conditions into consideration (e.g., globalization, the Rio Declaration and the new administrative division of the country). Its main goal is to provide environmental integrity for current and future generations by the adoption of a development model that will guarantee effective regulation and rational use of the environment. It seeks to ensure that the type and scale of its use will not pose any hazard to the quality of natural resources or to environmental sustainability (Napiorkowski et al. 2002: 8–12).

However, Poland's domestic environmental policy is still more to do with the Ministry of Environment than the policy of the whole government. Environmental projects are discussed only by small groups of specialists, and business representatives do not participate in the debate at all. Quite often contradictions between the decisions of the Ministry of Environment and other ministries occur, as each of them acts independently. The units of central administration responsible for management of natural resources and their protection are separated from the ones responsible for the economy. Therefore, the links between state policy and environmental policy are rather weak (Alberski et al. 2002: 31). Moreover, while investments in environmental protection amounted to 1.6 per cent of GNP in 1996–98, they decreased to 0.9 per cent in 2000 and 0.8 per cent in 2001, mainly due to decreased environmental investment by industry. Air quality protection took 35 per cent of the mentioned funds (Internet Service of the Central Statistical Office 2001).

Despite Poland having ratified the Framework Convention on Climate Change (FCCC) in 1994, no formal or legal framework has been created to attract private business to undertake actions that would result in decreased GHGs emissions (Kassenberg 2002: 47). The government claims to has realized programs of GHG emissions reductions in the state-owned sectors of the economy (ie energy production, steel production and coal mining), but those achievements are a byproduct of economic restructuring. This means that during the transformation from centrally planned economy to free-market economy, Poland had to first increase the effectiveness of energy usage, and that resulted in much lower emissions. This tendency continues and so there is no need to undertake any additional efforts. In other words, emission reductions are a result of economic changes, not of projects especially designed for that purpose (Olecka et al. 2001: 34). However, the largest Polish steelworks were sold to foreign private investors in 2004, and the increasing demand for steel in the world market may lead to more production and a return to larger emissions in the future.

At the end of 2003, the Polish government accepted 'The Climatic Policy of Poland' (Internet Service of the Ministry of Environment 2003b). The document was prepared by the Ministry of Environment as a requirement of Poland's participation in the FCCC. It summarizes the legal framework of Polish climatic policy (international agreements signed), and presents a wide variety of postulates concerning the actions that should be undertaken. However, the document is of only declarative and general character. The next few years will show whether those declarations will be turned into action supported by new legislative acts needed for its realization.

Following the typology of Barkdull and Harris (2002), a conclusion can be formulated. The ideas concerning environmental problems are widely held in some elite groups of Polish society, but they are hardly reflected in the activity of Poland's political and business elites, as far as domestic environmental policy is concerned. As Poland does not have a political environmental lobby, there is no group to seriously influence interests of the business sector on environmental issues. Therefore, Poland's climate foreign policy is not rooted in a domestic *environmental* policy so much as outside political stimuli. This situation is additionally connected with the fact that, under the FCCC, the Polish government does not have to undertake any actions where the main motivation would be decreasing GHG emissions. When economic decisions are realized, emissions are reduced anyway. It is a comfortable situation for the decision makers, on one hand, resulting from the conditions of the pre-1989 period, but there is also a danger, on the other hand, that Poland's GCC policy will remain a matter of external conditions instead of national debate based on public awareness of the issue.

MEMBERSHIP IN THE EUROPEAN UNION

In the 1990s, the Polish parliament began to modify Poland's legal system so as to adjust it to the new political reality, but also to make it compatible with the legal system of the EU in preparation for joining the organization. Twenty-eight new acts concerning environmental protection were passed (most of them were amended later, as often as 20 times). Unfortunately, Poland does not follow the tendency observed in Europe to consolidate the law and pass complex acts; on the contrary, one issue is sometimes regulated by four different documents (Alberski et al. 2002: 89–94). In the EU, the general approach to environmental legislation is different than in Poland because EU regulations should not only protect the environment but also contribute to the functioning of the common market.

Poland's membership in the EU demands that air pollution standards in EU Directive 96/62/EC are not exceeded. For Poland, the standards concerning tropospheric ozone and their particulates will be the most difficult ones to apply. Moreover, it will be necessary to improve the quality of gas used in Poland, according to the standards in EU Directives 93/12/EC and 98/07/EC. It means that some of Poland's refineries will have to be modernized; others will only be able to blend gas, and not to produce it. Additionally, Poland will have to reduce emissions of volatile organic compounds produced during the storage and distribution of gas, in accordance with EU Directive 94/63/EC. Therefore, significant investments will be needed in the fuel production and distribution sector (Internet Service of the Ministry of Environment 2003a).

The Polish government did not profit from the fact that the GHGs emissions decreased significantly in Poland in the 1990s. It could be an important argument in negotiations concerning conditions of Poland's accession because it is much more difficult for the West European countries to achieve such good results in GHG emissions reductions. Unfortunately, that opportunity was wasted (Kassenberg 2002: 49).

The total cost of adopting EU ecological law in Poland will be very high. For example, realization of Directive 94/63/EC will cost €411 million (Internet Service of the Ministry of Environment 2004b). Therefore, harmonization programs may be realized only if new national financial resources are generated – if Poland obtains financial help from the EU – and if the present system of financing environmental protection is maintained. That system is based on ecological funds at the national and regional level. New national financial resources should be generated by, for example, introducing product fees, wider usage of commercial credits by investors, and shifting a part of investment costs to the goods and services recipients. Financial help from the EU is realized within the following programs: the Instrument for Structural Policies for Pre-Accession (ISPA), the Special Accession Program for Agriculture and Rural Development, Labor Sponsored Investment Funds, and 'Phare' (*Pologne, Hongrie Assistance à la Reconstruction Economique*) program to aid new member states' accession to the EU. Foreign funds will cover only about 10–15 per cent of all costs, but they will be an important element of the whole system. Ecological funds at the national and regional level are not financed from the national state budget, but from (for example) payments for using natural resources and fines for exceeding pollution standards.

An important issue is the social costs of realizing the adaptation of the EU environmental law in Poland. Most probably the majority of costs (or all of them) will be shifted to users. The costs of the EU directives' application may be included in electricity or gas prices. Those payments are already very high for most Polish families, and therefore Poland has asked for quite long transitional periods for complete application of a few EU directives, even up to 10 years regarding selected kinds of air pollution (Internet Service of the Ministry of Environment 2003a, 2004a).

INTERNATIONAL FINANCIAL SUPPORT FOR POLISH CLIMATE-RELATED INVESTMENTS

At the beginning of the 1990s, Central Europe was a highly polluted region. Other European countries could not improve their environmental quality without decreasing pollution in Poland, the former Czechoslovakia and the GDR. On the other hand, it has become obvious that Central European countries do not have sufficient funds for performing those tasks. In the case of Poland, the problem has been solved by so-called 'eco-conversion'. On April 21, 1991, the Paris Club decided that 10 per cent of Polish foreign debt could be converted to environmental-protection objectives. (The Paris Club of creditor nations had lent Poland about US$18 billion in the period 1971–78.) Those debts were not repaid, and amounted to more than US$32 billion (including interest) in 1991. Poland was incapable of repaying this amount. Therefore, in 1991 the Paris Club decided to write off 50 per cent of Poland's debt, on condition that the remainder would be paid off by 2010. The agreement also allowed for a further 10 per cent of the debt to be converted for purposes defined in bilateral agreements.

One of those purposes was for environmental protection. An 'Ecofund' was established with the budget of US$570 million to be spent within 18 years. The creditors that joined the project included Finland, France, Italy, Norway, Sweden, Switzerland and the United States (US). The money is used, first of all, to reduce Poland's pollution of fresh water, which leads to improvement in the Baltic Sea's ecological conditions. The second goal is to decrease the emissions of SO_2 and NO_x from the so-called high emission sources (high chimneys of power plants and factories). Those international efforts turned out to be very efficient, especially in decreasing SO_2 emissions.

In recent years, Poland also received EU subsidies from the Phare program, and since 2000 it has been entitled to aid from the pre-accession ISPA fund, which is designated to adjust technical infrastructure of the candidate countries to EU standards. Foreign assistance made up from 4–7 per cent of the total amount of expenditures for environmental protection in Poland (Kozlowski 2000: 147–53; Napiorkowski et al. 2002: 150–57).

On February 16, 2005, the Kyoto Protocol finally came into force, which means that Poland gained the possibility of introducing new, clean technologies using joint implementation (JI) mechanisms. Many concrete actions have already been undertaken, such as a Polish-Dutch project on using biomass as a fuel for the municipal heating system in Jelenia Góra. Poland may also benefit from technological achievements abroad thanks to the Clean Development Mechanism, and it can profit from emissions trading (Olecka 2005).

POLISH INVOLVEMENT IN INTERNATIONAL ENVIRONMENTAL AGREEMENTS

Following the collapse of communism in Central Europe at the end of the 1980s, the 1990s brought new global environmental threats. Poland faced economic crisis and began its preparations for joining the EU. Poland regained its freedom and independence, but even today Poland's GCC foreign policy has been burdened with the socialist heritage. Nevertheless, Poland has signed the GCC agreements. Poland participates in important international initiatives concerning sustainable development, mainly those organized by the United Nations (UN), and concerning cooperation with (and since May 1, 2004, within) the EU. However, these international activities are not followed up by the incorporation of experience and solutions developed in other countries into Poland's policy, science or NGO activities. There is no system of knowledge and information transfer from experts and politicians to the society (Kassenberg 2002: 107–11).

The Geneva and Vienna Conventions

Poland participated in the negotiation of the Geneva Convention on Long-Range Transboundary Air Pollution and ratified it in 1985 (Official Journal of Laws 1985). Since 1988, Poland has also been a party to the Protocol to the Geneva Convention regarding long-term financing of the Cooperative Program for Monitoring and Evaluation of the Long-Range Transmission of Air Pollutants in Europe (EMEP). In 1988, Poland became a signatory of the Protocol to the Geneva Convention regarding control of emissions of nitrogen oxides and their transboundary fluxes, and in 1994 it signed the Protocol to the Geneva Convention concerning further reduction of sulfur emissions (the so-called Second Sulfur Protocol). In 1998, Poland signed two other protocols to that convention: one concerning control of persistent organic pollutants and the other regarding control of heavy-metal emissions. In 2000, Poland signed the protocol on the prevention of acidification, eutrophication and creation of near-ground ozone (Sommer 1995, Napiorkowski et al. 2002: 128–39).

These actions were possible because emissions in Poland decreased significantly after 1989 due to economic crisis, lowered production, improved efficiency and changes in the structure of the economy. In the period 1992–2000, SO_2 emissions decreased by 46 per cent, NO_x emissions by 26 per cent, and particulates by 70 per cent (Napiorkowski et al. 2002: 128–39). Thus this time, unlike in the 1980s, the coal-electricity lobby did not block the decision simply because Poland was able to fulfill the obligations of agreements without affecting the lobby's interests. In 1990, Poland issued a regulation in which, for the very first time, norms for emissions of SO_2, NO_x and particulates from fuel combustion processes in energy production were specified. Taking into account European legislation, the government introduced EU requirements for new facilities.

Still, emissions in Poland are one of the highest in OECD countries: 0.5 mt of particulates, 1.5 mt of SO_2 and 0.8 mt of NO_2 (Internet Service of the Central Statistical Office 2001). Increased concentrations of SO_2 and particulates, sometimes exceeding the norms, were recorded mainly in industrialized regions of southern Poland, especially in the cities of Silesia. (However, in that region a significant overall reduction in the concentration of pollutants has been observed.) Due to rapidly increasing traffic, admissible values of NO_x, CO_2 and particulates in urban areas are exceeded more and more often, which may lead to the origin of tropospheric ozone (Napiorkowski et al. 2002: 128–39). In the 1990s, the air pollution decreased significantly in many areas considered to be ecologically endangered, but 20 new endangered areas emerged, mainly in large city centers. Almost all investments in air quality improvement are addressed to the industry, while transportation and municipal management pose new threats. Moreover, there is no ecological policy for that sector; on the contrary, private transport is promoted while public transport is rather neglected (Kozlowski 1997: 9–32). Therefore, any tendency toward decreasing air pollution may have a fading character.

Since 1990, Poland has been a party to the 1985 Vienna Convention on the protection of the stratospheric ozone layer, which it ratified on May 25, 1992 (Official Journal of Laws 1992a). Next it ratified the Montreal Protocol on Substances that Deplete the Ozone Layer (Official Journal of Laws 1992b). In 2000, Poland ratified the 1990 London amendments to the Montreal Protocol (Official Journal of Laws 2001a), and in 2001 it did the same for the 1992 Copenhagen amendments (Official Journal of Laws 2001b). Poland has employed the regulations on the import and possible re-export of ozone-depleting substances (ODSs). In 1997, a system of permits for import and export was introduced, together with a ban on the importation and export of ODSs. The use ODSs has decreased; in 1986 the quantity of Freon (an ODS) consumed was equal to 6,656 tons, while in

1999 it had fallen to 187 tons. Exploitation of halon ODSs was completely eliminated in 1995. Of all substances controlled by the Montreal Protocol, only carbon tetrachloride is produced in Poland, which is now manufactured in small quantities (and treated by the Protocol as exceptional because it is part of the process that facilitates production of chlorine) (Napiorkowski et al. 2002: 128–39).

The Framework Convention on Climate Change

The FCCC was signed in 1992 and Poland ratified it in 1994 (Official Journal of Laws 1996). Poland actively participated in the work of FCCC from the beginning. During the December 1997 third conference of the parties (COP3) in Kyoto, Poland supported the position of the EU and was ready to reduce its emissions by 10 per cent by 2010. However, if emission trading and usage of the emissions reductions achieved in the past were to be included, Poland's position was closer to that of US, Canada, Japan and Australia (Sadowski and Olecka 1998). During COP5 in Bonn in 1999, Poland became responsible for coordinating the debate (Sadowski 2000). During COP7 in Marrakech in 2001, the Polish delegation managed to include in the 'Marrakech Accords' the provision that countries like Poland, going through systemic transformations (economies in transition), would qualify for financial and technological support for the construction of infrastructure serving development and technology transfer. At the same time, such states were not required to support developing countries, although they may do so. That is a great advantage that leaves more funds for improving the situation in Poland (Sadowski and Olecka 2002). The Marrakech Accords were adopted during the COP11/MOP1 meeting in Montreal, Canada (November to December 2005). Poland participated in the meeting and continuously supported the position of the EU, which was in accordance with Polish national interests (Internet Service of the Ministry of Environment 2006).

Generally, Poland fulfils its obligations under the convention in spite of this requiring comprehensive action throughout the economy and it being associated with the establishment of new legislative, economic and financial instruments. Poland was mentioned in Annex I of the FCCC as a developed country. This meant that the emissions of GHGs were supposed to be stabilized by the year 2000. However, as a country going through systemic transformation, Poland took advantage of the possibility offered in article 4.6, choosing 1988 as the base year, even though 1990 was recommended by the convention. That made Poland's task much easier, as CO_2 emissions in Poland in 1990 were 20 per cent lower than in 1988. This was the result of the economic crisis that followed the end of socialism in 1989, including reduced production as well as the introduction of new and more efficient technologies (Kassenberg 2002: 45). In 1998, Poland signed the Kyoto Protocol, agreeing to reduce its emissions of GHGs by 6 per cent in the years 2008–12, and it ratified the protocol in July 2002 (Napiorkowski et al. 2002: 128–39; Official Journal of Laws 2002).

As noted by Olecka (1999), 97 per cent of CO_2 emissions in Poland come from energy production. Since 1992, emissions have stabilized at a level 20 per cent below the base year of 1988. This means that Poland is able to follow the regulations of the Kyoto Protocol. The emissions stabilization after 1992 has been a result of increasing efficiency of the whole economy. Poland has undertaken various economic policies that have led to the reduction of GHG emissions, such as the restructuring of heavy industry and the energy production sector, and modernization and replacement of outdated technologies (Narodowy Fundusz Ochrony Srodowiska i Gospodarki Wodnej 1998: 6–7). Even though Poland has reduced emissions of GHGs significantly, their level is relatively high on the European scale. In 2000, Poland emitted 315 mt of CO_2, 2.2 mt of CH_4 and 77,000 tons of N_2O. Poland emits 1.6 per cent of global CO_2 input to the atmosphere (Internet Service of the Central Statistical Office 2001). At present, Poland fulfils agreements regarding the stabilization of the GHGs emissions, but it does not have an official, governmental strategy for the GHGs reduction. Therefore, the strategies of developing particular sectors of the national economy do not meet the FCCC's requirements. There are no systemic GCC actions (Kassenberg 2002: 45–9).

Cooperation with Central European Countries

At a 1991 meeting in the Czech town of Dobris, European ministers of environmental protection decided to prepare an environmental program for Central and Eastern Europe. The program was accepted during the next ministerial conference in Lucerne in April 1993. It was titled 'For Clean European Environment' and applied to Albania, Bulgaria, the Czech Republic, Croatia, Estonia, Latvia, Lithuania, Poland, Russia, Romania, Slovenia, Slovakia, Ukraine and Hungary. Those were the countries that suffered the most from the communist period. Their environment was polluted to such an extent that it constituted a threat to human health. Additionally, environmental degradation was causing significant losses in those countries' economies and their biodiversity. Each country was supposed to prepare a national program of environmental policy, focusing mainly on local problems. In each country, the responsibility for past actions harmful to the environment was to be defined, and the transfer of 'clean technologies' was to be organized between the European countries and institutions. The role of environmental ministries was supposed to be strengthened.

The next meeting was in Sophia in 1995. The participants accepted the 'Pan-European Strategy of Biodiversity and Landscape Diversity', for the years 1996–2000, as a contribution to the Convention on Biodiversity signed at the 1992 Earth Summit in Rio de Janeiro. The main aim was to achieve, within 20 years, protection of European biodiversity and landscape diversity from potential hazards, involving the whole of society in the process. The ministers met again in 1998 in Aarhus, where they discussed access to information, participation of society in decision-making processes, and environmental justice. However, at the end of the 1990s, cooperation among Central European countries grew weaker. This mainly resulted from the fact that many of them began negotiations with the EU concerning future membership, and environmental protection was one of the issues being negotiated. The tendency to create specific forms of cooperation for Central Europe decreased (Alberski et al. 2002: 144–7; Kozlowski 2000: 147–51).

CONCLUSION

Polish foreign policy on climatic issues is a complex result of many factors. It is a part of Poland's environmental policy. In the typology of approaches to environmental foreign policy proposed by Barkdull and Harris (2002), the systemic theories seem to be most useful in interpreting Poland's EFP on GCC. The demands placed on Poland in the run up to membership in the EU created a new state identity, determining Poland's interests and shaping its behavior on environmental (and other) issues. Even though different political parties had enough power to form governments in Poland after 1990, joining the EU was the main goal of foreign policy for all of them. Thus the EU may be treated as a 'hegemon' that made Poland follow its environmental requirements to have a clean environment. Since Europe is quite a small continent, what is done in one country strongly influences all of its neighbors. Poland's position in the international hierarchy depends strongly on its membership in the EU, and, again, following the EU environmental requirements is a prerequisite for that. Therefore, even though the vulnerability of Poland's environment to damage is rather low, and the cost of abatement very high, Poland is no longer a 'dragger' in international environmental policy.

The typology proposed by Barkdull and Harris (2002) is a complex one and allows multi-level analysis of GCC policy. However, it is also two-dimensional. In the case of Poland and other post-communistic countries of Central and Eastern Europe, a third dimension – time – is of crucial importance. The changes in Poland's and other former communist countries' political systems at the beginning of the 1990s had a great impact on their environmental policies. The past, however, is not simply embedded in the minds of these societies. Some political, administrative and economic structures, inherited from the pre-1989 period, still exist and continue to shape these countries' GCC policies, as in the case of Poland. Therefore, instead of solely defining the impacts and relations using the Barkdull and Harris typology, it is necessary to include a dynamic factor: temporal changes. With time the importance of some factors diminishes, while other factors become more decisive.

We can first follow the changes using a cognitive (ideas-based) approach. In the period before 1989, environmental problems in Poland were excluded from public debate; they were prohibited, ignored by the state as insignificant, and recognized only by a small group of experts. At present, environmental issues are open to public debate (but not often discussed), promoted by the state as a significant element of Poland's international position, and better recognized by a gradually increasing segment of society. From the perspective of theory, this may look like a shift from a cognition-systemic to a cognition-societal pattern, but this is a false impression. Poland remains in a cognition-systemic pattern; it is the geopolitical situation and the superpower that have changed. As far as interests are concerned, the interest-societal approach can hardly be discussed, as Poland does not have a green lobby; there is no social power to balance the coal-energy lobby inherited from socialist times. The interest-systemic approach allows highlights that Poland shifted from a 'dragger' to a 'bystander' in GCC affairs because the reduction of its GHGs since the 1990s is a byproduct of economic changes. Also, the interests-state-centric approach reveals that the interest of the state – becoming a member of the EU – was the only factor strong enough to counterbalance with the coal-mining and energy-production sectors to force significant changes in GHG emissions (although this was not the main aim).

Application of a power-based perspective, as described in the Barkdull and Harris typology, shows that Poland's foreign policy concerning GCC has been highly dependent on the international system. First Poland was forced to accept Marxism interpreted by Stalin and to ignore environmental issues, then it decided to adjust to the demands of the EU (power-systemic approach). Looking for power in society, we can see that despite the shift from socialism to democracy little has changed. Before 1989, the coal-energy lobbies were also bureaucratic state institutions. Now they are officially distinguished from the state structure, but they are still state owned, which on one hand is an element of Poland's national energy security, but on the other hand is the source of their power. As far as a power-based state-centric approach is concerned, it can be stated that Poland's domestic environmental policy is still more the policy of the Ministry of Environment than a policy of the whole government. The links between central state policy and environmental policy are rather weak (Alberski et al. 2002).

Generally, there is a huge difference between Polish domestic and foreign policy on environmental issues. The national indices of pollution, emissions and so forth have shown continuous improvement since 1989, but this was at first largely the result of economic crisis and reduced production. After 1991, however, a significant shift in technology has occurred as many old factories have been modernized, resulting in reduced polluting emissions and increased energy efficiency. Those tendencies were accompanied by rapidly increasing unemployment (to raise efficiency and allow for technical modernization) and a dramatic drop in the standard of living of many families. Still, the permanently unstable political situation, continuously changing legislation, the lack of a political body representing the environment (such as a green party), and constant pauperization of Polish science and education must be contrasted with the improving environmental indices.

What is done within environmental domestic policy has been mainly the result of Poland's foreign policy: first the preparation for EU membership and now adjusting Poland's system to EU requirements. Such a situation can be explained, to a large extent, by the mentality of Poland's decision makers at all administrative levels. Many of them operated most of their lives under the communist system, which did not allow for any freedom in thinking or acting. The only possibility was to follow orders obtained from the higher-ranking officials. In the past those officials were, ultimately, in Moscow. Today they are in Brussels. Democracy, to be successful in Poland, requires a huge change in the social mentality, which most probably can be achieved only by the next generation. Until then, we can expect Poland's foreign policy, on global climate change and other environmental issues, to largely mirror that of the EU.

NOTE

1. For a discussion of global climate change and broader EU enlargement, see Chapter 13.

REFERENCES

Alberski, Robert (1996), *Polityka ochrony srodowiska w Polsce* [*Environmental protection policy in Poland*], Wroclaw: Towarzystwo Naukowe Prawa Ochrony Srodowiska [Scientific Association for the Environmental Law].

Alberski, Robert, Halina Lisicka and Jerzy Sommer (2002), *Polityka ochrony srodowiska* [*The environmental protection policy*], Wroclaw: Wroclaw University Publishing House.

Barkdull, John and Paul G Harris (2002), 'Environmental change and foreign policy: a survey of theory', *Global Environmental Politics* **2**(2) (May): 63–91.

Bywalec, Czesław (1990), 'Industrializacja Krakowa: przeslanki, przebieg, skutki' ['Cracow's industrialization: causes, performance, consequences'], in Maria Guminska and Andrzej Delorme, *Kleska ekologiczna Krakowa* [*Ecological disaster in Cracow*], Cracow: Polish Ecological Club.

Delorme, Andrzej (1990), 'Stalinowska industrializacja przyczyna kleski ekologicznej Krakowa' ['Stalin's industrialization as the cause of the ecological disaster in Cracow'], in Maria Guminska and Andrzej Delorme, *Kleska ekologiczna Krakowa* [*Ecological disaster in Cracow*], Cracow: Polish Ecological Club.

Domanski, Boleslaw (2001), *Kapital zagraniczny w przemysle Polski* [*Foreign capital invested in the Polish industry*], Cracow: Institute of Geography and Spatial Management, Jagiellonian University.

Internet Service of the Central Statistical Office (2001), *Ochrona srodowiska w 2001 r.* [Environmental protection in 2001], available at www.stat.gov.pl.

Internet Service of the Ministry of Environment (2003a), *Integracja europejska* [*European Integration*], available at www.mos.gov.pl/integracja_europejska.

Internet Service of the Ministry of Environment (2003b), *Polityka klimatyczna Polski. Strategie redukcji emisji gazow cieplarnianych w Polsce do roku 2020* [*The climatic policy of Poland. Strategies of greenhouse gases emission reductions in Poland up to 2020*], available at www.mos.gov.pl/1strony_tematyczne/ochrona_powietrza/ dokumenty_opracowania_raporty/polityka_klimatyczna.pdf.

Internet Service of the Ministry of Environment (2004a), *Wykaz okresow przejsciowych w obszarze negocjacyjnym 'Srodowisko* [*List of transitional periods in the negotiation position 'Environment'*], available at www.mos.gov.pl/integracja_europejska/index.shtml.

Internet Service of the Ministry of Environment (2004b), *Wyniki negocjacji akcesyjnych o członkostwo Polski w Unii Europejskiej w obszarze 'Srodowisko'* [Results of accession negotiations concerning Poland's membership in the European Union in the area 'Environment'], available at www.mos.gov.pl/integracja_europejska/index.shtml.

Internet Service of the Ministry of Environment (2006), *Notatka prasowa z przebiegu i wyników XI Konferencji Stron Ramowej Konwencji Narodów Zjednoczonych w sprawie zmian klimatu oraz I Spotkania Stron Protokółu z Kioto do tej Konwencji (Montreal, 28 listopada – 9 grudnia 2005)* [*Press release on the course and effects of the 11th Conference of the Parties to the United Nations Framework Convention on Climate Change and 1st meeting of the Parties to the Kyoto Protocol (Montreal, 28th November–9th December, 2005)*], available at www.mos.gov.pl/1aktualnosci/informacje_rp/13.12.2005.shtml.

Kassenberg, Andrzej (ed.) (2002), *Kompas Rio+10* [*Compass Rio+10*], Report No. 2, Warszawa: Institute for Sustainable Development.

Kocik, Lucjan (2000), *Miedzy przyroda, zagroda i spoleczenstwem* [*Among nature, farms and the society*], Cracow: Jagiellonian University Publishing House.

Kortus, Bronislaw (1990), 'Ocena powojennego uprzemyslowienia Krakowa w aspektach przestrzennym i ekologicznym', [Evaluation of Cracow's industrialization after the Second World War; Spatial and ecological aspects], in Maria

Guminska and Andrzej Delorme (eds), *Kleska ekologiczna Krakowa* [*Ecological disaster in Cracow*], Cracow: Polish Ecological Club.
Kozlowski, Stefan (1983), *Przyrodnicze uwarunkowania gospodarki przestrzennej Polski* [*Environmental conditions for spatial economy in Poland*], Wroclaw: Ossolineum.
Kozlowski, Stefan (ed.) (1997), *Kierunki zmian zachodzace w srodowisku czlowieka w Polsce. Lepiej czy gorzej?* [*Directions of Environmental Changes in Poland. Better or Worse?*], Warszawa: Polish Academy of Sciences.
Kozlowski, Stefan (2000), *Ekorozwoj: Wyzwanie XXI wieku* [*Eco-development: The Challenge for the 21st Century*], Warszawa: Polish Scientific Publishers PWN.
Leszczycki, Stanisław (1965), 'Problems of the post-war industrial concentration and decentralization in Poland', *Geographia Polonica* **7**: 29–47.
Napiorkowski, Krzysztof, Anna Palusinska and Tomasz Podgajniak (eds), (2002), *Agenda 21 in Poland: 10 years after Rio 1992–2002*, Warsaw: National Foundation for Environmental Protection.
Narodowy Fundusz Ochrony Srodowiska i Gospodarki Wodnej [National Fund for Environmental Protection and Water Management] (1998), *Drugi raport rzadowy dla konferencji stron Ramowej Konwencji Narodow Zjednoczonych w sprawie zmian klimatu* [*Second Governmental Report for the United Nations Framework Convention on Climate Change*], Warszawa.
Official Journal of Laws [Dziennik Ustaw] (1985), 'Governmental statement of 18th October, 1985, concerning the ratification of the Convention on Long-range Transboundary Air Pollution, prepared in Geneva on 13th November, 1979, by the People's Republic of Poland ' ['Oswiadczenie Rzadowe z dnia 18 pazdziernika 1985 r. w sprawie ratyfikacji przez Polska Rzeczpospolita Ludowa Konwencji w sprawie transgranicznego zanieczyszczania powietrza na dalekie odległosci, sporzadzonej w Genewie dnia 13 listopada 1979']. *Official Journal of Laws of the Republic of Poland* **60**: 312.
Official Journal of Laws [Dziennik Ustaw] (1992a), 'Governmental statement of 25th May, 1992, concerning the accession of the Republic of Poland to the Vienna convention on ozone layer protection, prepared in Vienna on 22nd March, 1985' ['Oswiadczenie Rzadowe z dnia 25 maja 1992 r. w sprawie przystąpienia Rzeczypospolitej Polskiej do Konwencji wiedenskiej o ochronie warstwy ozonowej, sporzadzonej w Wiedniu dnia 22 marca 1985 r.'], *Official Journal of Laws of the Republic of Poland* **98**: 489.
Official Journal of Laws [Dziennik Ustaw] (1992b), 'Governmental statement of 26th May, 1992, concerning the accession of the Republic of Poland to the Montreal Protocol on substances depleting ozone layer, prepared in Montreal on 16th September, 1987' ['Oswiadczenie Rzadowe z dnia 26 maja 1992 r. w sprawie przystapienia Rzeczypospolitej Polskiej do Protokolu montrealskiego w sprawie substancji zubozajacych warstwe ozonowa, sporzadzonego w Montrealu dnia 16 wrzesnia 1987 r.'], *Official Journal of Laws of the Republic of Poland* **98**: 491.
Official Journal of Laws [Dziennik Ustaw] (1996), 'Governmental statement of 15th September, 1996, concerning the ratification of the United Nations Framework Convention on Climate Change, prepared in New York on 9th May, 1992, by the Republic of Poland' ['Oswiadczenie Rzadowe z dnia 15 wrzesnia 1996 r. w sprawie ratyfikacji przez Rzeczpospolita Polska Ramowej konwencji Narodow Zjednoczonych w sprawie zmian klimatu, sporzadzonej w Nowym Jorku dnia 9 maja 1992 r.'], *Official Journal of Laws of the Republic of Poland* **53**: 239.

Official Journal of Laws [Dziennik Ustaw] (2001a), 'Governmental statement of 27 December, 2000, concerning the validity of the amendments to the Montreal Protocol on substances depleting ozone layer, prepared in London on 29th June, 1990' ['Oswiadczenie Rzadowe z dnia 27 grudnia 2000 r. w sprawie mocy obowiazujacej Poprawek do Protokolu montrealskiego w sprawie substancji zubozajacych warstwe ozonowa, sporzadzonych w Londynie dnia 29 czerwca 1990 r.'], *Official Journal of Laws of the Republic of Poland* **44**: 491.

Official Journal of Laws [Dziennik Ustaw] (2001b), 'Governmental statement of 16 March, 2001, concerning the validity of the amendments to the Montreal Protocol on substances depleting ozone layer, prepared in Copenhagen on 25th November, 1992' ['Oswiadczenie Rzadowe z dnia 16 marca 2001 r. w sprawie mocy obowiazujacej Poprawek do Protokołu montrealskiego w sprawie substancji zubozająych warstwe ozonowa, sporzadzonych w Kopenhadze dnia 25 listopada 1992 r.'], *Official Journal of Laws of the Republic of Poland* **44**: 493.

Official Journal of Laws [Dziennik Ustaw] (2002), 'The act of 26th July, 2002, on the ratification of the Kyoto Protocol to the United Nations Framework Convention on Climate Change' ['Ustawa z dnia 26 lipca 2002 r. o ratyfikacji Protokołu z Kioto do Ramowej konwencji Narodow Zjednoczonych w sprawie zmian klimatu.'], *Official Journal of Laws of the Republic of Poland* **144**: 1207.

Olecka, Anna, Marta Radwan-Röhrenschef, Anna Romanczak, Maciej Sadowski and Alicja Sienkiewicz, (ed.) (2001). *Trzeci raport rzadowy dla konferencji stron Ramowej Konwencji Narodow Zjednoczonych w sprawie zmian klimatu* [*Third Governmental Report for the United Nations Framework Convention on Climate Change*], Warszawa: National Fund for Environmental Protection and Water Management.

Olecka, Anna (1999), 'Strategie ochrony klimatu' ['Strategies of climate protection'], in Alfred Dubicki, Malgorzata Gutry-Korycka, Krzysztof Kozuchowski, Halina Lorenc, Leszek Starkel (eds), *Zmiany i zmiennosc klimatu Polski* [*Changes and variability of the climate in Poland*], Lodz: Publishing House of the University of Lodz.

Olecka, Anna (2005), 'Protokol z Kioto wszedl w zycie' ['The Kyoto Protocol came into force']. *Gazeta Obserwatora IMGW* [*Journal of the Observers of the Institute of Meteorology and Water Management*] (forthcoming publication).

Pietras, Marek (1996), *Bezpieczenstwo ekologiczne w Europie* [*Ecological Safety in Europe*], Lublin: Publishing House of the Maria Curie Sklodowska University.

Raczynski, Maciej (1990), 'O dziejach lokalizacji Nowej Huty pod Krakowem' ['On the location of Nowa Huta near Cracow], in Maria Guminska, Andrzej Delorme, *Kleska ekologiczna Krakowa* [Ecological Disaster in *Cracow*], Cracow: Polish Ecological Club.

Sadowski, Maciej (2000), 'Problemy zwiazane z ratyfikacja protokolu z Kioto dotyczace zmian klimatu' ['Problems in the area of climatic changes relating to the Kyoto Protocol ratification'], *Przeglad Geofizyczny* [*Geophysical Review*] **1**: 7–15.

Sadowski, Maciej and Anna Olecka (1998), 'Kolejne fazy naukowej i politycznej ewolucji pogladow na problem zmian klimatu' ['Further development of scientific and political views on climate change issues']. *Przeglad Geofizyczny* [*Review of Geophysics*] **3–4**: 133–46.

Sadowski, Maciej and Anna Olecka (2002), 'Przelom w negocjacjach Ramowej Konwencji Narodow Zjednoczonych w sprawie zmian klimatu i protokolu z

Kioto' ['Turn in negotiations of the United Nations Framework Convention on Climate Change'], *Przeglad Geofizyczny* [*Geophysical Review*] **1–2**: 5–14.

Sommer, Jerzy (1995), 'Zobowiazania traktatowe Polski w zakresie ochrony srodowiska' [Poland's treaty obligations in the area of natural environmental protection'], *Rocznik Polskiej Polityki Zagranicznej* [*Yearbook of the Polish Foreign Policy*], available at www.sprawymiedzynarodowe.pl/rocznik/1995/jerzy_sommer_zobowiazania_ traktatowe.html.

Sowa, Kazimierz Zbigniew (1990), 'Nowa Huta – modelowy przypadek stalinowskiego kreacjonizmu' ['Nowa Huta model case of Stalin's creationism'], in M. Guminska and A. Delorme, *Kleska ekologiczna Krakowa* [*Ecological Disaster in Cracow*], Cracow: Polish Ecological Club.

6. Sweden, climate change and the EU context

Kate E Marshall

INTRODUCTION

The 1995 accession of Sweden to the European Union (EU) was largely viewed with promise, as perhaps ushering in a new era of more environmentally sensitive behavior by the EU. The effects that EU membership would have on Sweden's domestic policies were hotly debated within Sweden, and the left and green parties took an anti-accession stance (Miles 1997: 26). However, outside of Sweden the environmental benefits of Swedish membership in the EU were less contested, as Sweden was seen as bolstering the general level of environmental awareness among EU member states. In concert with Finland and Austria, who also joined in 1995, Sweden was a potential coalition partner with Denmark, thereby forming a 'Nordic block' to advance environmental issues within the EU (Barnes 1996: 217–18). This was a well-founded expectation, given the strong history of environmental cooperation among the Nordic states in the Nordic Council (Barnes 1996: 218), as well as the high environmental standards each state maintained individually. A perfect alliance with perfect outcomes was not expected; the ability of states to influence EU policy is subject to limits, and the Nordic EU members have national rivalries, as well as differing national policies and national political preferences on some issues (Barnes 1996: 217–18). All the same, the accession of a state with a proven history of progressive international environmental cooperation was cause for increased environmental expectations.

While the environmental record of Sweden is certainly laudable, Sweden has not always made environmentally optimal decisions within the EU context. Molin and Wurzel note that Sweden 'has exhibited a somewhat ambiguous climate change strategy' (Molin and Wurzel 2000: 174). In 1999, Sweden nearly unraveled the EU's global climate change (GCC) policy because of its preference for unlimited emissions trading (ET) instead of a ceiling on emissions trading, the option preferred by a majority of EU states

(Molin and Wurzel 2000: 175). The roots of this stance lie deep in domestic politics and relate to Sweden's high reliance on nuclear power, a form of energy production that does not generate greenhouse gas (GHG) emissions. Nuclear power remains an important issue in Swedish environmental policy, and the results of a 1985 referendum barring the expansion of nuclear power in Sweden place it in a difficult position with respect to its domestic (nuclear-energy-reducing) and international (GHG-reducing) commitments. This chapter will argue that, past successes notwithstanding, this tension between energy production methods and environmental goals augurs ill for the future of Swedish GCC diplomacy, both within and outside the EU.

I will develop this argument using a leaders-laggards dynamic and parts of the theoretical framework advanced by Barkdull and Harris (2002) regarding types of environmental foreign policy (EFP). In particular, this analysis of Sweden's past failures and uncertain future with respect to GCC diplomacy will draw upon the roles of power, interests and ideas at the systemic, societal and state levels, as outlined in Barkdull and Harris's typology. Particular attention is paid to norms and ideas, since this is the manner in which Sweden traditionally engages in international environmental diplomacy.

The next section outlines theoretical considerations. The third section has two parts: a background discussion on nuclear power in Sweden, and an analysis of a 1999 incident that followed international adoption of the Kyoto Protocol. This analysis unites the societal, state and systemic levels in a manner similar to that of Robert Putnam's two-level games (Putnam 1988). The fourth section discusses examples of Sweden's past successes with respect to GCC diplomacy, noting the decline of such actions since EU membership. Finally, I explore issues related to sovereignty and future Swedish actions on GCC, noting the implications of Sweden's reliance on nuclear power for national, EU and international outcomes.

THEORETICAL CONSIDERATIONS

Leaders and Laggards

The leaders and laggards dynamic is an oft-debated concept in EU environmental policy. This chapter will use McCormick's description of the concept: 'member states have a mixed record on different issues at different times, with *tendencies* either to lead, or to lag, or to come somewhere in between the two' (McCormick 2001: 90; emphasis in original). Sweden, in concert with other wealthier Northern European states, has earned an overall reputation as an environmental leader (McCormick 2001: 90; Börzel 2002; Liefferink and Anderson 1998a; Liefferink and Anderson 1998b). I will argue that Sweden's general leader designation is subject to important caveats that have potentially far-reaching effects.

Few scholars engage with these aspects of Swedish environmental performance, preferring to examine the dynamics that animate pro-environmental action, and dismissing un-environmental actions with unexplored generalities. For example, Kronsell notes that 'Swedish environmental policy is not always perfect but has its own set of flaws and weaknesses' (Kronsell 2002: 291). Molin and Wurzel write that Sweden's position with respect to GCC demonstrates 'that even environmental leader states occasionally fail to live up to their reputation' (Molin and Wurzel 2000: 174). That simply will not do. These deviations from Sweden's near-automatic position of environmental leadership make its international positions on GCC theoretically interesting. Regarding environmental policy, the fact that Sweden rarely behaves poorly makes these cases of anomalous behavior all the more interesting: what factors cause states that are environmental leaders to behave in a manner more befitting environmental laggards?

Foreign Policy

To understand why a given state adopts a particular policy or orientation with respect to international environmental concerns, we turn to theories of foreign policy. Barkdull and Harris advance a typology of EFP that relies on three levels of analysis – systemic, societal and state – and three paths of influence – power, interests and ideas (Barkdull and Harris 2002: 69). In the case of Sweden, power, ideas and interests are intertwined.

Power, ideas, norms and interests

How is it that Sweden has emerged as a powerful state in environmental politics? Kronsell's argument regarding Swedish success in the EU environmental arena as a result of norms-based influence (Kronsell 2001; Kronsell 2002) can be broadened to explain Swedish success on the world stage. In particular, Kronsell identifies four specific ways in which Sweden has been able to influence the EU environmental agenda (Table 6.1): expectations of other states regarding the type of policy it will advocate, knowledge of environmental processes, expertise with environmental management at the national level, and an ability to present a unified national position based on a shared understanding of the national interest among all Swedish government representatives (Kronsell 2001: 5–9).

As a result of these domestic normative factors, Sweden is able to exert a perhaps disproportional amount of influence and persuade other states to trust its analysis and assume its valuation and causal interpretation of the issue; this is, in essence, a type of 'soft power', which is 'the ability to get desired outcomes because others want what you want; it is the ability to achieve desired outcomes through attraction rather than coercion' (Keohane and Nye 2001: 220). Attention to soft power thus refines the way in which power matters in approaches to the study of EFP, as will be discussed below. The promulgation of norms is a key part of soft power; Sweden's power and efficacy result from its ability to influence the norms of the larger community. While attempting to directly influence the norms of the larger world community is one route, the potential for Sweden's global influence is enhanced once the larger EU community adopts a Swedish policy or a compromise policy based on Swedish national laws. In this scenario, adoption of the common EU policy position, in essence, magnifies the Swedish position, projecting it forcefully beyond Europe.

Table 6.1 Kronsell's avenues of normative influence

- Expectations of other states regarding the type of policy likely to be advanced
- Knowledge of environmental processes
- Expertise with environmental management at the national level
- Ability to present an unified national position based on a shared understanding of national interest

Source: Kronsell 2001: 5–9.

Normative power is heavily reliant on ideas, turning our attention to the role of ideas in foreign policy. It is helpful to view norms as principled meanings that guide, rationalize, justify, or express expectations about actions or events (Ruggie 1998: 97). Understanding norms as 'meanings that guide' fits the dynamic process of normative influence because of the stress placed on self-reflection and, thereby, the potential for internalization. Ruggie also argues that norms are causal beliefs without the actual power of causality (Ruggie 1998: 97). This is especially applicable in the area of environmental issues, where scientific approaches and uncertainty frequently create conflicts. Under conditions of uncertainty, norms can be conceptualized as descriptive and cause-laden stories that states use to persuade one another. Thus, norms condition the larger context of interstate influence through their descriptive and causal power.

Finlayson and Zacher argue that the importance of norms comes from 'the importance [that] the most influential members [in a regime or system] attach to them' (Finlayson and Zacher 1983: 305). In international environmental policy debates Sweden is frequently able to exercise enough influence – drawing upon its leadership in the environmental domain, which is fed by its domestic normative strengths – to sway more traditionally powerful actors. Mitchell suggests that within a given system, norm dominance may come as a result of a response to specific pressures: economic, political or ecological (Mitchell 1998). All three types of pressures have the potential to push the EU toward harmonized environmental action, and to push international environmental agreements onto the world stage.

Norms-based leadership is well suited to the style of EU policy making, which relies heavily on negotiation, bargaining and informational networks in the absence of a single, powerful co-ordinating body (Kronsell 2002: 292). These policy-making approaches allow new and non-traditional actors – states and non-governmental organizations (NGOs) alike – access to and influence upon the intricacies of EU policy making. With increasing popular pressure and institutional support for transparency and democratization, an increase in the importance of norms-based leadership is a reasonable expectation.

As in the EU, international politics is also characterized by the absence of a sole central government. Thus, the factors that allow Sweden to act as a successful actor on the EU environmental stage, employing normative persuasion, also predispose Sweden favorably on the significantly larger world stage. Sweden's interests, of course, serve to knit the role of power and the role of ideas together. For it is the content of these ideas, which represent Swedish interests – those interests deemed salient by the national public and expedient by political actors – which are then sold to other political actors in a process that empowers Sweden.

The systemic, societal and state levels

What is the appropriate level at which to understand the formation of Sweden's GCC diplomacy? First, we might note that systemic theories, which 'emphasize the importance of the international system in creating state identities, determining what a state's interests are and shaping state behavior' (Barkdull and Harris 2002: 68), do not allot causal importance to domestic politics. Domestic politics, as will be explained below, is key for understanding Sweden's GCC diplomacy. This does not mean, however, that we ought to discard the systemic level of analysis. Rather, we might note that Sweden's approach to international affairs has been deeply influenced by the distribution of power in the international system. Sweden is a small state whose national economy and military – sources of power according to more conventional readings – do not currently compare to those of the larger western states, foreclosing those avenues of influence. Thus, for

Sweden, the structural characteristics of the international system reinforce a focus on soft power and norm-based diplomacy.

Although norms are articulated by the state, they have roots within the state; the power that Sweden is able to exercise, the interests that Sweden is able to represent and the ideas/norms that Sweden is able to 'sell' at the international level all have an intimate relationship with domestic political processes. For example, Kronsell's factors of normative influence, cited above, all draw heavily on national policy experiences or repertoires. Both state and society share these experiences and repertoires. In the case of Sweden, with its tradition of corporatist politics, it is particularly difficult to disentangle the state from society; insights from both levels of analysis are needed to make sense of politics at the systemic or international level. The applicability of multiple levels of analysis is not unique to Sweden: Barkdull and Harris state that 'no one theoretical approach will give a complete picture' (Barkdull and Harris 2002: 84), and thus plea for synthesis.

A corporatist state, one in which societal interests are explicitly recognized by the state and receive privileged, non-adversarial access in the policymaking process, is excellent ground for the work of synthesis. From the societal level, we might take elements of the idea approach that stress the role of social movements and public opinion, factors whose importance is evidenced below in the national referendum. However, this analysis of Sweden and GCC diplomacy does not accept the view of the state as a neutral arbiter or arena, as many societal views of the state do (Barkdull and Harris 2002: 74–5).

Rather, for an understanding of the Swedish state's actions in response to societal pressures, we must turn also to the state level of analysis, which explores the way in which states and state agencies act to protect a multitude of perceived interests. The state – even the Swedish state – has interests beyond environmental protection. Namely, the state wishes to protect *itself*. Environmental protection may be one way in which state survival is assured, but this is not the only way. Sovereignty concerns also play an important role in how the state evaluates international environmental cooperation. Here, I wish to argue that, under the proper conditions, even to a leader state like Sweden, 'the environment is simply insignificant in comparison to other foreign policy concerns' (Barkdull and Harris 2002: 70).

Norms have their limits. Norms have a qualitative character: they reflect beliefs, understandings and perceptions. As such, they may be limited by other norms or by non-normative factors. This chapter takes a 'least-likely' case in which a state with a high level of national environmental knowledge, a strong environmental reputation, a proven environmental record, and a well co-ordinated national interest, failed to act in an environmentally sustainable way. In the case of global climate change (GCC), Kronsell's normative factors were all present. However, in her work Kronsell considers only successful cases in which Sweden was able to persuade the EU to take the environmental high road and improve EU standards in air pollution and chemicals regulation (Kronsell 2002; Kronsell 2001). Thus, this chapter uses theories of EFP and the GCC case, in which Sweden failed to live up to its leader reputation, to clarify arguments about norm-based influence.

The puzzle, then, is this: factors predisposing Sweden toward an environmentally friendly norm-based GCC foreign policy were present, yet such a policy was absent at a crucial moment. Sweden could have just as easily engaged in its traditional *modus operandi* of EFP: norms-based influence. I will argue that the choice to engage in a foreign policy of normative influence can be circumscribed by sovereignty concerns, even for Sweden. In order to explore this dynamic in depth, this chapter will examine the case of Sweden's Kyoto Protocol-related behavior. Before doing so, however, a review of Swedish EFP successes is in order.

SWEDISH SUCCESS IN ENVIRONMENTAL DIPLOMACY

Sweden has had a long history as an active participant in international environmental politics. Notably, the United Nations (UN) Conference on the Human Environment was held in Stockholm in 1972. The symbolic value of playing host to the first significant global environmental conference established Sweden as a state on the forefront of environmental politics. In addition, though this was long before GCC appeared as an international political issue, Sweden did manage to use this conference to push issues related to acid rain and international air pollution (Haigh 1992: 237). Sweden played an important role in the framing of the acid rain problem as stemming from transnational pollution (Zito 2000: 51). The actions of Sweden, in concert with other similarly harmed northern states, led to the adoption of a management regime for long-range transboundary air pollution in 1979 (Haigh 1992: 238). At a 1982 Stockholm Conference, Jan Nilsson of the Swedish Environment Ministry synthesized and presented air pollution data in ways that were easily understood by policymakers and that led to international cooperation on the problem (Zito 2000: 66). Acidification remained an important priority for Sweden upon EU accession, and Swedish pressure resulted in the adoption of a strengthened preventative EU strategy in 1997 (Zito 2000: 76). The case of acid rain demonstrates the early and consistent ability of Swedish policymakers to engage in norm- and interest-driven environmental diplomacy related to atmospheric environmental issues.

With respect to GCC, it was a Swede, Svante Arrenhius, who in 1886 published the first paper warning of human-induced global warming (Nordhaus 1997: 42). In more recent times, Sweden has been an active participant in combating GCC. Anderson and Liefferink write that:

> 'Sweden's role has been to act as a catalyst for international developments, and in this process Swedish experts and diplomats have come to play a significant role in providing scientifically based advice and international leadership. Sweden's role in the UN's International Panel on Climate Change, an advisory body with its secretariat in Stockholm [then] headed by the Swede Bert Bolin, is a prominent example of this role but so are the negotiations at the Rio Convention at which the work of Swedish diplomats stood out' (Anderson and Liefferink 1997: 22).

Swedes are, indeed, sensitive to issues surrounding GCC. In the early days of the GCC issue, prior to the founding of the Intergovernmental Panel on Climate Change (IPCC) in 1988, the Swede Bert Bolin worked tirelessly to draw international attention to the scientific findings (Bodansky 2001: 26).

Additionally, 'Sweden has long made considerable efforts to cut its "greenhouse gases" of which carbon dioxide is the most important' (Molin and Wurzel 2000: 174). Prior to the EU's formal program requirements (to be discussed below) Sweden engaged in cooperative GHG abatement (joint implementation) programs abroad, such as those in Poland (see Chapter 5), where the costs of abatement are lower than they are in Sweden (Beckerman 1992: 283). Sweden also attempted to link its vote in the EU ratification of the UN Climate Change Convention to the (unsuccessful) adoption of a European carbon tax (McCormick 2001: 282). Carbon taxes are important to Sweden as it is also one of the six EU states that have a national carbon tax (McCormick 2001: 283). This tax has helped Sweden to maintain 'a low profile with regard to CO_2 emissions. ... 1990 CO_2 emissions from industry and energy had been cut by 33 per cent compared with the 1970 levels' (Kronsell 1997: 67).

However, Sweden's low CO_2 profile is not simply the result of environmentally friendly taxes. Swedish reliance on nuclear power is key. Yet, the triumph over CO_2 emissions in this way might be a Pyrrhic victory, as dependence on nuclear power contributed to the difficult intra-EU negotiations over the European position on GCC, in addition to preventing opportunities for Swedish influence in other international forums.

As we will see, some Swedish actions with respect to GCC contrast with Swedish successes in framing the air pollution issue. We now turn to an examination of a specific instance of Sweden's behavior with respect to international GCC politics. Sweden makes for an excellent case study because of its generally environmentally friendly behavior and the expectations that are associated with Sweden, a fact highlighted by Kronsell's model of normative influence.

SWEDEN AND KYOTO IMPLEMENTATION: THE ROLE OF NUCLEAR POWER

Following the adoption of the Kyoto Protocol in December 1997, signatories were faced with drawing up detailed plans to reduce their GHG emissions. While the EU member states signed the Kyoto Protocol individually, the European Community (EC) is also a signatory. Under Article 4 of the Protocol, which established the 'EU Bubble', the EU member states are permitted to meet the total EU GHG reduction commitment through differentiated commitments among themselves (McCormick 2001: 288). With this differentiated commitment approach, Sweden is actually permitted to increase its CO_2 emissions by 4 per cent (McCormick 2001: 288). Meeting the 4 per cent goal, however, is unlikely to be easy for Sweden; this situation will be greatly exacerbated should Sweden cease using nuclear power. As Nordhaus notes, '[e]ven with the current nuclear power production, it will clearly be difficult for Sweden to keep its commitments to stabilize carbon dioxide emissions. If Sweden chooses to phase out its nuclear power plants, meeting that commitment will go from difficult to near impossible' (Nordhaus 1997: 44). More broadly, in 2004 the then-EU Energy Commissioner Loyola de Palacio, commenting on the energy-environment tradeoff, said that the EU cannot both close nuclear power plants and meet Kyoto targets (*The Economist* 2004).

Descending to the societal level, we will now explore the pressures on Sweden to reduce its reliance on nuclear power plants and then, subsequently, examine the state's response in EU negotiations.

Background

During Sweden's accession to the EU in 1995, the government insisted on a special provision attached to its accession treaty stating that 'the right to choose sources of energy would not be affected by EU membership' (Molin and Wurzel 2000: 167). This provision ensured that Sweden retained its right to use nuclear power. In the wake of the 1973–74 oil embargo, Swedish policy sought to reduce dependency on foreign energy suppliers (Kronsell 1997: 45; Nordhaus 1997: 12). In this context, nuclear energy was seen as a clean and cheap source of energy, and it was designed to keep the entire nuclear fuel cycle within Sweden (Nordhaus 1997: 12). Despite mild controversy, these factors eased the introduction of nuclear energy in Sweden, and large-scale development of nuclear power began in Sweden in 1975 (Kronsell 1997: 45; Nordhaus 1997: 3). The organized Swedish anti-nuclear movement is very small, counting only 3,000 members in 2000 (Molin and Wurzel 2000: 169) out of a population of 8.8 million.

However, the small size of the anti-nuclear movement in Sweden is deceiving: the use of nuclear power in Sweden has been a highly divisive issue. In 1980, Sweden held a controversial national referendum in which the majority favored the reduction and eventual phasing out of nuclear power within its borders (Matlary 1996b: 238). All three options on the referendum had in common the view that there should be no further construction of nuclear power plants; at issue was the phase-out date for existing nuclear power plants (Nordhaus 1997: 35). While referenda in Sweden are advisory, in earlier referenda the Swedish Parliament followed popular sentiment on two out of three votes (Nordhaus 1997: 35). Although the 1980 referendum was criticized for not offering voters an option to continue the nuclear program and the option selected by the largest number of voters was the least restrictive of nuclear power, subsequent governments have remained at least rhetorically committed to the eventual goal of a nuclear phase-out.

During the time preceding the referendum, ardent anti-nuclear activists were able to advance their cause on the grounds of the potentially environmentally devastating effects of nuclear power, particularly in the event of an accident (Downie 1980). The 1979 Three Mile Island nuclear accident in the United States (US) cast 'doubts on both the reliability of nuclear power and, perhaps even more importantly, the credibility of the nuclear power industry' (Nordhaus 1997: 34). The Social Democratic Party, which had already been wavering on the issue of nuclear power, withdrew its support for nuclear power in Sweden and called for a national referendum.

Although the outcome of the referendum was initially cited as a victory for the preservation of nuclear power, the winning provision contained a poison pill: it allowed for the development of six new nuclear reactors, but it also limited future expansion by stating that no further reactor construction would take place after those six were built (Carey 1980). Thus, the approval of this provision potentially meant 'the end of nuclear energy in Sweden after the 25- to 30-year lifespan of the new reactors' (Carey 1980). That time period was to allow Sweden to develop alternative energy systems based on renewable resources (Downie 1980). From the 'apocalyptic vision' (Downie 1980) that dominated the campaign to the outcome of the final vote, themes of environmental degradation related to nuclear power represented salient and real concerns for the majority of the Swedish polity.

However, during the time since the 1980 referendum, 'nuclear energy has more than doubled its share of total energy production' (Kronsell 1997: 46). Additionally, since 1980, scientific awareness about the negative effects of other energy sources has grown, making nuclear power relatively more attractive. In particular, the Swedish Parliament granted protected status to the four major rivers in the north of the country, thus diminishing the likelihood that increased hydropower could compensate for a shutdown of nuclear power plants.

In 1991, as a response to the increasing reliance on nuclear power, an agreement was reached between 'various corporate and political actors in the energy sector' (Matlary 1996b: 238) which held that the phasing out of nuclear power would be postponed until 2010, the expected lifetime of the youngest reactor (Nordhaus 1997: 3). In 2000, however, nuclear power represented 46 per cent of Sweden's gross electricity production, second only to hydropower (Nordhaus 1997: 27). Sweden's energy consumption per capita is high, and, even including taxes, electricity prices for industry and households remain among the lowest in comparison with other industrialized nations (Nordhaus 1997: 29; Kronsell 1997: 67).

In March of 2002 the Swedish government set forth the conditions that must be fulfilled before the decommissioning of nuclear reactors commences. Namely, Sweden's electricity consumption must be reduced and a secure, environmentally responsible alternative source of energy must be in place (Quinlan 2002). However, few serious investments in alternative energy sources have been made (Kronsell 1997: 67). The above circumstances have contributed to an 'informal reversal of the referendum' (Matlary 1996b: 238).

Yet, the results from the 1980 referendum suggest that the danger posed by nuclear power was a salient and long-lasting issue with the Swedish polity:

> 'Nuclear power has been a hot potato in Sweden for 20 years. The issue has nagged away with varying degrees of intensity in the political debate, even helping to decide the outcome of one general election. It has gained added urgency every time there is a nuclear spillage beyond Sweden's borders, such as the 1979 Three Mile Island disaster in the US and the Chernobyl disaster in the Ukraine in 1986' (Brown-Humes 1994: 4).

The results of the referendum combined with the national imperative for economic growth leave the Swedish state in a difficult place. Although the referendum was nonbinding, the Swedish government will not easily be able to abandon the referendum results as they reflect a deep split in society (Molin and Wurzel 2000: 175). Postponing definitive action with respect to nuclear power has worked thus far, but what will happen when the day of reckoning arrives and the choice must be made between GHG-reduction goals and reasonably priced efficient-energy sources?

Environmental policy-making is frequently a conflict-ridden process, a tendency likely to be exacerbated in this instance. Perhaps more important than the outcome of the choices concerning the continued use of nuclear power is *who* will make these choices. Will the Swedish state retain enough flexibility and autonomy or, as an EU member, will Sweden be subject to rules and policies that restrict its scope for action on this delicate issue? The capabilities referenced above are no trivial matter. Rather, this policy flexibility corresponds closely to a key element of statehood: sovereignty. Kenneth Waltz defines sovereignty this way: 'to state that a state is sovereign means that it decides for itself how it will cope with its internal and external problems' (Waltz 1979: 96). The events depicted below must be interpreted with these concerns in mind.

Swedish Actions in May 1999

The Swedish government nearly derailed the EU GCC strategy over the question of 'flexible mechanism' (Molin and Wurzel 2000: 175). This provision allowed a state to substitute some of its mandated national emissions reductions with either ET, sustainable joint development projects with developing countries, or facilitating CO_2-reducing activities, such as planting trees (Molin and Wurzel 2000: 175; McCormick 2001: 289; see Chapters 9–10). At issue was the extent of any future ET regime: most EU member state governments sought to place a ceiling on potential emissions trading, as those states felt that at least 50 per cent of the mandated national emissions reductions should come from the targeted state itself (Molin and Wurzel 2000: 175). Sweden, along with the Netherlands, initially threatened to block any efforts to limit ET, an outlook that put them closer to the US position than that of the EU, essentially driving a wedge into the European position. Eventually, a compromize position was found, in which further study would take into account Swedish and Dutch reservations about the manner in which the Kyoto Protocol process would take shape.

Why might Sweden seek to block proposal that seemed to be in line with many of its international and national priorities and past behaviors? A clue can be found in the low CO_2 emissions rates of Sweden. Unlike the US, Sweden had already cut many of its emissions. Sweden's behavior regarding the EU position was much more like a laggard than a leader, but the Swedish position reflects earlier leader behavior on GCC issues. The member-state sovereignty concerns surrounding nuclear power explain Sweden's position in this case. Indeed, it is significant that the two European states that objected to the EU proposal were states that are both facing a potential phase-out of their nuclear power generating capacity (on the Netherlands see Groubert 2000). Molin and Wurzel argue that Swedish opposition to this aspect of the EU GCC strategy should be understood in light of the slow progress Sweden has made on finding alternatives to nuclear power: 'the Swedish decision in the 1980s to phase out nuclear energy was not supported by a complimentary strategy aimed at introducing alternative energy sources which would not bring about increased carbon dioxide emissions' (Molin and Wurzel 2000: 175). Were the Swedes to phase out nuclear power, coal would be the most likely substitute, given the paucity, cost and logistical and transport difficulties of other appropriate energy sources. This predisposes the Swedes to favor unlimited ET, in the event that a nuclear phase-out were ever to become reality.

This is no doubt true. However, to see this event with such a narrow view misses the most interesting questions. Sweden had exercised leadership on GCC issues in the past and Sweden possessed the normative factors to exercise its leadership at the very moment it opted to be obstructionist. This invites an examination beyond economics. Kronsell's norm-based influence factors – knowledge, experience and expertise – were present here. As discussed above, Sweden had been an active participant in international GCC discussions. Additionally, Sweden also possessed national experts and national policy experience in responding to GCC concerns and, more generally, environmental policymaking with respect to atmospheric concerns. This set the stage for other states to expect environmentally friendly behavior from Sweden.

Nearly unraveling follow-up negotiations to the Kyoto Protocol is hardly an act with which one would be most likely to associate Sweden. Nevertheless, member-state sovereignty concerns over-rode Sweden's environmentally positive norm-based influence approach. The issue was not that Sweden was unwilling to pay the price for increased environmentally sensitive practices – its past attests to its willingness to do so – it was that Sweden wanted to make the decisions regarding the use of nuclear power, the balance of nuclear power substitutes, and the location of any required emissions reductions for itself, as a sovereign nation.

Sweden entered the detailed commitments phase of the GCC negotiations with a desire to protect its domestic situation. The problem posed by the EU institutions and negotiating forums was that they restricted potential Swedish policies on this important national issue. Not only were such actions initially constraining, but they had the potential to become more so in the future. Matlary notes that the EU Commission was aggressively seeking 'to enhance its role as an actor in international energy policy' (Matlary 1996a: 267). The ultimate success of this endeavor is unknown, but such a goal furthered concerns about EU decision-making power versus that of member-state decision-making power.

Additionally, the particularities of the energy sector make sovereignty with respect to energy a paramount concern, even given the sovereignty-infringing potential of the entire EU project. Sovereignty, defined in this context as the ability of the Swedish state to choose an energy policy free from the imposed constraints of others, was at stake. Barkdull and Harris

(2002: 81) cite control over the energy sector as an example of a guiding interest of the state.

Moving to a richer definition of sovereignty is to understand it as a bundle of concepts: authority, control and autonomy (Litfin 1998: 7). The scope of EU actions menaced the Swedish state's monopoly in each of these domains. Stephen Padgett argues that 'the strategic economic importance of the energy sector meant that policy autonomy was closely guarded by national governments' (quoted in Matlary 1996a: 261). Yet, autonomy in energy decisions is valued not just for its instrumental economic benefits, but also for its symbolic and strategic value. Energy equals political power; acting as an independent actor in this policy area carries with it a certain cachet that independent decision-making in other areas may not. Environmental lead states, regardless of the presence of normative influence mechanisms, may be equally susceptible to this.

The potential changes in state autonomy, understood as a right to mandate the outcome of energy decisions, are evident. Yet control and authority were also at stake. The EU, as a governing structure, provides forums that enhance the efficacy of citizen involvement and transnational activism. Namely, EU institutions offered a potential space in which the legitimacy of the actions of the Swedish state with respect to the 1980 referendum could be called into question; this allowed for the possibility of challenges to Swedish authority as a result of EU commitments. These challenges to the Swedish state's authority and control could come from either environmentalists or from industry on issues related to GHG emissions reduction commitments, nuclear power use, the relative price or availability of power, or the value of key economic indicators. Thus, the Swedish position on ET was much less about weighing and balancing the various components of sovereignty – autonomy, control and authority – in some sort of sovereignty bargain (Litfin 1997) than it was a wholesale defense of the entire sovereignty package as it applied to energy choice.

Sweden certainly cared about the environment and had made progressive moves on the international scene to this effect in the past, moves which this chapter has stressed. However, norm-based behavior has its limits; it would be simplistic to expect that the Swedes' environmentally friendly norms would guide their international and domestic policy choices and responses all the time. It is worthwhile, however, to understand the conditions under which these norms fail to dominate. In this case, examination reveals the way in which all three policymaking levels interact. The societal level, where public opinion provided a key enduring, if flexible, constraint. The state, however, was also a powerful actor, seeking to mediate between societal feelings on nuclear power, economic stability and GHG reduction commitments. These tensions played out at the systemic level where Sweden deviated from its usual norm-based foreign policy, akin to the focus on ideas in Barkdull and Harris' (2002) typology, to a foreign policy that protected sovereignty, akin to the focus on interests in the typology.

A LOOK TOWARD THE FUTURE: ENLARGEMENT AND RATIFICATION OF THE KYOTO PROTOCOL

Enlargement

The 2004 EU enlargement (see Chapter 13) suggests mixed outcomes for Sweden's GCC diplomacy. First, as one of the six net donor states to the EU's budget, Sweden has recently signaled its reluctance to donate more than 1 per cent of GDP annually (*The Economist*, 2004). The environmental effect of denying the EU extra funds is unclear. How these funds would have been spent and the degree to which environmental concerns, particularly those regarding clean energy, would have been integrated into the planning and execution of any projects, would have determined the eventual environmental outcome. However, with less money at its disposal, the EU is clearly not as able to aid its newest members as generously in all projects, including those related to sustainable development.

Secondly, as voting weights in the Council of Ministers and the total number of commissioners per state are adjusted in light of enlargement, Sweden may lose some of its institutional power with respect to environmental politics. Particular institutional positions have been vehicles through which Sweden advanced its environmental goals in the past. For example, from 1999 to 2004, the EU Commissioner for the environment was a Swede, Margot Wallström. In 2005, Wallström became one of several vice presidents in an enlarged European Commission, removing the environmental portfolio from her direct purview. In her 2005 position as Vice President of Institutional Relations and Communication, she may continue to advocate for environmentally friendly policies, but her primary duties will not include this. Thus, Sweden and Swedes may be able to ascend up the institutional hierarchy of the EU, but only at the cost of giving up control over particular issue areas. However, enlargement also brings a greater number of small states to the table, so the possibilities for alliances, tradeoffs and 'small state solidarity' may be enhanced, though important differences will, of course, remain.

Thirdly, and with respect to nuclear power, many of the new member states also rely on nuclear power and do not wish to become reliant on energy imports from Russia, a likely alternative to nuclear power (*The Economist* 2004). From a commercial perspective, British and French firms interested in maintaining and upgrading these nuclear power plants are likely to oppose

EU actions eliminating reliance on nuclear power (ibid 2004). This advocacy, combined with the legally binding Kyoto obligations, will give Sweden more breathing room, though the conflict between the environmental implications of continued nuclear power use and the alternative of increased GHG emissions as reliance on nuclear power is diminished will not disappear.

This tension underscores the fact that determining the 'good' environmental option, or even the option that represents the lesser of two evils, is a tricky process. In that sense, Sweden's troubles are a microcosm of the EU's troubles. Should Sweden develop a solution, it will no doubt become a very influential state, in both EU and international forums. Barring that, Sweden's international advocacy on GCC may be weakened.

Entry into Force of the Kyoto Protocol

With Russia's November 2004 decision to ratify the Kyoto Protocol, the treaty entered into force in February 2005. That year also marked the official start of the EU's Emissions Trading Scheme (ETS). As discussed above, under the ETS the EU sets limits for CO_2 emissions and allows EU member states to trade their allowances among themselves. States indicate to the Commission how they intend to meet their national goals in a National Allocation Plan. Sweden's plan, although requiring minor changes from the first draft submitted, was approved in August 2004 (Whitehead 2004). With respect to its 1990 GHG levels, Sweden has announced a domestic reduction target of 2 per cent in the period 2008–12, with the goal of an eventual 50 per cent decrease by 2050 (Ministry of Sustainable Development 2005). The last figure is likely to be unattainable without reliance upon emissions trading. To that end, Sweden promoted both the EU's ETS scheme as well as flexible mechanisms at the November-December 2005 conference of the FCCC parties in Montreal (ENB 2005).

While Russia is not included in the ETS, as it is directed at intra-EU emissions, the Russian ratification suggested a larger future for the emissions trading embodied in the ETS. The ETS system requires national registries of GHG emissions. For domestic trading between individual firms, Sweden has purchased a national registry system developed by the UK (Whitehead 2004). Although the market facilitated by national registries will be co-ordinated through a central EU hub, member states will still be held responsible for national excesses of CO_2 emissions. They will be able to purchase emissions credits from Russia and other states under the joint implementation plan if they have provided for this in their National Allocation Plan (Whitehead 2004). In early 2005 Sweden was negotiating with Bulgaria regarding the development of a joint implementation project (Velikova 2005). Subject to the Commission-approved National Allocation Plan, member states will also be able to make use of the Clean Development Mechanism.

CONCLUSION

The Swedish example of intra-EU Kyoto protocol negotiations provides a valuable counterweight to Kronsell's persuasive illustrations of how small states might exert power in the EU and on the world stage through normative influence.[1] What at first seems like a clear argument is complicated by the fact that small states, such as Sweden, do not always avail themselves of their normative power over others. While small states can exert power via normative influence, small states are also powerful when they retain their right to choose their own policies. This can be a means to facilitate the acquisition of a different, and perhaps more recognizable form of power: economic growth. Environmentally friendly norms do not vanish, nor do their tangible manifestations – state reputation, national expertise, national policy experience and co-ordinated interests – wither away. Rather, other concerns enter the policy-making domain and trump environmental ones, making international environmental norm influence a non-event. Normative foreign policy is not always operative, even among its best practitioners.

Do the results of this study imply that norm influence has a constrained sphere – that norm influence is not the power mechanism Kronsell envisions it to be? Perhaps. If a state, particularly an environmental leader, finds itself constrained by sovereignty and economic concerns, might this bode ill for the global environment? If environmental lead states can experience moments in which their environmental commitments are eclipsed by other factors, what will the future hold, particularly when tough choices – choices no doubt concerning questions of sovereignty and economic sacrifice – need to be made? Such a question might expect too much of lead states and might overstate the power of, and expectations associated with, environmentally friendly norms. To expect perfect performance all the time is a standard even an environmental lead state may not be able to meet. If, however, a norms- or idea-based foreign policy is likely to be the route by which the society of states engages with the GCC issue, then we ought to bear this constraint in mind.

We also need to look at the situations in which environmental leaders stumble and the incentives that contribute to their doing so. For Sweden, the larger incentive structure was the EU and the global energy supply. The effect of these factors is not particular to this instance: 'since joining the EU, the Swedish government has not taken on a leadership role with regard to combating climate change' (Molin and Wurzel 2000: 174).

What does this mean for future GCC policies? First, as an EU member and a state committed to cooperative environmental remediation, Sweden is unlikely to strike out alone; Swedish policies and EU policies are likely to be coterminous. While Sweden can be a difficult coalition partner in some instances of intra-EU politics, it usually adopts the common EU position in the end. Secondly, GCC diplomacy places Sweden in an awkward position with respect to both the international GHG reduction commitments it has made and the environmentally friendly norms it traditionally espouses, versus the domestic constraints placed on energy policies by Swedish voters. This is an anomalous role for Sweden, given its general proclivity toward international and EU environmental leadership. This leaves room for other states, perhaps even those that have not traditionally been seen as environmental leaders, to come to the fore.

What does this mean for studies of environmental foreign policy? First, an idea-based foreign policy can be inconsistent, just as a power- or interest-based foreign policy can be. Norm influence is the primary method for Sweden's exercise of international power, but even when all the factors that support norm influence were present, a state can fail to engage in an idea-based EFP. Sovereignty concerns and domestic constituency concerns are reasons for this, but they are unlikely to be the only reasons. In effect, there are situations when the realist description of state behavior is apt, even for states whose international identity and power may be based very much on a rejection of the realist worldview.

This case study of Sweden suggests that the actions of environmental leaders can be constrained by these identifiable factors. This, in turn, suggests that arguments about norm influence are more applicable in some situations than in others, and that there might be, in fact, a more complex picture than the linear relationship suggested by factors supporting norm influence leading to actual norm influence. This can make theory-based prediction difficult. A potential direction for further research is therefore to take this framework and examine individuals' political will, or lack thereof, as a further explanation for policy-making choices concerning environmental issues. This proposal corresponds to Barkdull and Harris's (2002: 84) call for studies on the psychology of individual environmental policy decision-makers.

Additionally, this chapter has responded to Barkdull and Harris's (2002: 65) call for more empirical work. Sweden is an important state in international environmental policy-making. It behooves us to understand the complex pressures and distinctive role that characterize Sweden's GCC diplomacy. The focus on Sweden also allows an examination of the role of small states' foreign policies in international politics and supra-national governance. This chapter further draws attention to an under-studied class of events as well as providing a new way to examine the leader-laggard dynamic in European environmental politics. Further study is of course justified because global climate change will only grow in prominence in both European foreign policy and world politics.

NOTE

1. For a discussion of another 'small' state's influence in the GCC negotiations, see Chapter 4.

REFERENCES

Anderson, Mikael Skou and Duncan Liefferink (eds) (1997), in Lee Miles (ed.), *European Environmental Policy: The Pioneers*, New York: Manchester University Press.

Barkdull, John and Paul G Harris (2002), 'Environmental change and foreign policy: a survey of theory', *Global Environmental Politics* **2**(2) (May), 63–91.

Barnes, Pamela (1996), 'The Nordic countries and European Union environmental policy', in Lee Miles (ed.), *The European Union and the Nordic Countries*, New York: Routledge.

Beckerman, Wilfred (1992), 'Global warming and international action: an economic perspective', in Andrew Hurrell and Benedict Kingsbury (eds), *The International Politics of the Environment*, New York: Oxford University Press.

Bodansky, Daniel (2001), 'The history of the global climate change regime', in Urs Luterbacher and Detlef Sprinz (eds), *International Relations and Global Climate Change*, Cambridge, MA: MIT Press.

Börzel, Tanja A (2002), 'Pace-setting, foot-dragging, and fence-sitting: member-state responses to Europeanization', *Journal of Common Market Studies* **40**(2): 193–214.

Brown-Humes, Christopher (1994), 'Survey of Sweden', *The Financial Times* (19 December): 4.

Carey, Elaine T (1980), 'Swedes: sticking with the cheaper way to keep warm: vote nuclear', *The Christian Science Monitor* (25 March): 1.

Downie, Leonard Jr (1980), 'Voters in Sweden reject phase-out of nuclear plants; Swedish plebiscite rejects phase-out of nuclear power', *The Washington Post* (23 March): A1.

Earth Negotiations Bulletin (ENB) (2005), 'COP11 and COP/MOP1 highlights: Thursday 8 December 2005', *Earth Negotiations Bulletin* **12**: 290 (9 December), available at www.iisd.ca/vol12/enb12290e.html.

The Economist (2004), 'Nuclear power: out of Chernobyl's shadow', *The Economist* (8 May): 57–58.

The Economist (2004), 'How to blow a trillion euros', *The Economist* (24 January), available at: www.economist.com/displaystory.cfm?story_id=2367432.

Finlayson, Jock A and Mark W Zacher (1983), 'The GATT and the regulation of trade barriers: regime dynamics and functions', in Stephen D Krasner (ed.), *International Regimes*, Ithaca: Cornell University Press.

Groubert, Jonathan (2000), 'Mean or green: nuclear power in the Netherlands', *Radio Netherlands* (6 November), available at: www2.rnw.nl/rnw/en/features/cultureandhistory/nuclearpower001114.html?view=Standard&version=1.

Haigh, Nigel (1992), 'The European Community and international environmental policy', in Andrew Hurrell and Benedict Kingsbury (eds), *The International Politics of the Environment*, New York: Oxford University Press.

Keohane, Robert and Joseph Nye (2001), *Power and Interdependence*, San Francisco: Longman.

Kronsell, Annica (1997), 'Sweden: setting a good example', in Mikael Skou Anderson, and Duncan Liefferink (eds), *European Environmental Policy: the Pioneers*, New York: Manchester University Press.

Kronsell, Annica (2001), 'Can small states influence EU norms? Insights from Sweden's participation in the field of environmental politics', *EU Center of Seattle Conference: The EU and Scandinavia Today*, Seattle (24–25 February).

Kronsell, Annica (2002), 'Can small states influence EU norms? Insights from Sweden's participation in the field of environmental politics', *Scandinavian Studies* **74**(3): 287–304.

Liefferink, Duncan and Mikael Skou Anderson (1998a), 'Greening the EU: national positions in the run-up to the Amsterdam Treaty', *Environmental Politics* **7**(3): 66–93.

Liefferink, Duncan and Mikael Skou Anderson (1998b), 'Strategies of the "Green": member states in EU environmental policy-making', *Journal of European Public Policy* **5**(2): 254–70.

Litfin, Karen (1997), 'Sovereignty in world ecopolitics', *Mershon International Studies Review* **41**: 167–204.

Litfin, Karen (ed.), (1998), 'The greening of sovereignty: an introduction.' In *The Greening of Sovereignty*, Cambridge, MA: MIT Press.

Matlary, Janne Haaland (1996a), 'Energy Policy: from a national to a European framework?', in Helen Wallace and William Wallace (eds), *Policy-Making in the European Union*, New York: Oxford University Press.

Matlary, Janne Haaland (1996b), 'The Nordic countries and EU membership: the energy factor', in Lee Miles (ed.), *The European Union and the Nordic Countries*, New York: Routledge.

McCormick, John (2001), *Environmental Policy in the European Union*, New York: Palgrave.

Miles, Lee (1997), *Sweden and European Integration*, Brookfield, VT: Ashgate.

Mitchell, Ronald B (1998), 'Forms of Discourse/norms of sovereignty: interests, science, and morality in the regulation of whaling', in Karen Litfin (ed.), *The Greening of Sovereignty in World Politics*, Cambridge, MA: MIT Press.

Molin, Katarina and Rüdiger KW Wurzel (2000), 'Swedish environmental policy', in Lee Miles (ed.), *Sweden and the European Union Evaluated*, New York: Continuum.

Ministry of Sustainable Development (2005), *Government of Sweden: Climate Policy*, available at: www.sweden.gov.se/sb/d/2125/a/21787.

Nordhaus, William D (1997), *The Swedish Nuclear Dilemma: Energy and the Environment*, Washington DC: Resources for the Future.

Putnam, Robert (1988), 'Diplomacy and domestic politics: the logic of two-level games', *International Organization* **42**(3): 427–60.

Quinlan, Martin (2002), 'Strange northern lands: to continental European eyes, the Nordic countries present a strange model of energy supply', *Petroleum Economist*, (1 December): 15–16.
Ruggie, John G (1998), *Constructing the World Polity: Essays on International Institutionalization*, New York: Routledge.
Velikova, Iva (2005), 'Bulgaria aims to beat Russia to global CO_2 market', *See News* (31 January).
Waltz, Kenneth N (1979), *Theory of International Politics*, Reading, MA: Addison-Wesley.
Whitehead, Paul (2004), 'Poland pedals for its place in the EU', *Platts EU Energy* **75**: 18.
Zito, Anthony (2000), *Creating Environmental Policy in the European Union*, New York: St Martins Press.

7. A new climate for Spain: accommodating environmental foreign policy in a federal state*

J David Tàbara

INTRODUCTION

Global climate change (GCC) has come to the forefront of international environmental policy discussions. Within the field of international relations, GCC is triggering a large amount of research aimed at understanding the factors that shape related foreign policy. In addition to the Intergovernmental Panel on Climate Change (IPCC) and the United Nations (UN) Framework on Climate Change (FCCC) reports and monitoring system, extensive and valuable reviews are now available describing the different arenas and actors, as well as explaining the growing climate regime architecture and the opportunities and pitfalls of current negotiations and mechanisms.[1] However, this chapter is not aimed at reviewing the existing literature in this matter. Rather, it will concentrate on describing and analyzing the general and current situation of GCC politics in Spain, particularly with regard to the domestic and foreign causes and effects of its implementation of the Kyoto protocol and the European Climate Change Programmes (ECCP) (see EC 2001)[2]. In order to understand the following lines it is important to underline that Spain is now a highly decentralized and federally organized state composed by 17 Autonomous Communities (ACs), some of them (e.g., Andalusia, the Basque Country, Catalonia and Galicia) with very distinct cultural traditions and political arrangements. These distinctions explain many of the different domestic responses and organizational policy arrangements to environmental issues, including climate change, in Spain.

This chapter first describes Spain's greenhouse gas (GHG) emissions and related policies. It then reviews theories of environmental foreign policy (EFP) as portrayed in Barkdull and Harris (2002), before explaining the main goals, actors and forces, as well as the underlying political culture and processes which have shaped Spain's GCC foreign policy. Although GCC issues have been on the Spanish political agenda since the early seventies, it was only after 2001 that some new, albeit very weak, administrative structures and organizational changes began to come into shape in a more visible way. However, such new arrangements just started to produce some decisive effects in 2004 with the publication of the Allocation Plan (AP), which limited the permits for emissions at national level for a selection of key economic sectors. As it will be argued next, the new action on the climate policy and its related institutional setting emerged as a result of the interplay of increasing pressures from *outside* Spain, mainly from the European Union (EU) and the FCCC, the growing role taken by regional actors, and finally as a result of a change of government in March 2004. The new context made it possible to start working in more notorious way on a large portfolio of climate measures, which had been left aside during the previous 8 years of conservative rule.

Spain is a good example of the contradictions that come with the incorporation to the EU of those less-developed countries whose main goal is to catch up with the economic standards and integration of the rest of the EU members states *but*, at the same time, trying to attain their EFP commitments. European market integration devotes large sums of public investments to the development on new infrastructures aimed at enhancing national economic competitiveness, such as investment in transport, a sector that is one of the largest GHG emitters. Furthermore, environmental policy is limited by proposing measures which are mostly non-inflationary (thus avoiding substantive taxing) and which do not create market disequilibria.

In this context of support for growth and ambivalent environmental protection, in 2001 Spain was the fifth largest emitter of GHGs in the EU, and indeed it went from emitting 6.86 per cent of all European Community GHGs in 1990 to 9.32 per cent in 2001. Furthermore, in 2001 Spain was one of only two EU member states (barely surpassed by Ireland) furthest away from attaining its GHG emissions target (EEA 2003a). Thus it is simply not possible that full effective domestic changes can be put in place to reduce the growth of GHG emissions in order to fulfill Spain's Kyoto commitments and for it to participate proactively in the ECCP. For a long time, Spanish politics framed GCC either as a non-issue or as a burden on economic growth and competitiveness. As a result, the accumulated passivity and inertia are very difficult to overcome, but (as will be argued below) it is not impossible. One of the main lessons to be learned from the Spanish case is that, to be successful, global environmental agreements, and the GCC regime in particular, need to take into account and accommodate intrastate regional actors and related federal institutional arrangements. The building of institutions of a multi-level collaboration is a necessary condition for the success of the EFP, and in particular for the implementation of very specific policy goals with notorious effects on the key economic sectors such as those included within the national APs. This parallels the philosophy and implementation of the EU's subsidiarity principle, which inspires a great deal of the current environmental policy making in Europe. However, this research shows that the role of regional actors and governments may be greater at the earlier stages of the climate policy development but once in motion, national agencies and large corporate interests may overtake them in leading and implementing the attendant measures and policies which creates questions with regard to equity of the whole climate policy process.

THE EVOLUTION OF SPAIN'S GREENHOUSE GAS EMISSIONS

Greenhouse gas emissions in Spain increased 40.48 per cent between 1990 and 2002 (see Table 7.1).[3] This means that emissions were 2.7 times over the 15 per cent FCCC limit (by 2008–2012) to which Spain committed itself. According to European Environment Agency projections (EEA 2003b), based on existing measures emissions in 2010 could be 48.3 per cent higher than in the base year (1990), although official trends (BOE, 2004) put this figure over 60 per cent. Furthermore, the gap between the burden-sharing target of Spain in 2010 and project emissions, based on existing domestic policies and measures, are to reach 33.3 percent, although if (unlikely) additional measures are implemented this figure would be 13 per cent (EEA 2003b). Greenhouse gas emissions fell in 2001 (–1.1 per cent) for the first time since 1990, but this is unlikely to be the result of active policy making but instead an outcome of the slowing down of the economic activity (GNP growth fell to 2.8 per cent in 2001 compared to 4 per cent and more during the four previous years). The distribution of GHG emissions by sector, as represented in Table 7.2, shows that energy industries and energy use accounted for 53.7 per cent of the total emissions in 2002, followed by transport (23.6 per cent) and agriculture (10.7 per cent).

Looking at the evolution of some key environmental and monetary indicators, Table 7.3 shows how carbon and energy efficiencies of the Spanish economy did not improve during the period of 1990–2002, instead having worsened to some. While gross domestic product (GDP) grew at a yearly rate of 2.6 per cent during this period, for a total increase of 36.6 per cent, Spain's GHG emissions (excluding land-use change and forestry) grew at a rate of 2.9 per cent, a total increase of 40.4 per cent. With regard to energy, a similar pattern can be observed: total primary energy supply

Table 7.1 Evolution of GHG emissions in Spain (1990–2002)

GHG	Base year	1990	1991	1992	1993	1994	1995	1996	1997	1998	1999	2000	2001	2002	Change 2002–1990 (%)
Net CO_2 emissions/ removals	215,295.23	215,295.23	220,119.35	226,233.14	214,715.49	225,413.66	235,549.32	221,863.83	238,468.90	244,904.87	268,502.62	275,083.56	276,801.91	290,146.68	**34.77**
CO_2 emissions (without LUCF)	224,751.23	224,751.23	231,316.08	239,387.99	230,165.91	241,295.56	251,902.10	240,649.24	260,074.46	268,775.66	295,260.27	306,829.65	308,278.41	325,448.17	**44.8**
CH_4	30,244.07	30,244.07	30,517.38	31,567.48	31,963.01	33,039.06	33,728.16	35,399.48	36,434.07	37,682.86	38,064.81	39,265.40	40,312.18	41,136.13	36.01
N_2O	26,273.22	26,273.22	26,034.70	25,330.38	23,438.32	25,649.26	25,311.69	27,645.45	26,971.07	27,665.96	29,007.40	30,320.52	29,140.11	28,755.38	9.45
HFCs	4,645.44	2,403.18	2,179.01	2,762.60	2,258.39	3,458.21	4,645.44	5,196.84	6,125.88	5,809.01	7,163.91	8,171.44	5,287.84	3,896.11	62.12
PFCs	790.37	828.41	787.15	781.84	793.76	785.14	790.37	758.93	784.32	749.69	695.53	404.84	228.82	257.05	–68.97
SF_6	93.58	55.75	61.30	63.80	67.50	75.70	93.58	101.34	121.87	140.57	185.46	211.02	212.31	238.69	428.14
Total (with net CO_2 emissions/removals)	**277,341.91**	**275,099.86**	**279,698.87**	**286,739.25**	**273,236.46**	**288,421.03**	**300,118.56**	**290,965.88**	**308,906.12**	**316,952.96**	**343,619.74**	**353,456.77**	**351,983.17**	**364,430.03**	**32.47**
Total (without CO_2 from LUCF)	**286,797.91**	**284,555.86**	**290,895.61**	**299,894.10**	**288,686.88**	**304,302.93**	**316,471.34**	**309,751.28**	**330,511.68**	**340,823.76**	**370,377.38**	**385,202.87**	**383,459.67**	**399,731.53**	**40.48**
Annual Index	–	99.22	101.43	104.57	100.66	106.10	110.35	108.00	115.24	118.84	129.14	134.31	133.70	139.38	–
GNP variation at 1995 constant prices	–	3.8	2.5	0.9	–1.0	2.4	2.8	2.4	4.0	4.3	4.2	4.4	2.8	2.2	–

Sources: MIMAM (2004), INE (2002) and UNFCCC (2005).

grew at at a yearly rate of 3.1 per cent, a total growth of 44.3 per cent between 1990 and 2002, which exceeded the growth of GDP. This may partly explain some of the difficulties of adapting the Spanish economy to the challenge of GHG reductions, although the final explanation must also be found in the particular aims and arrangements of related policy structures during this time.

Table 7.2 Distribution of GHG emissions per sector in Spain (2002)

Sector	**GHG Emissions (%)**
Energy industries	28.8
Energy use in industries	15.9
Energy use in other sectors	9.0
Transport	23.6
Agriculture	10.7
Industrial processes	6.9
Waste	3.9
Fugitive emissions	1.1

Source: UNFCCC (2005) and MIMAM (2004).

We can say, therefore, that from a physical accounting point of view, Spain has been one of the EU countries that have performed most ineffectively in curbing domestic GHG emissions growth on the basis of its own foreign policy commitments. Based on existing measures, its projected GHG emissions are by far the worst of the entire EU. (The same pattern can be observed with respect to some other countries listed under Annex I of the FCCC [see Figure 7.1].) However, at the same time, it must be said that Spanish GHG in 2002 had not yet reached the EU average GHG emissions. The lack of compliance with international commitments will have important negative effects on EU efforts to meet its emissions targets under the Kyoto protocol.

Despite such trends are difficult to be substantially reversed in the short or medium term, the second half of the year 2004 seems to have marked the start of a whole new period for climate policy in Spain. In the face of the imminent European Emission Trade Scheme in 2005, the newly elected Socialist Party (*Partido Socialista Obrero Español*, PSOE), which came into power in March 2004, had to rapidly put together a whole new set of actions to comply with the unavoidable economic and environmental integration with EU regulations. Up to that point, however, Spain proved incapable of cutting – or insufficiently interested in cutting – its own GHG emissions.

Table 7.3 GHG evolution in Spain 1990–2002: relation to population and selected economic indicators

	1990	2001	2002	Change 1990–2002 %	Growth rate 1990–2002 %/year	Change 2001–2002 %
Population (millions)	38.9	40.3	40.5	4.1	0.4	0.5
GDP (US$ billions 1995 PPP)	573	767	783	36.6	2.6	2.1
TPES	91.2	127.8	131.6	44.3	3.1	3.0
GDP capita (US$ thousands 1995 PPP)	14.8	19.1	19.3	30.4	2.3	1.0
TPES capita (kgoe)	2.3	3.2	3.2	39.1	2.7	0.0
CO_2 emissions without LUCF (Tg)	224.8	308.3	325.4	44.8	3.1	5.5
GHG emissions without LUCF (Tg CO_2 eq)	284.6	383.5	325.4	40.4	2.9	4.2
GHG emissions/removals by LUCF (Tg CO_2 eq)	–9.5	–31.5	–35.3	271.6	11.6	12.1
GHG emissions with LUCF (Tg CO_2 eq)	275.1	352.0	364.4	32.5	2.4	3.5
CO_2/capita (Mg)	5.8	7.7	8.0	37.9	2.8	3.9
CO_2/GDP (kg CO_2 eq per USD 1995 PPP).	0.39	0.40	0.42	7.7	0.5	5.0
GHG/capita (Mg CO_2 eq)	7.3	9.5	9.9	35.6	2.5	4.2
GHG/GDP (kg CO_2 eq per USD 1995 PPP)	0.50	0.50	0.51	2.0	0.2	2.0

Source: UNFCCC (2005).

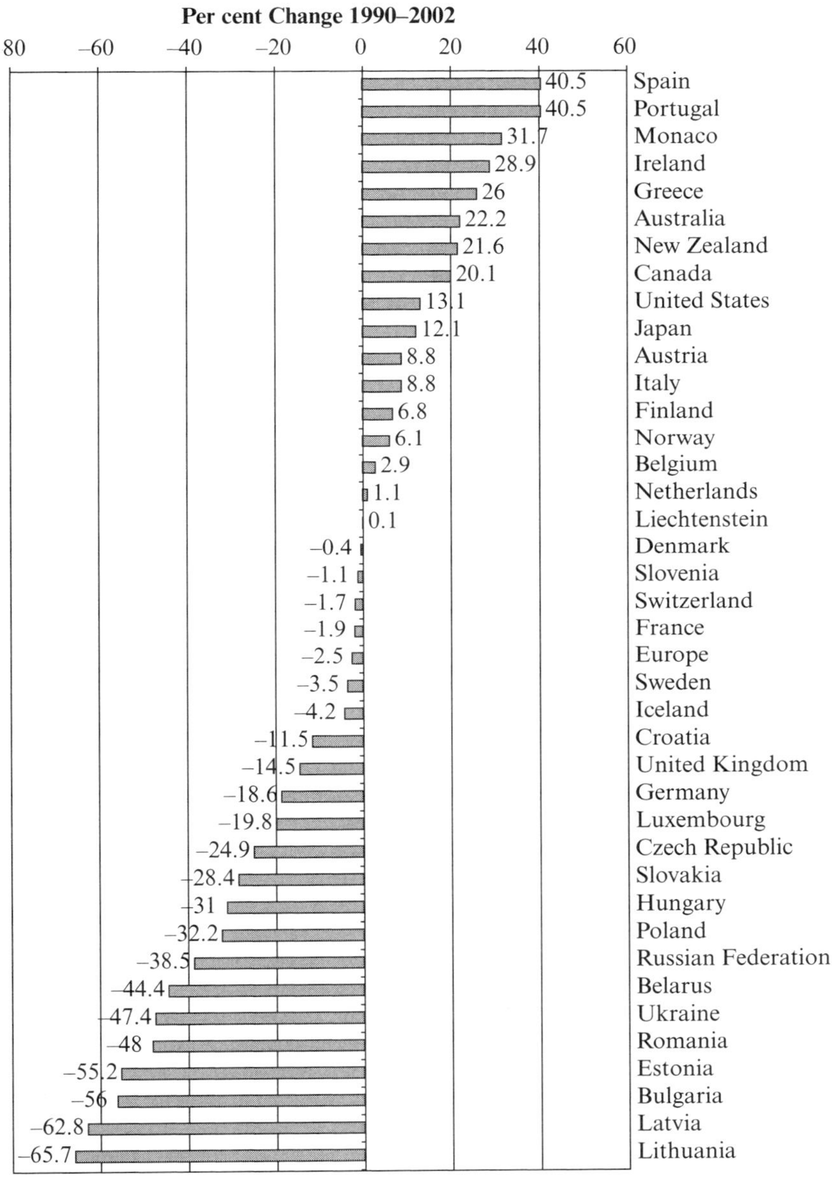

Source: UNFCCC (2004a).

Figure 7.1 Change in total aggregate GHG emissions in individual Annex I Parties, 1990–2002

WORDS ARE NOT ENOUGH TO CUT EMISSIONS GROWTH

For almost two decades and until the year 2004 Spanish rhetorical discourses on GCC occupied a large amount of room in official documents and plans presented both in national and international arenas (see MIMAM 1997, 2004, 2002a, 2002b; UNFCCC 1994; and, for reviews, UNFCCC 1996, 2000, 2002). However, even though GCC was present in domestic policy talks during this time, Spain was characterized by a lack of a global, direct, explicit and transparent strategy on this matter. The existence of very few options for public accountability made it very difficult for social actors to influence the formulation of the climate measures at the national or regional level or to help improve the mechanisms necessary to comply with Spain's international GCC-related obligations.

The peculiar socio-economic structure and geographic situation of Spain provides a high potential for contributing to curbing GHG emissions growth and for taking a significant role in the use of the ECCP's flexibility mechanisms. This is the case, for instance, if Spain were to take advantage of its potential for renewable sources of energy, mainly wind and sun. Such potential, framed under the GCC policy discourse was only to come fully recognized in the late 1990s. By 2005, Spain became the second largest producer of wind energy (with 9.947 MW) in the world, production being mostly concentrated in the regions of Galicia, Castilla-La Mancha and Castilla-Leon (SdeO comunicación 2005). However, instead, GCC had for a long time been framed as a threat to growth and not as an opportunity for social, economic and ecological development and modernization.

To a large extent, GCC politics epitomize a recurrent problem of environmental and sustainability politics in general. They are often understood mainly as symbolic demands with little attention needed to make real changes in economic, institutional and environmental conditions from which such demands depart (see Edelman 1977). This was certainly the case for Spain, at least until the approval in September 2004 of the national Allocation Plan (BOE 2004), a decisive moment when the affected industries started taking seriously into account the new regulatory framework in their corporate strategies. Nevertheless, for a decade, Spain had been missing out on many of the opportunities that the GCC arena could offer in terms of developing new research and development strategies, reducing the costs of its European and international integration and cooperation, and expanding its international environmental service markets.

Despite a period of almost two decades of being oblivious to the core of the international climate politics and debate, Spain seems to have finally awakened. Spain developed some institutional arrangements and innovation to deal with GCC, mostly as a consequence of foreign pressures, in particular from the EU. As early as 1992, and to avoid embarrassment at the UN Conference on Environment and Development held in Rio de Janeiro, the Spanish government created the National Commission for Climate Change (*Comisión Nacional del Clima*) just a few days before the conference began. The Commission was unable to complete a final version of a national plan for climate change, let alone present it before the Council of Ministers. However, Spain signed the FCCC in December 1993 and the Kyoto Protocol in April 1998.

Since the National Commission for Climate Change proved to be largely inoperative, in 1998 it was replaced by the National Council on Climate Change (*Consejo Nacional del Clima*) (NCCC) (Oroza 1999; BOE, 2001a). Originally, the NCCC was presided over by the Minister of the Environment, and was composed of five working groups together with a technical secretary. The working groups were in the areas of industry and energy, agriculture and environment, inter-departmental co-ordination, transport and institutional relations (Garcia 2001). At that time, there was no direct representation of the 17 ACs, and there were several overlapping functions and tasks. (For example, the Ministry of the Environment was both at the presidency, in the working group on agriculture and environment and in the working group on institutional relations, while the National Institute of Meteorology initially was involved in both the technical secretary working group and the inter-departmental working group.)

Given such circumstances, in July 2001 the functions and structure of the Office of Climate Change were regulated and assimilated into the functions of the technical secretary of the Council. In November 2001, the Spanish government decided to reform the entire composition and all functions of the NCCC. Two types of membership of the new Council were established: the plenary and 'vocal' members. The plenary was headed by the Minister of the Environment, together with a Vice-President (the Secretary General of the Environment), the secretary of the Spanish Office of Climate Change, and a representative of 12 ministries. The vocal consultancy group consisted of five general directors of research institutes and other governmental agencies, a representative of each AC, three representatives of local municipalities, two representatives of business organizations, a representative of chambers of commerce, two experts designated by the Minister of the Environment, three representatives of ecological organizations, two members of trade unions and two representatives of consumer associations. The aim was that the Council should meet at least in plenary once a year and in permanent commission once per semester in order to elaborate, assesses and monitor the highly elusive Spanish Strategy on Climate Change.

The number of human resources involved in climate politics was expanded and reoriented towards both new agencies, but the legislation that regulated them explicitly stated that their functioning would not include an increase in funding beyond that already available from the Ministry of the Environment or other public agencies which compose them (BOE 2001a; BOE 2001b; MIMAM 2002a). The official mandate of the NCCC was to prepare the national Strategy on Climate Change, but its first complete draft did not appear until February 2004. The EU Environment Commissioner, Margot Wallström, warned that the EU would take actions against Spain for not having presented, as late as May 2004, any plan for the allocation of emissions trading rights under the EU emissions trading system that was to start in January 2005 (Cañas 2004).

Before Spain signed the Kyoto Protocol and joined the GHG 'bubble' allocation system of the EU, it was allowed to increase its emissions by 17 per cent to 2010 (compared to 1990) due to its differential economic development. However, following agreements reached at the Council of Luxemburg in June 1998, this percentage was reduced to 15 per cent. Spain's initial target was to increase GHG up to 25 per cent (UNFCCC 2004b), more in accordance with other less developed EU countries such as Portugal and Greece, which themselves had negotiated increases of 27 per cent and 25 per cent respectively. In 1995, the Spanish government announced that the economic slowdown in 1992–94 would enable Spain to lower its CO_2 related emissions, although (as shown in the figures above) this did not happened. It is clear that Spain underestimated the real costs and necessary institutional changes required to achieve the 15 per cent target. It is difficult to understand why such a strict and unrealistic figure was chosen, but it is possible that Spain wanted to portray an image of modernity, solvency and European institutional capacity, even though at that time it did not have an assessment of the domestic policy and economic implications of meeting that target.

In June 2001, the EU agreed that it would ratify the Kyoto Protocol. Spain was to hold the presidency of the EU the following year. On 1 February 2002, the Spanish government decided to send the Protocol for approval to the Spanish Parliament, and Spain ratified the Kyoto Protocol on 31 May 2002. However, Spain had not yet sent the Third National Communication (MIMAM 2002b), in which GHG emissions were to be inventoried and a strategy specifying different measures to be taken was to be detailed, to the FCCC secretariat (Rivera 2002). Most of the GHG-related actions taken in Spain, although slowing the growing rate of the country's GHG emissions, have not been the result of any explicit national, integrated strategy on GCC, but instead derive from the application of EU directives or are the result of sectoral plans.

The indirect mitigation 'measures' described in the Third National Communication submitted to the FCCC in Spring 2002 did not presuppose a constriction on economic growth, but rather constituted a further stimulus to it. The Communication did not give a detailed plan quantifying the impact of the different measures on curbing CO_2 emissions by 2010 (MIMAM 2002a, 2002b). With regard to adaptation measures, the finally withdrawn National Hydrological Plan, which aimed to further exploit hydrological resources (and which had been highly contested by a wide range of non-governmental organizations [NGOs] and academic communities), was mentioned

In spite of this historical record, recently there have been some signs of change. Only few weeks before the Socialist Party (PSOE) came into power in March 2004, the National Council on Climate Change approved on 5 February a document containing 440 measures to abate GHG emissions. In September 2004, as a result of mounting pressure from the EU and a greater receptiveness to GCC concerns by the new government – and in particular by the new environment minister, Cristina Narbona – the GHG allocation plan was refined. The plan, which was further specified in January 2005, selected 957 industrial facilities, allowing total emissions of 513.6 mt of CO_2 equivalent. The highest quotas were allocated to the electricity sector (256.2 mt) and to the cement industry (82.5 mt). Despite some early opposition within the industrial sector, the plan was finally being fully implemented. In March 2005, the new Designated National Authority (*Autoridad Nacional Designada* (AND)) was created to deal with Clean Development and Joint Implementation Projects and also the Commission for the Coordination of Climate Change Policies (*Comisión de Coordinación de Políticas para el Cambio Climático*) to harmonize the actions taken by ACs and the central government.

Despite the new actions, the case of Spain shows how non-policy on GCC finally leads to many inefficiencies and inequities that forgo many opportunities and benefits that would arise from early action on climate change. Spain has had to recourse to the more expensive – and perhaps less efficient and equitable – flexible mechanisms as the inescapable way to meet its Kyoto commitments for the total amount of 100 mt of CO_2 equivalent for the period 2008–12, which equals the 7 per cent of its baseline emissions (almost half of its original commitments).

EXPLAINING THE LATE CLIMATE POLICY IN SPAIN

The typology of EFP provided by Barkdull and Harris (2002) helps us to identify some of the main factors that explain why GCC foreign policy in Spain was taken into account so late and in a way which had negative consequences both for the international and domestic arenas. On the one hand, explanations for the Spanish case are closely related to the interpretations of EFP given by state-centric theories, bearing in mind the emphasis on institutional changes that occurred in March 2004 with the election of the Socialist Party. Such political changes have opened a new window of opportunity for Spain's GCC policy, although the pervasive reluctance to do more with regard to GCC over the last decade cannot be fully understood unless other factors are taken into account. This is why we must also underline the role of the weak environmental institutional culture in Spain, which has been more interested in growth and security, as a factor that provides a more in-depth explanation of the failure of the GCC foreign policy in Spain. We must also look at the impossibility for regional governments and civic actors to actively participate in the development of state-wide climate policy. In this sense, the interpretation of the Spanish case would support the societal type of EFP theories, given the interplay of the main domestic policy goals with a peculiar political culture for dealing with environmental issues. That political culture is characterized by the strength and predominance of the state over regional governments, individual actors and private interests. Table 7.4 summarizes the main goals, actors and forces that have driven GCC policy in Spain, as well as the processes and results of making EFP in this issue area.

Another element that can help to explain what occurred in Spain is the fact that the environment traditionally had very low salience for the Spanish political agenda. Spain shows some of the lowest percentages of green voting and environmental civic membership in Europe, the former being below 5 per cent in almost all types of elections and the latter around 1 per cent of the whole population. In contrast to the political cultures of many of the Anglo-Saxon countries, where environmental problems, measures and conflicts tend to be discussed and oriented towards civic society (including large voluntary public campaigns to set aside private land) in Spain, as has happened in other Mediterranean and some central European countries, environmental issues tend to be oriented towards the state. It is within the limited circles of government and the different levels of administration where interested parties expect to obtain regulations and enforcement measures, which then will oblige private parties to act accordingly and respect the new institutional framework for individual action. This environmental political culture, which confers a stronger authority to the state and to the community above the individual, at the same time means there is less demand for commitment from the citizen in terms of individual participation and reflection. This in turn has resulted

Table 7.4 Key dimensions of climate policy in Spain, 1992–2005

Key Dimensions	**Spain's Climate Foreign Policy, 1992–2005**
GOALS	Mainly to participate and implement EU climate policy to the extent it did not substantially affect Spain's commitment to economic growth and the goal of 'catching up' with EU standards of living. EU economic integration also supported the emphasis on growth and initially also on the development of high intensive energy and carbon development, e.g., in the transport sector
DRIVING FORCES	EU being the main driver of domestic climate change policy and institutional change, but failing to generate effective domestic institutional changes to significantly curb GHG emissions growth. A change of government in March 2004 opened a new window of opportunity; new policies were implemented to comply with EU commitments (mainly the GHG allocation plan).
CONSTRAINTS AND OPPORTUNITIES	Strongly decentralized state, with primary competencies in environmental, industry and energy issues given to the Autonomous Communities, entailing high costs of co-ordination but offering greater potential for actors' engagement.
ACTORS	Central government placed climate issues at the very end of the environmental political agenda during most of the period. The Conservative Party, which ruled from March 1996 to March 2004, resisted taking action on climate change. NGOs, epistemic communities and the media did not participate substantially in creating public awareness that might have activated effective institutional change. However, after 2000 regional governments increasingly took on major roles in climate politics, together with major affected industries, mainly as a result of the 'threat' that the EU emissions trading system posed (the EU directive on emissions trading was approved in 2003).
PROCESSES	Characterized by a lack of participation of civil society in climate politics together with an embedded political culture of dealing with environmental issues on the basis of powerful state intervention and regulation. For the first time, a GHG allocation plan for the reduction of GHG emissions was approved by the Socialist Party (approved in September 2004 and revized in January 2005).
RESULTS	A series of continuous institutional reforms and innovations, but failure to deliver a throughout national strategy on climate change until February 2004. Spain's non-compliance with GHG emissions-limitation commitments adversely affects the EU's performance with regard to the Kyoto Protocol commitments and may also negatively affect aspirations for larger GHG limitations in the post-Kyoto negotiations.

in a peculiar approach to the way environmental questions are dealt with in Spain. On the one hand, in certain areas it has allowed for greater levels of environmental protection by means of regulatory measures than in other European countries, as is the case with nature and land conservation. However, on the other hand, it has led to a situation where many of those improvements in environmental quality that require strong *individual* pro-environmental action and private coordination have been very low, if not completely paralyzed.

Government officials undoubtedly played a major influence, particularly during the 8 years of conservative government. Climate change was perceived either as a non-issue or as a threat to economic growth and competitiveness.[4] When something was argued for protecting climate, it was to enhance particular industrial interests. To give an example, one of the former Conservative ministers of the environment, Isabel Tocino, an allegedly very keen supporter of the nuclear lobby, pressed – together with Spanish power companies – for more nuclear energy as a means to meet Kyoto commitments. Corporate demands were also receiving serious consideration at the EU level by the then EC vice-president and transport commissioner, Loyola de Palacio, who was from the same governing party, despite the ECCP explicitly advocating *non*-nuclear renewable sources of energy. After over a decade of inaction in curbing GHG emissions growth, Spain, together with its echoing and amplifying conservative voice at the EU, was still arguing that there was no other way to meet Kyoto goals than to push harder, in a high-risk manner, for more nuclear energy (Carcar 2002).

Nevertheless, as suggested above, the approaching of the date for the entering into force of the European emission trade scheme (ETS) in 2005 led to the Spanish government to begin taking a new position toward GCC. The more intensive external pressures were combined with a more open and decisive attitude of the newly elected government on these issues. The new political situation offered now potential and more possibilities to fulfill Spain's EU and international climate obligations. Albeit very slowly and in a very peculiar way, traditional negative perceptions also seem to be changing and are being reframed. To a large degree, it must be admitted that the new situation is the result of the EU market-environment integration regulations, and in this regard the role of the market in 'socializing' the problem of climate change should not be underestimated. Increasingly, a new awareness is trickling down, which shows how climate change can also offer new and very attractive windows of opportunity for Spanish corporate interests, which originally may have little to do with the goal of protecting the environment, improving the safety or sustainability standards.

THE GROWING ROLE OF REGIONAL AND LOCAL ACTORS

Because GCC policies are increasingly regarded as having an important effect on economic growth, and given the lack of decisive action in these affairs by the central government, some of Spain's ACs, and even some local municipalities, began to take a lead in proposing specific climate policy plans. In Spain, local authorities and ACs are responsible for developing and implementing housing programs and have many responsibilities with regard to environmental issues. In some ACs they have full responsibility in these matters. This allows the ACs notable opportunities for intervention in GCC policy at the local and regional level and it is particularly true for sectors such as waste and energy, as is the case of Navarre, an AC where wind power generation for electricity is very developed. Some ACs also began to produce their own GHG inventories well in advance, to prepare to the new own strategies on GCC.

One of the most notable examples of these new actions taken by ACs is provided by the regional government of Andalusia (the *Junta de Andalusia*). In September 2002, Andalusia AC prepared its own *Regional Strategy for Climate Change*, which was approved in February 2003. Among the proposed measures were the following: the creation of a regional panel for assessment and monitoring of the implementation of the regional strategy; a new set of legal instruments to curb GHG emissions; the elaboration of regional GHG inventories; a specific plan for forest management to enhance the function of forests as sinks and as mitigation and adaptation instruments; and several educational and information actions. Included in their climate strategy were plans to improve energy efficiency, as well as actions in agricultural, the fishing sector, science, education and monitoring (Junta de Andalucía 2002).

Furthermore, on 24 November 2003, a coalition of left-wing parties from Catalonia (*Entesa Catalana del Progrés*, which a month later was actually ruling the Catalan Parliament), presented a proposal to the Spanish Senate (which represents regional governments) for a law aimed at putting the Spanish Strategy on Climate Change in motion and providing it with the necessary resources and intergovernmental support to do so. However, given the Conservative Party majority at that time, the proposal was rejected. In Catalonia an assessment, carried out by 45 experts, was commissioned to evaluate the possible impact of climate change at the regional level to support regional climate policy (Llebot 2005).

Other concerns about the posture of the state toward GCC, and demands to incorporate the growing number of regional climate policy stakeholders, appeared in other several AC governments, such as the Canary Islands – for obvious reasons (tourism and water resources) concerned about GCC – and the Basc Country, which already has its own GHG inventory. Indeed a new 'regional climate diplomacy' within Spain was emerging. As early as the Milan Conference of the Parties to the 2003 conference of the parties to the FCCC (COP9), at the 2004 Buenos Aires COP10 meetings and at COP11 in Montreal, several regional and federal representatives, including some from Spanish AC governments, met to discuss their roles, difficulties and opportunities for participating directly in this area of foreign policy.

There was also a debate in Spain about who should have responsibility, and at what level, to assign GHG emissions permits. At present, Spanish ACs have responsibility for emissions authorizations under the Integrated Pollution Prevention and Control Directive, so it would seem logical that they would do so in the case of GHG permits. However, NGOs and trade unions who were participating in the NCCC claimed that the government was acting without enough transparency in making the national permit plan for GHGs, in opposition to what was called for in the EU's emissions trading directive.

In addition, at the local level, several cities were also early in taking an active role in GCC politics. For instance, in the Castilla-La Mancha AC, a network of over 150 small towns and municipalities aiming to curb GHG emissions growth was established. Similar actions were taking place in other ACs, such as Aragon and the Basc Country. Barcelona was the first city in Spain to have compulsory legislation on solar power. It requires that all new buildings, and those that consume more than 2,000 liters of hot water daily, install solar heat systems. Barcelona, which signed the Heidelberg Declaration of commitments on climate change, has also approved a plan to improve the energy efficiency of the city. By 2010 the capacity of urban photovoltaic panels is expected to increase to 14.14 MW in the city and several public buses in Barcelona are now running on hydrogen (Ajuntament de Barcelona 2002). In June 2005, a new Spanish network of Cities for Climate (*Red de ciudades por el Clima*) was created, which included over 80 large cities from most of the Spanish ACs.

CONCLUSION

Spain is one of the EU countries that has most failed to fulfill its international obligations with regard to GHG emissions reductions. This prompted a new set of the country's regional political actors, both private and public, to become increasingly organized in order to take action on this problem. The failure of Spanish climate policy until the imminent penetration of the European ETS in 2005 can be explained to a large extent by the lack of participation of civil society and the difficulties regional political actors and institutions have faced in becoming involved in the climate policy process. A conservative government that was reluctant to take proactive action in this domain exacerbated this situation. As a result, Spain had to take recourse in the flexible mechanisms by buying reduction credits from abroad and, in 2006, by creating their own new ones. Yet at the same time, Spain had to find new ways to reduce its domestic GHGs in a more decentralized manner, in tune with its current regional political organization (Autonomous Communities). The strongly decentralized political structure of Spain with regard to environment, industry and responsibilities for energy use poses special difficulties for implementation of the GCC regime, but it does not explain Spain's failure to comply with its international GCC commitments. On the contrary, in the context of the growing role taken by intermediate and regional actors in addressing GCC, Spain's AC governments provide differentiated and more flexible opportunities for the successful implementation of international environmental policies.

By late 2004, a new set of GCC policies were initiated in Spain as the result of a change of government and greater pressure and policy integration with the EU, as well as a consequence of greater involvement of new actors, such as local municipalities, regional governments and private companies. However, the renewed involvement of these new actors, mainly the private ones, in the GCC policy should be taken with caution and not automatically be seen as a greater 'democratization' or improvement of the equity content of its related decisions. A large amount of these new actors are of a corporate and transnational nature and therefore they are difficult to be accountable by the existing local or regional political agencies. For instance, in Spain, in 2005 only three large companies dominated 58 per cent of the whole wind energy production sector, this sector having placed a strategic role in fulfilling the Spanish objectives in meeting its GCC commitments. Before such large corporate interests, which often operate on a global scale, regional governments have found a limited scope for action.

Spanish GHG emissions will likely be around 33 per cent larger than the amount that Spain committed itself to (for the period 2008–12) under the Kyoto Protocol (EEA 2003a, 2003b, 2003d). After pushing Spanish climate policy far away from the political agenda for almost two decades, external pressures have finally brought the problem back into the domestic economy but with more expensive and stronger effects than if it had been taken earlier and more decisive action. This in turn, has made it even harder for effective or more equitable measures and instruments to be adequately implemented in time.

It was particularly remarkable that the mounting external forces on Spain in the climate arena for over a decade did not substantially affect the existing administrative arrangements of the state as a whole, particularly with regard to key areas such as industry and energy production and use, but are instead creating differentiated responses at lower levels of governance. Spain resisted for a long time to the foreign policy pressures and it was the combined result of change of government and EU integration that the country became more permeable to international cooperation. However, such responses still vary according to institutional capacities, cultures and perceptions of GCC opportunities and risks, which exist in the different Spanish ACs and at the municipal (local) level. The allocation plan for emissions of GHG in Spain, despite provoking some initial concerns from industry and trade unions in sectors such as energy, cement and paper production, was fully implemented by early 2006. Consequently, a new and real window of opportunity for climate politics was opened in Spain for the first time.

In terms of its international and European consequences, it is clear that the burden of non-compliance and inaction of Spain during such a long period may have had a noticeable negative impact on the EU's ability to reach its common GHG reductions targets, and indirectly added further difficulties to the feasibility and credibility of the international Kyoto process as whole. This argument was reinforced by the fact that European Environment Agency projections (EEA 2003c) estimated that the EU would only achieve a 0.5 per cent GHG reduction in 2010 (using existing measures[5]), compared with its Kyoto commitment of 8 per cent, because Spain was not alone among EU member states in failing to meet its commitments.

Most of the societal actors that usually contribute to the perception of GCC as a social and political problem in other Western contexts have not played a significant role in Spain. The lack of an environmental movement able to influence action by ACs and even less to deal with global problems, as well as the weakness of national epistemic communities and the low level of importance placed on environmental issues by the mass media, can partly explain the lack of collective action in this area. The inexistence of a green party with representation at the Spanish Parliament is indicative of political fragmentation in relation to environmental issues. However, above all, it seems that Spain, having emerged from dictatorship and relative underdevelopment in the 1970s and trying to catch up with the living standards and economic integration of its EU partners, had grossly relegated environmental issues to the very bottom of the political portfolio.[6] When these issues penetrated political discourses, it was only as a reaction to external pressures, mainly from EU directives and regulations. This is why Spanish environmental administrative capacity was far behind its needs, reactive and sluggish, and unable to incorporate complex and global issues like GCC that went beyond the boundaries of its perceived national priorities and responsibilities.

One of the main lessons to be learned from the Spanish experience is that in highly decentralized countries like Spain, a general state (i.e., centralized) framework for the implementation of international environmental protections, such as those related to GCC, may be inadequate. Indeed, the traits that characterize such federal policy settings – high co-ordination costs or the resistance of regions that do not wish to lose their ability to take decisions – can explain the reasons for Spain's failures, particularly at the early stages of the process. In these early developments of the policies, the most active and interested actors at the regional level do not have the capacity to operate abroad, and the difficulties of creating a unifying voice between the different regional levels may be too high for an issue that has not been regarded as central to the country's foreign policy agenda. This does not mean that the implementation of international environmental agreements implies a reorganization of environmental responsibilities at home so as to give the central government an opportunity to recentralization policy. This is a process that is no longer feasible or adequate if Spain is to also implement the EU's subsidiarity principle. On the contrary, it seems sounder to believe that international and EU agreements should also be able to accommodate regional institutional diversity and give regional governments an opportunity to take a more active role, particularly when central governments have proven to be highly ineffective, as in the case of Spain.

For Spain this would mean allowing greater opportunities for different ACs to participate in some of the existing GCC flexibility mechanisms, such as emissions trading, as if they had the status of states. Creating a domestic ETS within Spain may increase the feasibility and flexibility of existing policy instruments, and it may help to adapt them more closely to current administrative reality. This could also immediately have several positive consequences, such as increasing regional capacity to monitor and account for GHG emissions, create more cost-efficient specific plans and strategies to meet each region's policy objectives and to follow their own measures within a national plan, and to avoid what have been recurrent problems of environmental co-ordination among Spain's 17 ACs. At present, however, given that Spain is now trying to catch up very fast with its foreign GCC commitments, there is the risk that a greater role be given to the central and large corporate private actors which may improve the efficiency of their domestic policies but at the costs of meeting standards with regard to the equity of the related processes and decisions.

Indeed, as a result of the administrative reform that followed the end of dictatorship in the late 1970s, most environmental responsibilities were transferred to the ACs. In those days, environmental issues may have seemed of little importance in relation to other issues that were reserved to the state. However, according to Aguilar (1997: 120–26), the transfer of environmental functions to the ACs was done without a legal framework able to provide homogeneous state criteria for its adequate management. This process yielded confusion, disorganization and a lack of vertical co-ordination in environmental administration. In fact, when the Spanish Environmental Ministry was first constituted in 1996, environmental responsibilities were already distributed among 61 regional sub-ministries and agencies. Nevertheless, some AC governments are adopting certain EU environmental directives more quickly than the state. Despite some pressures and tensions, it seems unrealistic to believe that these regional governments will give back to the central government those administrative gains achieved over the two decades of democracy, particularly in those areas like the environment that have increasing implications for development.

A more important role for the ACs as GCC actors has important economic and political consequences. It is no surprise that some of the most proactive regional governments on environmental issues in Spain are also the most industrialized ones that enjoy higher levels of per capita income. These regions are also some of the higher GHG emitters. If the criterion of common but differentiated responsibility (see Chapter 15) were to be applied, these regions should pay more or do more to compensate for their emissions, which in turn would make it even more necessary to develop some kind of system for emissions trading among regions. In this way, inaction from some regions or from the state as a whole could be partly compensated for by a greater involvement of those regional actors that have traditionally been more active and now have administrative responsibilities in dealing with the environment in Spain.

In sum, the low-profile history of GCC politics in Spain has been characterized by a series of administrative failures when confronted with unavoidable international obligations, such as the 1992 Rio conference, Spain's presidency of the EU, or the EU ETS implementation (which entailed having a national allocation plan ready so that required GHG reductions could be estimated). The poor environmental political culture, a lack of efficient environmental diplomacy, the obliviousness to the GCC discourse by the ruling parties, and the impossibility of local and regional actors participating in (a non-existent) single Spanish GCC strategy can explain the very late development of Spain's GCC policies. To what extent Spain's reactive, messy and diluted responses to GCC will be transformed by the new situation into a more effective foreign policy, and what effect that will have on GHG emissions, cannot be fully assessed. Being unable to curb its GHG emissions at home as a single country, Spain is now in need of buying reductions abroad, supporting the creation of carbon funds (particularly in Latin America), and also finding collaborative ways to harmonize in a more regionally decentralized manner the abatement of its GCC pollution. Needless to say, additional, more proactive and stronger measures, perhaps among the more than 400 already approved by February 2004, will have to be adopted if Spain is to fulfill its commitments under the ECCPs, the Kyoto Protocol and future emissions-limitation agreements (MIMAM 2004b).[7]

NOTES

* Portions of this chapter appeared in Tàbara (2003).

1. Contrasting views on these matters can be found in works such as Downing et al. (1999), Grubb (1999), Jäger and O'Riordan (1996), Luterbacher and Sprinz (2001), Social Learning Group (2001), Victor (2001), and Yamin and Depledge (2004), see Chapter 1.
2. The second European Climate Change Programme (ECCP II) was launched in October 2005.
3. The latest figures for 2003 put this figure to a 40.6 per cent GHG emissions increase, with respect to those of 1990 (EEA 2005).
4. In March 2002, the Spanish Business Organization asked the government to postpone the transposition of the Emission Trade Directive (Mardones 2002).
5. See Chapter 15 for more optimistic EU emissions scenarios based on additional GHG-limitation measures that are being put into effect across the EU.
6. In this sense the Spanish incorporation to the EU had a double effect. On the one hand it initially stimulated fast economic growth, hence accelerating the increase in GHG emissions. However, after 2005, such emissions had to be harnessed as a result of a greater integration between market and environmental objectives.
7. At the time this chapter went to press, it was not possible to include the more recent figures of GHG emissions and policy developments in Spain. Those can be found in MIMAM (2006). These figures reinforce the arguments in this chapter.

REFERENCES

Aguilar, Susana (1997), '*El reto del medio ambiente: Conflictos e interesses en la política medioambiental europea*' ['*The challenge of the environment: Conflicts and interests in European environmental policy*'], Madrid: Alianza.

Ajuntament de Barcelona (2002), *Pla de millora energètica de Barcelona* [*Barcelona Energy Improvement Plan*], Barcelona: Ajuntament de Barcelona.

Barkdull, John and Paul G Harris (2002), 'Environmental change and foreign policy: a survey of theory', *Global Environmental Politics* **2**(2): 63–91.

Boletín Oficial del Estado (BOE) (2001a), 'Orden de 17 de julio de 2001 por la que se regula la composición y estructura de la Oficina Española de Cambio Climático' ['Order of 17 July 2001 to regulate the composition and structure of the Spanish Office of Climate Change'] *Boletín Oficial del Estado* [*Spanish Official Gazette*] **171**: 25888–25889.

Boletín Oficial del Estado (BOE) (2001b), 'Real decreto 1188/2001, de 2 de noviembre, por el que se regula la composición y funciones del Consejo Nacional del Clima' ['Royal decree 1188/2001 of 2 November to regulate the composition

and functions of the national council for climate change'], *Boletín Oficial del Estado* [*Spanish Official Gazette*] **264**: 40026–40027
Boletín Oficial del Estado (BOE) (2004), 'Real Decreto 1866/2004, de 6 de septiembre, por el que se aprueba el Plan nacional de asignación de derechos de emisión, 2005–2007 [Royal decree 1866/2004 of 6 September to approve the national allocation plan], *Boletín Oficial del Estado* [*Spanish Official Gazette*], **216**: 30616–30642
Carcar, Susana (2002), 'Nucleares bajo Pressión. Las grandes compañias reabren el debate sobre la energia atómica con el apoyode la comisaria Loyola de Palacio' ['Nuclear power station under pressure. Large companies reopen the debate on nuclear energy with the support of the commissary loyola de palacio']. *El País negocios* [*The Business País*] (31 March).
Cañas, Gabriela (2004), 'La UE apercibe a España por no hacer su plan para cumplir Kioto: Narbona presentará en julio el reparto de emisiones de CO2' ['The EU warns Spain of not making its plan to comply with Kyoto: Narbona will submit the CO_2 emissions allocation plan in July']. *El País* (19 July).
Downing, Thomas E, Alexander J Olsthoorm and Richard SJ Tol (eds) (1999), *Climate, Change and Risk*, London: Routledge.
Edelman, Murray (1977), *Political Language: Words that Succeed and Policies that Fail*, London: Academic Press.
European Commission (EC) (2001), *European Climate Change Programme (ECCP): Long Report.* Brussels: Environment DG.
European Environment Agency (EnEA) (2003a), *Annual European Community Greenhouse Gas Inventory 1990–2000 and Inventory Report 2003*, Technical Report No. 95, Copenhagen: European Environment Agency.
European Environment Agency (EnEA) (2003b), *Greenhouse Gas Emission Trends and Projections in Europe 2003*, Copenhagen, Denmark: European Environment Agency.
European Environment Agency (EnEA) (2003c), 'EU Greenhouse Gas Emissions Rise for Second Year Running', *EEA news release* (6 May), available at: http://org.eea.eu.int/documents/ newsreleases/ghg-2003-en.
European Environment Agency (EnEA) (2003d), 'Domestic measures taken or planned so far are insufficient to meet EU climate emissions targets, projections show', *EEA News Release* (2 December), available at: http://org.eea.eu.int/documents/newsreleases/ghg-en.
European Environment Agency (EnEA) (2005), *The European Environment. State and outlook 2005*, Copenhagen, Denmark: European Environment Agency.
Garcia, Mas (2001), *Desarrollo de un marco regulador para alcanzar los compromisos de Kioto en España.* [*Development of a Regulatory Framework to Achieve the Kyoto Commitments*], Madrid: Ministerio de Economía [Ministry of Finances].
Grubb, Michael (1999), *The Kyoto Protocol: A Guide and Assessment*, London: Earthscan.
Instituto Nacional de Estadística [Madrid Ministry of Finances] (INE) (2002), *Contabilidad Nacional de España* [*Spanish National Accounting*], Madrid: INE.
Jäger, Jill and Timothy O'Riordan (1996), *Politics of Climate Change*, London: Routledge.
Junta de Andalucía [Autonomous Government of Andalusia] (2002), *Estrategia del Gobierno Andaluz ante el Cambio Climático. Acuerdo del Consejo de Gobierno de la Junta de Andalucía* [*Andalusian Government Strategy on Climate Change*], Seville: Junta de Andalucía [Autonomous Government of Andalusia] (3 September).

Llebot, JE (2005), *Informe del canvi climàtic a Catalunya* [*Report on Climate Change in Catalonia*], Barcelona: Institut d'Estudis Catalans [Institute of Catalan Studies].

Luterbacher, Urs and Detlef Sprinz (2001), *International Relations and Global Climate Change*, Cambridge, MA: The MIT Press.

Mardones, Immaculada (2002), 'La "venta" del Protocolo de Kioto: La CEOE pide al Gobierno español que aplace la directiva sobre el comercio de emisiones' ['The "Selling" of the Kyoto Protocol: the Spanish business organization CEOE asks the government to postpone the emission trade directive'], *El País* (31 March).

Ministerio de Medio Ambiente (MIMAM) (1997), *Segunda Comunicación Nacional de España. Convención Marco de las Naciones Unidas sobre el Cambio Climático* [*Second Communication of Spain to the UNFCCC*], Madrid: Ministerio de Medio Ambiente [Ministry of the Environment].

Ministerio de Medio Ambiente (MIMAM) (2002a), 'Estrategia frente al Cambio Climático' "Strategy on Climate Change"', in Ministerio de Medio Ambiente [Ministry of the Environment] (ed.), *Actuaciones Publicas en Materia de Medio Ambiente* [*Public actions on Environmental Issues*], Madrid: Ministry of the Environment.

Ministerio de Medio Ambiente [Ministry of the Environment] (MIMAM) (2002b), *Tercera Comunicación Nacional de España. Convención Marco de las Naciones Unidas sobre el Cambio Climático* [*Third Communication of Spain to the UNFCCC*], Madrid: Ministerio de Medio Ambiente [Ministry of the Environment].

Ministerio de Medio Ambiente [Ministry of the Environment] (MIMAM) (2004a), *Greenhouse Gas Emissions Inventory Report From Spain 1990–2002: Communication to the European Commission.* Madrid: Directorate General for Environmental Quality and Assessment (March).

Ministerio de Medio Ambiente (MIMAM) (2004b), *Estrategia Española sobre Cambio Climático para el cumplimiento del Protocolo de Kioto. Borrador num 4.* Madrid: Ministerio de Medio Ambiente [Ministry of the Environment].

Ministerio de Medio Ambiente (MIMAM) (2006), *Cuarto Communicación Nacional de España. Convenio Marco de las Naciones Unidas sobre el Campio Climático.* [Fourth communication of Spain to the UNFCCC] Madrid: Ministerio de Medio Ambiente [Ministry of the Environment].

Oroza, Sergio Alonso (1999), 'El Cambio Climático y Global: ¿Qué opciones políticas, económicas, legales y competenciales pueden elaborarse para obtener un mayor bienestar en la sociedad española y a la vez cumplir con nuestra cuota de solidaridad ambiental?' ['Climate and Global Change: What policy, economic, legal and competence option can be developed to obtain greater welfare and comply with our share of environmental responsibility?'] in J Grau Rahola and JE Llebot (eds), *Política Ambiental y Desarrollo Sostenible* [*Environmental Policy and Sustainable Development*], Madrid: Institute of Ecology and Markets.

Rivera, Alicia (2002), 'El gobierno da luz verde al Protocolo de Kioto sin tener un plan para cumplirlo' ['The government gives the go ahead to the Kyoto Protocol without having a plan on how to comply with it'], *El Pais* (2 February): 24.

SdeO comunicación (2005), 'SdeO comunicación', available at www.sdeocom.com.

Social Learning Group (2001), *Learning to Manage Global Environmental Risks: A Comparative History of Social Responses to Climate Change, Ozone Depletion and Acid Rain* 2 vols, Cambridge, MA: MIT Press.

Tàbara, David (2003), 'Spain: words that succeed and climate policies that fail', *Climate Policy* **3**(1): 19–30.

United Nations Framework Convention on Climate Change (UNFCCC) (1994), *Resumen de la Primera Comunicación Nacional de España.* [*Summary of the First Communication of Spain to the UNFCCC*], (A/AC.237/NC/14), New York: United Nations.

United Nations Framework Convention on Climate Change (UNFCCC) (1996), *Spain: Report on the In-Depth Review of the National Communication of Spain*, (FCCC/IDR.1./ESP,. New York: United Nations.

United Nations Framework Convention on Climate Change (UNFCCC) (2000), *Spain: Report on the in-Depth Review on the Second National Communication of Spain*, (FCCC/IDR.2./ ESP), New York: United Nations.

United Nations Framework Convention on Climate Change (UNFCCC) (2002), *Report of the Individual Review of the Greenhouse Gas Inventory of Spain Submitted in the Year 2001*, (FCCC/WEB/IRI(3)/ESP), New York: United Nations.

United Nations Framework Convention on Climate Change (UNFCCC) (2004a), *Information on National Greenhouse Gas Inventory Data From Parties Included in Annex I to the Convention for the Period 1990–2002, Including the Status of Reporting*, (FCCC/CP/2004/5), Executive Summary, New York: United Nations.

United Nations Framework Convention on Climate Change (UNFCCC) (2004b), *Report on the In-Depth Review of the Third National Communication of Spain*, (FCCC/IDR.3/ESP), New York: United Nations.

United Nations Framework Convention on Climate Change (UNFCCC) (2005), *Country Profile: Spain* (28 February), New York: United Nations.

Victor, David G (2001), *The Collapse of the Kyoto Protocol and the Struggle to Slow Global Warming*, Princeton: Princeton University Press.

Yamin, F and J Depledge (2004), *The International Climate Change Regime: A Guide to Rules, Institutions and Procedures*, Cambridge, UK: Cambridge University Press.

8. Shaming and framing: Norwegian NGOs in the climate change negotiations*

Andreas Tjernshaugen and Ho-Ching Lee

INTRODUCTION

Since the late 1980s, environmental NGOs (ENGOs) have sought to make governments in Europe and the rest of the industrialized world commit to and implement substantial and early reductions in greenhouse gas (GHG) emissions. One of several political arenas where they have worked to promote this goal is during international negotiations of the United Nations (UN) Framework Convention on Climate Change (FCCC). The aim of this chapter is to throw light on the strategies ENGOs employ in seeking to influence international negotiations and foreign policy on global climate change (GCC). To illustrate the discussion, we specifically examine how ENGOs attempted to influence Norway in the FCCC negotiations from 1995 to 2001. We look at the activities of Norwegian ENGOs as well as their international counterparts in an international ENGO coalition, the Climate Action Network (CAN), during the main negotiating sessions. Not surprisingly, Norway's behavior was the main concern for the Norwegian groups, while multinational ENGOs such as Greenpeace International or WWF International gave attention to Norway more sporadically.

Our focus on *strategy choice* says little by itself about the environmental groups' actual results in terms of *influence* over outcomes (Betsill and Corell 2001). The NGOs' strategies could simply be a failure, and their actions irrelevant to outcomes. However, careful attention to the strategies actually chosen may help us to identify potentially important forms and channels of influence. Specifically, we point out that even when attending the UN conferences, ENGOs spent much of their energy seeking to influence domestic debates and decisions on foreign policy – the government's negotiating position, as well as the interpretation and implementation of existing commitments – rather than trying to influence negotiators directly. Observers who assume that NGO influence over the outcome of international negotiations results only from direct lobbying at the international conferences could easily overlook the impact of the more indirect strategies we discuss here.

In the following section we start by showing how scholars of international relations and social movements understand the role of NGOs in international negotiations and foreign policy, introducing some concepts useful for understanding the presentation of the case. The next section provides background on Norway's role in international climate policy and the NGOs in question. The main section then presents and discusses the NGOs' strategy choices at the international conferences. In concluding, we confront our findings with some assumptions frequently made in analyses of the influence of NGOs in international negotiations.

BACKGROUND: NGOs, FOREIGN POLICY AND INTERNATIONAL ENVIRONMENTAL NEGOTIATIONS

Thousands of individuals attend the annual conferences of the parties to the FCCC. Official representatives of the states that are parties to the convention and are thereby eligible to vote have frequently been outnumbered by registered observers who are there to gather information or seek some form of influence or attention. A majority of the observers represent NGOs, particularly environmental groups and business associations. Most of them are trying to influence the outcome of negotiations through one means or another. In the literature on international environmental politics, much attention has been directed at the role of such NGOs. Opinions about the scope and degree of their influence are widely diverging. This chapter seeks to illuminate one aspect of this discussion, namely the various *strategies* non-state actors use to influence the outcome of ongoing negotiations.

While representatives of sovereign states hold a monopoly on formal decision-making power at international negotiations, it is increasingly recognized that other actors may still play crucial roles in facilitating and shaping international environmental cooperation. Particularly when it comes to bringing attention to issues and thereby shaping the political agenda, the role of NGOs is widely recognized. The forms and degree of NGO influence during the actual negotiations are, however, currently subject to considerable debate and research by students of foreign policy and international relations.

A recently proposed framework for analysis of NGO influence in international negotiations assumes from the outset that 'tactics such as the granting of rewards or the threat of punishment ... generally are not viable options during a UN negotiation on an environmental issue', and that *information* therefore is 'the key currency for NGOs in exerting influence' (Betsill and Corell 2001). Indeed, NGO influence is *defined* as occurring 'when [NGOs] intentionally transmit information to negotiators that alters both the negotiating process and outcome from what would have occurred otherwise' (ibid). According to these authors, NGO influence is thus present when NGOs participate actively during the negotiations and have access to negotiators and the information or knowledge resources to make them listen, as well as when the information they provide can also be seen to have altered or shaped outcomes such as treaty texts.

The framework proposed by Betsill and Corell explicitly seeks to embrace both international and nationally based NGOs, but it is exclusively concerned with events on the political arena provided by international negotiations. Following the conventional distinction between international and sub-national levels of decision-making, any influence NGOs might have over decisions regarding the relevant domestic and foreign policies of national governments is excluded from the analysis.

In contrast to this approach, we will in the following discussion stress the interconnectedness of political processes at the international and national levels regarding policies to limit GHG emissions. In taking this perspective, we follow authors such as Skodvin and Andresen (2003), who entertain a broader understanding of NGO influence on negotiations, including influence through domestic politics. In other words, what Barkdull and Harris (2002) call *societal theories* of environmental foreign policy (EFP) is a necessary supplement to theories focusing on the international system. This also invites consideration of a broader set of means of influence than merely providing information. The case we present below provides support for the usefulness of this approach.

In Barkdull and Harris's typology of foreign policy theories, we find that society-oriented approaches stressing the role of interests and ideas are particularly helpful in throwing light on the strategies employed by ENGOs seeking to influence Norway (Barkdull and Harris 2002). Specifically, a few concepts borrowed from the literature on international regimes and negotiations, and from the literature on social movements, prove useful in understanding domestic-international linkages in this case. Students of international regimes have observed that the *shaming* activities of NGOs can strengthen regimes. 'Shaming' refers to an activity that 'exploits the symbolic legitimacy of foreign pressure and international institutions to unleash domestic moral opprobrium' (Moravcsik 1995). By pointing to controversial government practices, international norms that condemn such practices, or international reactions to the practices in question, shaming strategies are intended to induce changes in government policy. Such strategies are frequently pursued by NGOs to put pressure on governments to keep agreements or to modify controversial domestic policies (Rengger 1997). Shaming activities could equally well be directed at positions taken in international negotiations, as will be illustrated below.

Students of social movements have also stressed the importance of *framing* activities – the interpretive work that activists do in presenting issues to the public and decision makers. The framing that NGOs give their issues could simply reflect unconscious or taken-for-granted assumptions about the issues at stake. In many cases, however, activists use *strategic framing* as a tool to further their political or organizational goals (Zald 1996). When policy is made in a multi-level process, the interpretation of events at the international level of action becomes a key part of the efforts of NGOs to strategically frame the domestic debate.

A *two-level game* has been suggested as a metaphor for political processes where participants in intergovernmental negotiations also engage in negotiations with key sub-national actors at home regarding foreign policy. Analytically, one can distinguish between a *negotiation phase*, where governments bargain at the international level, and a *ratification phase*, where ratification (and thereafter implementation) of the outcome of the negotiations at the international level is considered and contested within each country. In practice, expectations about events at one level influence strategies at the other level (Putnam 1988).

Expanding on Putnam's (1988) perspective on two-level games in the context of social movement studies, della Porta and Kriesi observe that non-governmental actors do not limit themselves to mobilization at the domestic level. They offer a typology of social movement activities beyond the nation-state (della Porta and Kriesi 1998). Through *transnational* mobilization, a national movement facing a difficult situation domestically could seek alliance with movements in other countries that will either support the movement or direct international pressure at the government of the country in question. *Cross-level* mobilization refers to cooperation between a national movement and a foreign government. Finally, international levels of the political system represented by institutions such as the European Union (EU) and the UN offer *new political arenas* for national movements, as well as opportunities and incentives to establish genuinely *transnational mobilizing structures* that are often oriented towards influencing decision-making at the international level.

NORWAY, THE KYOTO PROTOCOL AND ENGOs

Compared to the situation in the United States (US), Russia and many developing countries, the Norwegian GCC policy debate has been characterized by a relatively high degree of consensus that climate change is a real threat worth addressing through mitigation policies. At the same time, Norway's reliance on petroleum and other emissions-intensive industries for export income, as well as on road traffic for keeping a large but sparsely populated country connected, makes policy instruments such as environmental taxes or tradable quotas controversial (see Chapter 12).

During the 1990s, much of the GCC policy debate in Norway – both regarding domestic policies and the Norwegian position in the FCCC negotiations – centered on the relationship between international cooperation and national action to limit emissions. Should the Norwegian government take ambitious measures to reduce domestic emissions regardless of what other countries did, or should such action be conditional upon international agreements? Was it crucial to reduce emissions domestically, or was financing of cheaper emissions reductions abroad through 'flexibility mechanisms' (after 1997 also known as 'Kyoto mechanisms'), such as international emissions trading or project-based cooperation, of equal value? The environmental movement favored early action at home over conditionality and flexible international solutions (Kasa et. al., 2001; Hovden and Lindseth 2002). This position taken by the Norwegian ENGOs mirrored the views of their international counterparts, which stressed each country's responsibility to start reducing its own emissions without delay, and viewed 'flexible' solutions with varying degrees of suspicion.

Our discussion in the following sections focuses on the years 1995–2001. Events in this period should, however, be understood in light of the earlier climate policy debates since the second half of the 1980s. In 1987, the World Commission on Environment and Development (WCED), headed by the Norwegian prime minister at the time, Gro Harlem Brundtland, published its report, *Our Common Future* (WCED 1987). Here, the question of human-induced GCC was highlighted as one of the main challenges facing the world. The report contributed strongly to attracting political attention to the issue internationally and even more so in Norway. Brundtland's prominent role in international environmental affairs made her government especially sensitive to embarrassment regarding its own environmental conduct. The environmental movement eagerly exploited this sensitivity.

An early example was seen in 1989, when activists from *Natur og Ungdom* (Nature and Youth (NU), a membership-based ENGO for people younger than 25 years old) obtained a draft plan for the Brundtland government's work on the issues raised in *Our Common Future*. The draft suggested far less restrictive limits for national CO_2 emissions than the environmentalists wanted. Nature and Youth activists took the draft to an international environmental conference in The Hague. Their 'goal was to draw international attention' to the unsatisfactory action taken by Brundtland's own government on the issues raised by the WCED (Persen and Ranum 1997). They received attention in Dutch and Danish newspapers. Several international NGOs signed petitions demanding stronger environmental policies from Norway and expressing disappointment with Brundtland. This international attention in turn led to publicity on the issue in the Norwegian news media – and a prolonged, heated argument between the Prime Minister's office and NU (Persen and Ranum 1997; Nilsen 2001). The affair placed the question of a national CO_2 target firmly on the political agenda, where environmental issues were already getting increasing attention. In the pro-environmental political climate at the time, a broad majority in Parliament soon passed a stricter CO_2 target than that proposed by the Brundtland government, namely to stabilize emissions at the 1989 level by 2000 (Nilsen 2001; Reitan 2001).

About the same time, negotiations concerning a global regime to limit human contributions to GCC began. This process led to the adoption of the FCCC at the UN 'Earth Summit' in 1992 (an event which again pitted Brundtland against domestic environmental groups). The Climate Convention set a non-binding target of stabilizing emissions from industrialized countries at 1990 levels by 2000. In Norway, a CO_2 tax was imposed on emissions from road transport, offshore oil and gas extraction, and several other sectors, while export industries to a large degree were exempted.

During the early 1990s, it gradually became clear that Norway would have a hard time fulfilling its national stabilization target. Especially unforeseen growth in oil and natural gas output from fields in the North Sea led to increases in expected emissions growth. A controversy in 1990–91 concerning industrial use of natural gas, the so-called *Heidrun* project, stemmed from the conflict between reaching the CO_2 target on one hand and industrial development on the other. The environmental movement and its political allies successfully opposed plans for a gas-fired power plant with considerable CO_2 emissions. During the summer of 1994, new plans for gas-fired power in Norway were launched. The company *Naturkraft* was established to develop and run two power plants on the western coast of Norway. The owners were the petroleum company *Statoil*, power company *Statkraft* (both owned by the Norwegian state) and *Norsk Hydro* (partially state-owned).

The period in focus here (1995–2001) started with several important developments. In Spring 1995, the first conference of the parties (COP1) of the FCCC was convened in Berlin and issued a mandate ('the Berlin

Mandate') to negotiate a treaty with legally binding commitments to reduce GHG emissions in industrialized countries. In June of that year, the Labor Government, led by Brundtland, officially announced – in a long-awaited action plan on GCC policy – that the stabilization target was unattainable and therefore abandoned in practice. During the same year, Framtiden i Våre Hender (FIVH – The Future in Our Hands), Greenpeace Norway (GPN), Naturvenrnforbundet (NNV – Norwegian Society for the Conservation of Nature) and NU joined forces under the umbrella *Klimakameratene* (Climate Friends). This alliance produced joint documents criticizing the government's action plan and negotiating position, and lobbied and campaigned both jointly and individually against the two gas-fired power plants planned by *Naturkraft* that would increase domestic CO_2 emissions by about 5 per cent.

During the winter of 1995–96, the gas-power issue was singled out by the Norwegian environmental movement as a main target for protest in order to raise awareness and political pressure around the broader question of rising domestic CO_2 emissions and Norway's contribution to GCC before the Kyoto conference (Nilsen 2001: 221). The decision of whether or not to build the two projected power plants was soon declared 'the most important environmental issue of the 1990s' (Engeland 1996a). The movement's framing of a single industrial project as the symbol for the whole GCC issue was a conscious choice that enabled activists to mobilize broadly around a problem that could otherwise have seemed too abstract and distant. This was recognized as a strategic choice not only by movement leaders, but also by activists at the grassroots level (Carlsson and Tjernshaugen 2000: 36). It allowed the complex, abstract and distant climate problem to be focused into an issue sufficiently tangible, close-to-home and simple to facilitate broad mobilization and crystallize political alliances and lines of conflict. This was clearly a case of *strategic framing* (Zald 1996).

In April 1996, *Klima-alliansen* (the Climate Alliance), a broader alliance against the *Naturkraft* project that included several opposition parties as well as the youth branches of the major pro-gas Labor and Conservative parties, was established. In May, it presented its counter-report to the government's white paper on gas-fired power. In June, a clear majority in Parliament approved of the project subject to certain conditions to be considered by the government. At the COP2 meeting in Geneva, also in June, a surprise announcement by the American delegation that the US supported binding targets for GHG emissions added new momentum to the negotiations. In September *Klima-alliansen* announced a campaign to collect 100,000 signatures on a petition demanding that the Norwegian authorities reconsider the project; the signatures were submitted in December.

During 1997, intense negotiations took place before the COP3 meeting in Kyoto, which was to finalize a treaty with binding emissions targets. In January that year the NU congress announced its intention to use civil disobedience against the *Naturkraft* project if necessary. The ad-hoc *Fellesaksjonen mot gasskraftverk* (joint action against gas-fired power plants (FMG)) was established on an individual membership basis to organize protests in the event of any construction work, which was expected to start any moment. But in May 1997, Prime Minister Thorbjørn Jagland (Labor) made statements that in effect postponed the project until after the parliamentary elections and the Kyoto conference, both of which were to take place during the fall of that year. After the elections, a minority coalition of three centrist parties that were all against the *Naturkraft* project and members of *Klima-alliansen* took over government offices. The majority in Parliament still supported the project. At COP3 in December, Norway was therefore represented by Minister of the Environment Guro Fjellanger (Liberal), who opposed the *Naturkraft* project but otherwise disappointed the activists.

The Kyoto Protocol was finalized as planned and set a target of maximum 1 per cent increase in total Norwegian emissions of six GHGs compared to 1990 within the period 2008–2012. However, crucial rules regarding implementation of the protocol were left open for later meetings. This also meant that the prospects for gas-fired power in Norway were not definitely settled by the outcome of the negotiations. An announcement during the Spring of 1998 by *Norsk Hydro*, one of the *Naturkraft* partners, that it would give priority to developing its own technology for gas-fired power with reduced CO_2 emissions represented a major setback, and low producer prices on electric power kept delaying the project. Environmental activists tried to keep the threat of massive mobilization alive during the next few years, while the formalities regarding *Naturkraft's* permission to build power plants and to emit CO_2 and NO_x was again haggled over by various government agencies after complaints from environmental groups. During the same period, other commercial actors proposed several new gas-fired power projects. In March 2000, the centrist coalition government resigned on grounds of being instructed by Parliament regarding the use of environmental legislation in this matter. Although Norwegian authorities soon granted the necessary permissions for the planned gas-fired power plants, it took until 2005 before Naturkraft started building one of them. The activities of Norwegian ENGOs clearly delayed the building of the *Naturkraft* project, in conjunction with economic factors. An even more important outcome of the controversy over gas-fired power in Norway was the commitment by a centre-left coalition government, which took office in 2005, to help fund a pioneering, full-scale CO_2 capture and storage facility at Naturkraft's plant, which will remove most of the CO_2 emissions from the plant.

After President George W Bush's withdrawal from the Kyoto Protocol in March 2001, Norway joined the EU and others in seeking to save the agreement (see Chapter 12). At COP7 in Marrakech in late 2001, Norway supported a compromise proposed by the EU and the developing country group, G77-and-China. Norway explicitly opposed demands from Japan, Russia, Canada and Australia for easing the rules on several accounts, including how to include 'forest sinks' in emissions budgets. This stance with respect to fellow members of the so-called Umbrella Group (an alliance of non-EU industrialized countries which also includes the US) gave the Norwegian government a somewhat improved standing with the ENGOs.

In 2002, the Norwegian Government ratified the Kyoto Protocol. A milder climate between ENGOs and the Norwegian Government still prevailed in at COP11 in Montreal towards the end of 2005.

THE ORGANIZATIONS

International as well as nationally based ENGOs cooperating under the umbrella of the Climate Action Network are a well integrated part of the community of officials, experts and lobbyists that gather regularly at FCCC negotiating meetings (Walk and Brunnengräber 2000; Gough and Shackley 2001). Below, we describe and discuss the political strategies used by CAN members – both nationally based groups and their international allies – in seeking to influence Norway's behavior. The four Norwegian ENGOs that participated actively in the CAN network during the relevant period were FIVH, GPN, NNV and NU, an independent youth branch of NNV. Since the early 1990s, NNV and NU have been affiliated with Friends of the Earth International [FoEI].

Representatives of these four groups followed the international negotiations as observers, and from 1995 established formalized cooperation among themselves on domestic GCC policy. The Norwegian Society for the Conservation of Nature, NU and FIVH are membership-based organizations with local branches and an internal democratic structure. They all receive considerable funding from the Norwegian government. Greenpeace Norway was the Norwegian branch of Greenpeace International until 1998, when the Nordic branches were joined in a common Nordic organization. Greenpeace has had remarkably little success in establishing a foothold in Norway, with a few hundred members compared to more than 70,000 in neighboring Sweden (Bortne et al. 2001). In the GCC issue, however,

GPN played an important role as a representative of the international environmental movement in the national debate. Generally, only a handful of representatives from the Norwegian ENGO's headquarters participated at the COPs. An exception to this rule was the COP1 meeting in Berlin, where a busload of NU activists participated in an alternative conference and demonstrated against the Norwegian environment minister (Carlsson and Tjernshaugen 2000; Walk and Brunnengräber 2000).

All four of the *Klimakameratene* organizations participated in CAN, a broad network of environmental NGOs primarily active in and around the negotiation process (Walk and Brunnengräber 2000). The network also includes key lobbyists from big international groups who are intensely involved in the negotiations, right down to the level of technical details. The most visible multinational CAN members were Greenpeace International, FoEI and WWF International. The CAN members at the COPs included activists from other countries, in addition to Norway, taking part in their country's official delegations. The daily CAN meetings that took place during COP meetings were a crucial source of information for the Norwegian activists. In addition, NNV and NU were members of FoEI. Perhaps more important was the linkage between GPN and Greenpeace International. Kalle Hesstvedt, the leader of the Norwegian branch's climate work during much of the 1990s, had worked in Greenpeace International for some time. For the rest of the Norwegian organizations, the GCC activists never delved deeply into the world of international negotiations, and did not occupy central roles in the CAN or FoEI networks.

The Norwegian environmental movement has traditionally been strongly oriented towards national institutions in its political strategies (Gundersen 1991; Seippel 1999). A study based on survey data concludes that Norwegian environmentalism – manifested both in the views and strategies of environmental movement organizations and in popular attitudes – is a unique case (Bortne et al. 2001). One defining trait is the inclusive political system shared by Scandinavian countries, which goes along with a 'state-friendly' society where NGOs can cooperate closely with the state and receive public funding without losing their credibility as critics of public policy, and where citizens have high levels of trust in government institutions, but where the state also tends to co-opt new social movements. The other characteristic trait is a cultural orientation toward local, rural communities, which goes along with a pragmatic, use-oriented view of nature, with little room for issues such as animal rights within the domestic environmental movement. Bortne et al. (2001) claim that not only is this combination unique, but that it also explains why attempts to establish a viable Greenpeace branch or green party in Norway have been unsuccessful.

They also stress that the 'state-friendliness' of Norwegian society makes it difficult for a movement or an organization to grow while keeping its distance from the state (Bortne et al. 2001). Norwegian GCC activism has certainly been shaped by state acceptance of the legitimacy of the movement and its activities. The Ministry of the Environment financed the participation of leading activists as observers at the COP meetings, and the police had advance meetings with activists planning civil disobedience against the building of new, gas-fired power plants, in order to exchange information and build mutual trust. It would, however, be mistaken to assume that there was a confluence of views on climate policy between the Norwegian government and the environmental movement, as Kellow has observed for certain European countries (Kellow 2000). Rather, the views on such crucial questions as CO_2 taxes and rules for burden sharing between countries and implementation of commitments under the Kyoto Protocol have tended to diverge strongly. In championing strict regulation of domestic CO_2 emissions both through international treaties and national legislation, the Norwegian environmental movement has met massive resistance from energy-intensive industries with strong ties to state bodies and the political establishment, especially in the Labor and Conservative parties (Kasa 2000; Nilsen 2001). The Norwegian delegations to the GCC negotiations have not included NGO representatives, as several other national delegations have done (Walk and Brunnengräber 2000). This may relate to the fact that Norway's role as major oil exporter and as self-proclaimed environmental leader since the 1970s are hard to reconcile in the GCC arena, and that the domestic environmental movement is certainly not making it any easier for the government.

In the following section we move on to an analysis of the strategy choices of the ENGOs.[1] Non-governmental organisation activists may seek to influence negotiation outcomes either through influencing diplomats during the negotiations (lobbying) or by using appeals to the domestic public through the news media (the 'shaming' and 'framing' strategies described below) to influence government officials responsive to public attitudes and perceptions.

STRATEGY CHOICE

We now turn to the strategies chosen by the ENGOs seeking to influence Norway's behavior in the GCC negotiations (see Figure 8.1). Their efforts may be categorized under three headings: lobbying (which in fact formed a minor part of their efforts), shaming and framing.

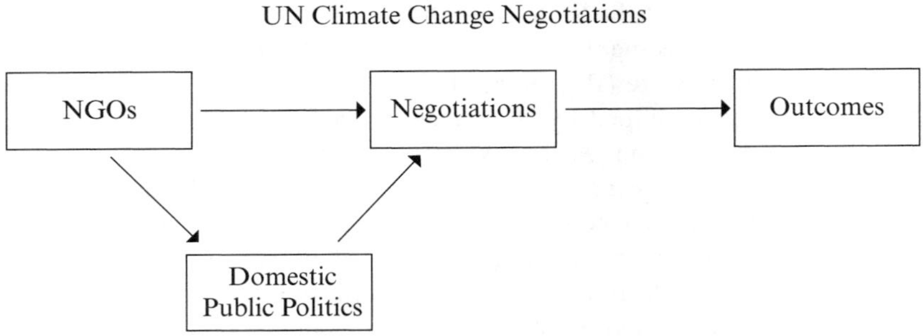

Figure 8.1 Channels of NGO influence in international negotiations

Lobbying

The ENGO representatives clearly had ideas about what they wanted the Norwegian negotiators to do, but in fact they rarely spent much energy during the meetings trying to influence them. Mostly the Norwegian NGOs used the formal briefings with the Norwegian delegation (which were open to Norwegian NGOs representing business, environmental and scientific interests that were present at the COP meetings) and even their more informal contacts with members of the delegation at and around the conference venue, mainly to gather information. A simple reason for this was that the government and the ENGOs were usually well aware of each other's points of view long before the COP took place. The ENGOs usually prepared and published lists of demands and proposals to the Norwegian delegation well ahead of the conferences, and attended advance briefings by the Ministry of the Environment in Oslo. At any rate, based on earlier experience, the ENGOs usually considered their opportunities for exercising direct influence to be very limited. There were exceptions. At the Kyoto conference, ENGO representatives tried intensely but unsuccessfully to lobby the Liberal environment minister Guro Fjellanger, who had once been employed by NNV, and who had previously expressed opinions closer to the ENGO's positions than the official Norwegian line.

If lobbying the Norwegian delegation was not the key preoccupation of the activists, neither was influencing other delegations or the overall agenda of the COPs. The Norwegians usually attended the CAN meetings and took part in discussions on common strategy. However, as will be elaborated below, even here they were to a large degree seeking support in their domestic struggles rather than trying to influence overall CAN strategy. Direct attempts to influence the COP agenda, such as organizing 'side events' (open seminars for delegates and observers), making statements at formal sessions or to the international press, or engaging in the detailed negotiations on sub-issues such as forestry practices or emissions trading rules, was for the most part left to CAN colleagues specializing in the FCCC process, usually representatives of the international offices of multinational organizations such as FoEI, Greenpeace and WWF. The general attitude of Norwegian NGO representatives was summed up by the president of NU in an interview in advance of COP3 in Kyoto: 'When we meet with … the Norwegian delegation, we don't really try to persuade them. They have their instructions from the Norwegian government. The point is to put as much pressure as possible on the politicians' (Tjernshaugen 1997, author's translation).

Shaming

If the Norwegian ENGOs were not trying to influence delegates, what exactly were they up to at the COP meetings? First and foremost, the Norwegian GCC activists sought – rather successfully – to influence the domestic public's perception of what was going on in the negotiations through their role as sources for the national news media. International ENGO representatives in the CAN network backed them in these efforts. The ENGOs' hope was that this would put pressure on the government to move its policies in the direction they favored. During each of the COP meetings, Norwegian newspapers reported the reactions of ENGOs to the latest developments, particularly the Norwegian negotiators' role. During COP3 in Kyoto, for instance, NNV alone was referred to as a source by the media more often than the Norwegian Confederation of Business and Industry (NHO) – an organization comprising most of Norway's private business sector – which was also represented at the meeting. Access to the media was one of the main resources of the activists.

The considerable knowledge gained though participation at the FCCC negotiating meetings as observers, and especially through participation in the international NGO networks, made leading activists very useful for journalists trying to make sense of confusing and technically complex negotiations. (As the Norwegian delegations to the FCCC meetings are well staffed and experienced, the same resource did not translate equally well into access and influence with respect to the delegates.) The environmental organizations' willingness to speak in media-friendly sound bites made them attractive media sources. They also enjoyed a high level of trust from the Norwegian population in environmental issues (Bortne et al. 2001). Similarities in background and outlook between activists and journalists covering the field may also have played a role in this connection.

Their position as news sources gave the activists some control over an important political resource: the public perception of Norway's role on the international scene, and thus the evaluation of the overall environmental performance of the government. The picture the environmental activists – Norwegian and foreign alike – painted of Norway was that of a petroleum-dependent economy protecting its narrow self-interest and thus endangering the prospects for international climate cooperation, as well as its own reputation in the international community. This portrayal of Norway as a 'climate villain' by environmentalists was at times linked to specific issues, such as the plans for gas-fired power or increased extraction of petroleum, or the insistence on certain positions in the international negotiations. The statements by environmentalists were obviously intended to put pressure on the government, to induce it to demonstrate willingness to contribute to climate protection by adjusting its negotiating position and domestic policies. Whether or not this strategy was able to influence GCC policy decisions, it probably had some bearing on Norwegian voters' perception of the government's environmental conduct. The hope of the Norwegian ENGOs was that this would influence the shifting governments to adjust some policies to improve their green image.

Moral support from the international NGO community gave the Norwegian GCC activists increased credibility at home. Both in terms of mobilizing for protest, and in terms of access to the media, the credibility gained by being seen as national representatives of a global environmental movement was important. The Norwegian activists contributed actively to pulling international ENGOs into the domestic debate. Throughout the 1990s, Norwegian environmentalists would repeatedly solicit statements from allied groups abroad, leading to reports in the Norwegian press about international environmental opposition to the government's domestic policies and negotiating position. The episode with the leaked draft from 1989 (see above) may have served as inspiration. It is likely that later activities in the international arena by Norwegian GCC activists were inspired by the success of this appeal to international opinion in grabbing media attention at home, embarrassing authorities and thereby influencing the national political agenda.

During COP1 in Berlin 1995, the Norwegian delegation was, for instance, presented with a statement from several international environmental groups criticizing the Norwegian position on GHG emissions reduction targets and so-called flexibility in the implementation of targets (Hjorthol 1995). At the Kyoto conference (COP3) in 1997, representatives of the World Council of Churches, criticized the position of the Norwegian government led by ordained priest Kjell Magne Bondevik, in line with the environmentalist framing of the debate (described in the next subsection below). One of the sources of the newspaper article was an observer from the Church of Norway, who was also a member of the board of FMG (Furuly 1997). Foreign environmental activists' characterization of oil-rich Norway as a nation of 'closet Arabs' was reported in the national press (e.g. Hegtun 1995). On several occasions, Norway was also nominated or selected for the 'Fossil of the Day Award', presented by CAN organizations during the negotiating sessions (see Climate Action Network 2006).

In della Porta and Kriesi's (1998) terms, these are instances of transnational mobilization of support, where a national movement appeals for help from movements abroad. However, in this case the transnational contacts in question were of a quite different character than in della Porta and Kriesi's example from the struggle against apartheid in South Africa. It does not seem that raising the pressure on the Norwegian government by influencing the actions of foreign governments (cross-level mobilization) was an important strategy for the Norwegian activists; the intended audience for their actions in the international arena was mainly the public *back home*. The support solicited from foreign NGOs by Norwegian GCC activists was purely symbolic, in the form of statements, with no material resources being provided or withheld by the international allies. The support given by members of the international NGO community cost those groups a negligible amount of effort, time and money. Apparently, activists from Norwegian organizations carried out work on joint statements regarding Norwegian policy.

On the other hand, one might well say that in putting pressure on the Norwegian government to accept strict targets, limited flexibility and strict rules in other regards, the Norwegian activists acted as local representatives of the international coalition of ENGOs cooperating under the CAN umbrella. The CAN organizations made similar demands to other industrialized governments. The ENGOs' strategies on the international and domestic arenas were in fact closely related.

Framing

Closely related to the shaming activities were activities to *strategically frame* the domestic debate on international commitments and their domestic implementation. While criticizing Norway's negotiating position and the outcomes of negotiations, the Norwegian ENGOs and their allies abroad also continuously sought to promote an interpretation of Norway's existing commitments under the FCCC, and later under the Kyoto Protocol, that fit their agenda of promoting early and substantial emissions reductions at the domestic level. At every turning point, including the COP meetings, they have argued that Norway is morally obliged to abide by the spirit and intentions of these agreements as well as their formal requirements, and highlighted passages that emphasize industrialized countries' responsibility to reduce emissions domestically (see Chapter 15).

Key individuals in the CAN network were at least occasionally briefed on domestic issues, such as gas-fired power, by Norwegian activists to ensure that they backed up their colleagues when interviewed by national news media. One important contact between the Norwegian and the international NGO community was between Greenpeace International and GPN. For instance, Bill Hare from Greenpeace International and Kalle Hesstvedt from the Norwegian branch co-authored a chapter on 'international perspectives' in the 'counter-report' against the *Naturkraft* project (Klima-alliansen 1996). During the second half of the 1990s, Bill Hare was the Climate Policy Director of Greenpeace International and the international environmentalist perhaps most frequently interviewed by the Norwegian media during international GCC meetings.

One particularly interesting form of strategic framing deserves closer scrutiny, namely the ENGOs' attempts to exploit their role as news sources to influence domestic expectations about the outcome of the negotiations. From 1995, negotiations under the FCCC had as their explicit goal to define legally binding limits for GHG emissions from individual countries. The Kyoto Protocol set such limits for industrialized countries in the years 2008–12, but it left crucial rules for implementation of the commitments to be decided at later meetings. The prospect of regulations on GHG emissions from 2008 was relevant for a number of domestic actors, including businesses considering investment in emission-intensive industries, government agencies considering policies for emission reduction, and various interest groups with a stake in the outcome of such policy decisions. At the same time, the lingering uncertainty about the exact nature of the international regulations, and about the measures the government would eventually take to implement those commitments, made it difficult to anticipate the consequences.

In this situation, actors who wanted to influence either domestic investment in emission-intensive (or emission-reducing) industry and infrastructure, the formation of domestic political coalitions around climate-relevant policy decisions, or even public opinion concerning such issues, could try to influence the expectations held by other actors about the emergent climate regime. This strategy was attempted by Norwegian GCC activists and their political allies. They repeatedly claimed that the Kyoto Protocol meant grave problems for the planned gas-fired power plants. Either the Protocol would simply make the plants unprofitable (because implementing the emissions target through CO_2 taxes or emissions trading would make CO_2 emissions costly), or building the plants with the Kyoto regime in place would impose extra costs or emissions restrictions on other sectors, such as road transport.

The premise for these claims was that the Norwegian economy would have to adapt to an absolute ceiling for national emissions, either because further international negotiations would lead to a quantitative limit to the use of international emissions trading and the other flexibility mechanisms in the protocol, or because of a decision by Norwegian authorities to limit the use of these mechanisms. This would, in turn, mean that the domestic allocation of emission rights to firms and households would become a zero-sum game, leading to considerable conflicts of interest between different sectors of the economy. Allowing one sector to increase emissions would imply forcing others to reduce theirs. This would be the case whether emissions were regulated through tradable quotas, non-transferable permits, technical standards, or emissions taxes. In other words, the expectation of an absolute ceiling on national GHG emissions would tend to split emission-intensive sectors in competing camps, and thereby counteract political alliances between the potential opponents of the environmental movement.

During the crucial meeting in Geneva in the summer of 1996 (COP6), the environmentalists tended to simply ignore the discussion about flexible implementation. As soon as the US delegation had made their historical statement that they supported binding emissions reduction targets, Norwegian GCC activists were quick to offer their view of the consequences. A spokesperson for GN said there would 'probably not' be room for emissions from the planned power plants if the American proposition of flat percentage cuts was carried out. The president of NU added that it would be embarrassing for Primer Minister Brundtland if an international treaty forced her to improve a failed environmental policy (Bjørkeng 1996; Hjorthol 1996; Dagbladet 1996). The environmentalists dominated the news reports after the American announcement, although a spokesman for the minister of the environment was quoted as questioning their conclusions and pointing to the American enthusiasm for emissions trading. In the following days, more representatives of the Labor government, as well as the CEO of *Naturkraft*, referred hopefully to the US support for flexibility mechanisms (Bonde 1996; Bull 1996; Stoltenberg 1996).

A couple of days after the news from Geneva, an economics professor explicitly presented a choice between losing jobs in heavy industry or canceling the plans for gas-fired power in order to fulfill emissions reduction targets (Engeland 1996b; NTB 1996). During 1997, this zero-sum argument (often referring to permanently parking a substantial number of private cars to make room for emissions from the power plants) would be used by GCC activists as well as by their political allies in the struggle against gas-fired power. It was used very actively both by environmental activists and oppositional candidates during the campaign for the parliamentary elections in the fall of 1997. After the new minority centrist government of gas-power opponents came into office, it became a mainstay of their rhetoric.

A newspaper report from the fall of 2000 gives a hint that this strategy at least created some uncertainty. The president of NHO and the energy spokesman of the right-wing Progress Party said that the gas-fired plants may have to be stopped if the Kyoto rules turned out to be very strict, in order to avoid placing too much of a burden on industry and car owners (Ellingsen 1997). During the Kyoto meeting, the new Liberal minister of the environment used every opportunity to warn against starting the tough job of complying with the Kyoto Protocol by building new gas-fired power plants that would put an increased burden on other sectors. Although bitterly disagreeing with the activists on almost all other issues in the negotiations, the minister took over their arguments in this respect. As the Kyoto Protocol did not provide exact rules for the use of flexibility mechanisms, it left the issue open for interpretation and speculation. The FMG used the zero-sum argument again immediately after the Kyoto conference. However, it was generally politicians that used it with greatest enthusiasm. The three parties of the centrist government stuck with it through their fight with the majority in parliament, and their eventual loss and resignation.

Naturkraft and their political allies, on the other hand, envisioned an international GCC regime where businesses could freely buy emissions quotas on the international market. The company itself was active in tree-planting projects in developing countries, with the expressed goal of eventually obtaining emission rights under the international climate regime (Alstadheim 1997). Labor politicians expressed confidence that it would be possible to buy the necessary quotas abroad. With domestic firms operating on international quota markets with little restrictions, increased emissions from one domestic activity would not have to be offset by corresponding domestic reductions. This would render the zero-sum argument more or less invalid.

It is important to note that the two competing conceptions of the restrictions Norway would face under the Kyoto Protocol were each compatible with negotiating positions held by major players in the international negotiations. The environmental movement favored the view, also championed by the EU and several developing countries, that the right to buy quotas from abroad should be restricted. The pro-gas coalition favored the American view that use of the mechanisms should be unlimited. Both sides tended to simply assume that their view would be the outcome of the negotiations. It is reasonable to interpret this debate at least partly as strategic argumentation from both sides. In Putnam's (1988) terms, the activists (just as their opponents) tried to use expectations of the outcome of the negotiation phase to influence positioning in preliminary rounds of the ratification phase. Again, we are seeing an instance of *strategic framing* (Zald 1996). Any ENGO influence that may have stemmed from such a strategy was clearly constrained by the fact that Norwegian media, government and business representatives were also present at the meetings, and that their organizations thus had independent access to information.[2]

CONCLUSION: WHAT ARE GREEN PLAYERS IN TWO-LEVEL GAMES UP TO?

In concluding, we will recapitulate the findings on ENGO strategies reported above, and then point to some implications for the understanding of, first, environmental foreign policy and, secondly, the role of ENGOs in international negotiations. Regarding strategy choice, we have seen that ENGOs pursue several different strategies to reach their political goals in ongoing international environmental negotiations. Members of the CAN coalition – including both nationally based and multinational environmental groups – to a large degree pursued joint goals with regard to the negotiation outcomes and the behavior of individual governments. What linked the joint CAN position papers and 'Fossil of the Day' awards (Climate Action Network 2006) at the COPs to the ENGO campaigns against gas-fired power plants in Western Norway was a goal of making each industrialized country government commit to and implement stringent targets for their domestic emissions of GHGs. While lobbying at the international conferences was one strategy to promote these goals (particularly pursued by multinational groups such as Greenpeace, WWF and FoEI), strategies directed at domestic publics and decision makers were also important.

In the ENGOs' efforts to change the behavior of the Norwegian delegation, attempts to influence delegates directly through lobbying at the international conferences only played a minor part. The ENGOs' main strategies at these meetings could be summed up as *shaming* and *framing*: shaming the Norwegian government into changing its policies, and framing the domestic debate over international commitments in terms that underpinned their own policies. In both these respects, NGO action at the international and domestic levels was intimately connected. The Norwegian ENGOs' activities at the international meetings were mostly geared toward influencing the domestic public and thereby, hopefully, the national government. In these activities, the activists were advancing viewpoints shared by the wider CAN network, and were actively supported by their fellow CAN members.

For our understanding of EFP, the lesson from the above discussion is, first, that ENGOs do engage in influencing foreign policy and could potentially be one of the actors shaping actual policies. In Norway, ENGO's opinions strongly influence public *debate* on EFP, and it is not unreasonable to assume that this could have some bearing on foreign policy *decisions*. By harshly criticizing Norwegian governments, ENGOs tried to make it politically costly to pursue policies they disagreed with. As in other countries in Northwestern Europe, a Norwegian government's green credentials usually matter to a considerable proportion of the electorate. The shared assessment of the Norwegian government by domestic organizations such as NNV, FIVH and NU, and multinational groups such as Greenpeace and WWF, presumably carried some weight for environmentally minded Norwegians. Our discussion underscores that the resources ENGOs have at their disposal are mostly symbolic rather than material. When foreign policy relates to international environmental agreements, the strategic interpretation and framing of existing or proposed international commitments is one available strategy, as well as influencing public perceptions of the government's performance in international arenas. Finally, both domestic groups and organizations based abroad could play a role in foreign policy-making, and they do sometimes (as in this case) form transnational alliances with the purpose of influencing domestic and foreign policy, as well as negotiation outcomes.

These strategies and their potential impact are unlikely to be limited to the Norwegian case. In any country where governments are at least moderately inclined to listen to the environmental movement, the influence of ENGOs is likely to be one of the many factors determining foreign policy regarding GCC. Generally, governments and public opinion in the Northwestern part of Europe have been relatively susceptible to arguments from environmentalists.

The preceding discussion of ENGO strategies also has implications for the understanding of their influence in international environmental negotiations, precisely because of the ENGOs' involvement in foreign policy processes. Our case suggests that the assumptions that information provided to negotiators is the predominant vehicle of NGO influence, and that influence at the international and domestic political levels can be viewed separately are both highly problematic. A *societal* understanding of EFP processes is in fact crucial to understand NGO influence even in international negotiations (Barkdull and Harris 2002).

In other words, the framework for analysis proposed by Betsill and Corell (2001) would not be suitable to assess the overall outcomes of the NGO activities discussed here and in similar NGO work directed at the governments of other countries. One 'key currency' in the Norwegian ENGOs' dealings was indeed information. The information was, however, rarely provided to negotiators. It was rather one of several resources which helped the ENGOs gain access to the national news media and other domestic players in their efforts to *strategically frame* the domestic debate on the government's performance and the content of Norway's international climate policy commitments. In this effort, the Norwegian ENGOs and their foreign allies cooperated. The stressing of Norway's obligation to reduce its domestic emissions was a part of CAN's wider effort to ensure that developed countries did commit to reduce their own GHG emissions substantially, instead of relying strongly on so-called flexible solutions such as financing emissions reductions abroad or planting trees.

While information was important, an equally important 'currency' for the ENGOs was *legitimacy*, which is the issue at stake in strategies of *shaming*. The environmental movement's popular credibility in defining who and what is helping the environment is perhaps its most important resource, which enables it to dispense and withhold environmental credibility to governments, politicians and other actors. In this symbolic sense, grants *are* continually rewarded and threats *are* made by NGOs during UN environmental negotiations – not in terms of money or physical resources, but in terms of governments' reputation among domestic constituencies.

While our case study of ENGO strategies is limited to the period 1995–2001, recent developments suggest that our findings remain relevant. A presentation in early 2006 by one Norwegian activist suggests that the basic repertoire of strategies is unchanged (Kismul 2006). However, reduced tensions between the ENGOs and Norwegian governments on the issues in the negotiations appear to have led to a somewhat stronger emphasis on dialogue with the Norwegian delegation during the COPs – in other words, on what we have termed lobbying strategies. While the shaming tactics have been employed more rarely (indicated by only one 'Fossil of the Day' nomination for Norway in the 4 years 2002–2005), using the COPs to frame the domestic debate is still important, and will likely remain so for some time to come.

NOTES

* An earlier draft for this chapter, by Andreas Tjernshaugen, was presented at the Research Network on Social Movements, ESA Fifth European Conference of Sociology, August 28–September 1, 2001, Helsinki, Finland. The authors thank Tora Skodvin and Paul G Harris for insightful comments, and Lynn Parker Nygaard for editorial assistance.

1. The work of two important Norwegian ENGOs – WWF Norway and the Bellona Foundation – will not be discussed here. The WWF International organisation is also a part of CAN, and has an active Norwegian branch. The WWF Norway group, however, did not engage actively in the international climate policy process during the time period in question. The Oslo-based Bellona Foundation did attend some of the meetings in question and were active in the domestic climate policy debate, but was not part of the international or other joint ENGO initiatives discussed here.

2. In the end, the rules for implementation of the Kyoto Protocol, set out in Marrakech in 2001 and confirmed in Montreal in 2005, placed no quantified cap on overall use of the Kyoto Mechanisms. However, since 2001, successive Norwegian governments have said that they wish to abide by less specific formulations, stressing that use of the mechanisms should only be a supplement to domestic action, and thus emphasizing substantial (but non-quantified) emissions reductions within Norway's borders as a policy goal.

REFERENCES

Alstadheim, Kjetil B (1997), 'Klimafloker til miljøtoppmøte' ['Climate troubles to environmental summit'], *Dagens Næringsliv newspaper* (30 October): 22.

Barkdull, John and Paul G Harris (2002), 'Environmental change and foreign policy: a survey of theory', *Global Environmental Politics* **2**(2): 63–91.

Betsill, Michele M and Elisabeth Corell (2001), 'NGO influence in international environmental negotiations: a framework for analysis', *Global Environmental Politics* **1**(4): 65–85.

Bjørkeng, Per Kristian (1996), 'Norske gasskraftverk i fare' ['Norwegian gas-fired power plants in peril'], *Aftenposten newspaper* (18 July): 6.

Bonde, Aslak (1996), 'Ja til gass, nei til bil!' ['Natural gas, yes; cars, no!'], *Aftenposten newspaper* (23 July): 2.

Bortne, Øystein, Gunnar Grendstad, Per Selle and Kristin Stroemsnes (2001), '*Norsk miljøvernorganisering mellom stat og samfunn*' ['Norwegian organizations for environmental protection: between state and society'], Oslo: Samlaget.

Bull, Bernt (1996), 'Kronikk: et positivt klimainitiativ' ['A positive initiative on climate'], *Aftenposten newspaper* (30 July): 9.

Carlsson, Camilla and Andreas Tjernshaugen (2000), 'Subpolitikk på norsk' ['Subpolitics in Norwegian']. *Arr – tidsskrift for idéhistorie* **2–3**(12): 33–38.

Climate Action Network (2006), Fossil of the Day, available at http://www.fossil-of-the-day.org.

Dagbladet (1996), 'USA viser vei' ['The US leads way'], *Dagbladet newspaper* (18 July): 13.

della Porta, Donatella and Hanspeter Kriesi (1998), 'Social movements in a globalizing world: an introduction', in D. della Porta, H. Kriesi and D. Rucht (eds), *Social Movements in a Globalizing World*, London: MacMillan.

Ellingsen, Lajla (1997), 'Heller bilisme enn gasskraft' ['Rather cars than gas-fired power'], *Dagbladet newspaper* (5 November): 15.

Engeland, Sveinung (1996a), 'Jubler for gasskraftverk' ['Cheering for gas-fired power plants'], *Bergens Tidende newspaper* (24 February): 9.

Engeland, Sveinung (1996b), 'Kan stanse gasskraftverk' ['Could stop gas-fired power plants'], *Bergens Tidende newspaper* (20 July): 4.

Furuly, Jan Gunnar (1997), 'Kirke-observatører kritiske til norsk linje' ['Church observers critical towards Norwegian line'], *Aftenposten newspaper* (9 December): 6.

Gough, Clair and Simon Shackley (2001), 'The respectable politics of climate change: the epistemic communities and NGOs', *International Affairs* **77**(2): 329–45.

Gundersen, Frode (1991), 'Utviklingstrekk ved miljøbevegelsen i Norge' ['Trends in the development of the environmental movement in Norway'], *Sosiologi i dag* **2**: 12–35.

Hegtun, Halvor (1995), 'Tyskland pådriver for CO_2-kutt' ['Germany a pusher for CO_2 cuts'], *Aftenposten newspaper* (6 April): 7.
Hjorthol, Lars Martin (1995), 'Miljøvernere refser Norge' ['Environmentalists scold Norway']. *NTB news agency* (3 April).
Hjorthol, Lars Martin (1996), 'Økt klimapress på Norge' ['Norway under increased pressure on climate'], *NTB news agency* (18 July).
Hovden, Eivind and Gard Lindseth (2002), *Discourses in Norwegian Climate Policy: National Action or Thinking Globally?* Oslo: ProSus.
Kasa, Sjur, Henrik Malvik and Arild Underdal (2001), 'Globale miljøproblemer og nasjonal maktfordeling' ['Global environmental problems and national distribution of power'], in BS Tranøy and Ø Østerud (eds), *Mot et globalisert Norge? Rettslige bindinger, økonomiske føringer og politisk handlingsrom*, Oslo: Gyldendal Norsk Forlag.
Kellow, Aynsley (2000), 'Norms, interests and environmental NGOs: the limits of cosmopolitanism', *Environmental Politics* **9**(3): 1–22.
Kismul, Ane H (2006), 'Hvilken rolle spiller uavhengige organisasjoner i det internasjonale klimaarbeidet og i møter under Klimakonvensjonen og Kyoto-protokollen?' ['The role of NGOs in the climate change negotiations'], Presentation at Klimaforum, Oslo, January 30 (cited with permission).
Klima-alliansen (1996), 'Motmelding til St. mela. nr. 38 (1995–96). Konsekvenser fa miljø, susselsetting og økonomi ved bygging argasstraftverk'. [Canter-report to Government's White Paper on Gas Fired Power Plants]. Oslo: Klima-alliansen.
Moravcsik, Andrew (1995), 'Explaining international human rights regimes: liberal theory and Western Europe', *European Journal of International Relations* **1**(2): 157–89.
Nilsen, Yngve S (2001), '*En felles plattform?: norsk oljeindustri og klimadebatten i Norge fram til 1998*' ['*A Shared Platform? Norwegian Oil Industry and The Climate Policy Debate in Norway up to 1998*'], Oslo: Senter for teknologi innovasjon og kultur Universitetet i Oslo/Unipub.
NTB (1996), 'Gasskraftverkene bør legges på is' ['Gas-fired power plants should be put on ice'], *NTB news agency* (20 July).
Persen, Åsne Berre and Nils Hermann Ranum (1997), '*Natur og Ungdom. 30 år i veien*' ['Nature and youth. Blocking the road for 30 Years']. *Natur og Ungdom* [*Nature and Youth*], A history of the Nature and Youth environmental group, published by the organization at its 30th anniversary.
Putnam, Robert D (1988), 'Diplomacy and domestic politics: the logic of two-level games', *International Organization* **42**(3): 427–60.
Reitan, Marit (2001), 'Den nye miljøpolitikken og de etablerte institusjonene' ['The new environmental policy and the established institutions'], in BS Tranøy and Ø Østerud (eds), *Den fragmenterte staten. Reformer, makt og styring*, Oslo: Gyldendal Akademisk.
Rengger, Nicholas (1997), 'The ethics of trust in world politics', *International Affairs* **73**(3): 469–87.
Seippel, Ørnulf (1999), 'Environmentalism, democracy and political opportunity structures: the case of the Norwegian environmental movement', *Environmental Politics* **8**(3): 49–76.
Skodvin, Tora and Steinar Andresen (2003), 'Non-state influence in the international whaling comission, 1970–1990', *Global Environmental Politics* **3**(4): 61–86
Stoltenberg, Jens (1996), 'Replikk: Gledelig USA-utspill om klima' ['Positive US initiative on climate'], *Dagbladet newspaper* (25 July): 2.

Tjernshaugen, Andreas (1997), 'Poenget er å presse' ['The point is to put pressure on the politicians'], *Natur og Miljø*, **4**:32.

Walk, Heike and Achim Brunnengräber (2000), '*Die Globalisierungswächter*' ['The Guardians of Globalization'], Münster: Westfählishes Dampfboot.

World Commission on Environment and Development (WCED) (1987), *Our Common Future*, Geneva: Oxford University Press.

Zald, Meyer N (1996), 'Culture, ideology, and strategic framing', in MZ McAdam (ed.), *Comparative Perspectives on Social Movements*, Cambridge: Cambridge University Press.

PART II

The European Union and global climate change

9. Articulating a consensus: the EU's position on climate change

Nuno S Lacasta, Suraje Dessai, Eva Kracht and Katharine Vincent*

INTRODUCTION

The issue of global climate change (GCC) ranks high on Europe's political agenda and continues to be a key area of foreign policy for the European Union (EU). In fact, the EU and its member states have, for over a decade, claimed domestic and international leadership with regard to the GCC challenge (Gupta and Grubb 2000; see Chapter 15). The EU has historically supported both the 1992 United Nations (UN) Framework Convention on Climate Change (FCCC), as well as its 1997 Kyoto Protocol. After the withdrawal of the United States (US) from the protocol in mid-2001, the EU ratified it in 2002 and actively pursued its entry into force. The aim of the FCCC is to 'achieve stabilization of greenhouse gas concentrations in the atmosphere at a level that would prevent dangerous anthropogenic interference with the climate system'. This is a simultaneously ambitious and ambiguous aim that began with unspecified targets (see Dessai et al. 2004). The first legal instrument for implementing the FCCC was the Kyoto Protocol, adopted at the third conference of the parties to the convention (COP3) in 1997. This landmark international agreement commits developed countries and economies in transition to reduce their overall emissions of six greenhouse gases (GHGs) to at least 5 per cent below 1990 levels during the 2008–12 commitment period. In order to meet their commitments in an economically efficient manner, parties to the protocol can make use of market-based instruments known as the Kyoto mechanisms (international emissions trading, joint implementation and the Clean Development Mechanism). Emissions reduction commitments can also be met through capitalizing on carbon sequestration options ('carbon sinks').

The Kyoto Protocol raised the issue of GCC into the realm of 'high politics', with the negotiations at times involving the US president, prime ministers of the United Kingdom (UK) and Japan, and the chancellor of Germany. The adoption of the protocol was seen as a major success for international environmental cooperation, even though it left many details unfinished and nearly broke down at critical stages (Grubb et al. 1999; Oberthür and Ott 1999). The subsequent path of negotiating rules for operationalizing the protocol has not been smooth, punctuated by the collapse of COP6 in the Hague in 2000 (Dessai 2001) and the withdrawal of the US in 2001 (Dessai et al. 2003). Nonetheless, agreement on how to operationalize the Kyoto Protocol was reached at COP7, with the adoption of the Marrakech Accords (Dessai and Schipper 2003). After substantial delays, the treaty entered into force on 16 February 2005.

This chapter aims to provide an overview of key processes and actors in the evolution of the EU's GCC policy from the early 1990s to 2005. It does so by considering theoretical frameworks of foreign policy that can provide insights into the development and current shape of the international GCC regime. The next section looks at the main features of the EU as an actor vis-à-vis its member states and the international community at large. The third section identifies key actors involved in negotiating European GCC policy. The fourth section explains, from an EU perspective, selected key topics in the international GCC negotiations. Finally, we debate future directions for EU climate policy and the role of the EU in shaping and developing the GCC regime in the future.

THE EU AS AN INTERNATIONAL ACTOR

The EU is an unusual international organization. Its characteristics are unique to international law in two main ways. First, European law has supremacy over national law in all member states. Secondly, the EU itself has the power to perform certain functions that are traditionally within the realm of sovereign states. Notably in a number of areas it has the authority to pass legislation which has the force of law in all member states (regulations). In this respect it functions as a supranational organization. However, not all sovereign powers have been surrendered, or at least not surrendered fully, and in these cases the EU effectively acts as an intergovernmental organization, where unanimity from members is required. European environmental law exhibits both characteristics. This section outlines the implications of this multi-level governance structure and the role of the various actors for GCC-related foreign policy.

Exclusive vs Shared Competence

One of the features of the EU that most clearly distinguishes it from a state relates to the degree of competence. Rather than the comprehensive competence of a sovereign state, it only has competency to the extent that member states have granted it (Craig and de Burca 1998; Hession 1995). In a few areas, the member states have decided that the EU should be solely responsible for dealing with issues that may arise pertaining to a particular subject matter ('exclusive' competence). This is arguably the case, for example, with the common commercial, agricultural and fisheries policies, but it is not the case for the field of environmental policy. According to the first paragraph of Article 174 of the European Community (EC) treaty, which determines the scope of the Community's environmental competence, EU environmental policy only 'contributes' to the conservation and improvement of the environment. This is a case of 'shared' or mixed competence between the Community and its member states. Both the organization and its members have the power to take action, legislative and non-legislative, in the field of environmental protection. The member states can act insofar as the EU has not done so, or are free to adopt more stringent measures.

It is important to note that for areas of shared competence, the EC Treaty contains a presumption that the member states, rather than the Community, should take necessary action. According to the 'subsidiarity' principle contained in Article 5, paragraph 2, the EU takes measures 'only if and insofar as the objectives of the proposed action cannot be sufficiently achieved by the member states and can therefore … be better achieved by the Community'. Subsidiarity restricts the Community's scope of action in the environmental field. In particular, the subsidiarity principle has been used to impede the surrender of further powers to the Community in the field of climate change (Manners 2000; Dahl 2000).

Of particular consequence for GCC policy is the limited EU competence in the energy sector. The Treaties do not contain any formal Community competence in this field. Energy legislation has therefore been based on the exclusive EU competence with regard to internal market issues, and attempts have been made to take energy-related measures on the basis of EU environmental powers. However, by invoking the subsidiarity principle, member states have frequently managed to retain their sovereignty in some all-important areas. Internal EU policy proposals in this field have therefore frequently failed or have been considerably weakened. The European Commission first proposed an EU-wide CO_2/energy tax in 1991. Following years of fruitless debate a modified version was proposed in 1997, allowing for the creation of an EU-wide energy taxation framework. This took a further 5 years of negotiation before coming into force in 2004 as a weak energy taxation directive. Similarly, the European energy efficiency program (SAVE) was considerably weakened as a direct result of member states bringing the subsidiarity principle into play.

Internal vs External Competence

In terms of foreign policy, the external powers of the EU reflect the degree of internal competence. Insofar as it is competent to promulgate regulations within the Community, the EU can also enter into negotiations and conclude treaties with other states and international organizations. As Community law preempts member state law, only the EU has the power to accept and implement international obligations in areas of EU competence. Where the Treaty does not provide explicitly for such external competences, they are therefore considered to be 'implied powers' (Macrory and Hession 1996; Craig and de Burca 1998). As the external competence corresponds to the internal powers, the split of competences described in the previous section also occurs in the external sphere. Article 174 expresses this by referring to the EC and its member states' 'respective spheres of competence'. This is the reason why both the EU and its member states are parties to a range of international agreements, including the FCCC and the Kyoto Protocol, and they both needed to ratify these agreements.

The FCCC and the Kyoto Protocol, as well as other environmental treaties, are therefore what can be termed 'mixed' agreements. They cover a variety of subject matters, including, for example, environmental protection and trade, so that neither the EU nor its members have the exclusive power to execute these accords (Jupille and Caporaso 1998). In theory, only the EU is entitled to participate in negotiations concerning matters of exclusive Community competence, while the member states are entitled to participate in negotiations in areas of exclusive member state competence. Both are entitled to participate in fields of shared competence. These areas cannot, however, be easily delineated from one another. This has at times caused the EU members and institutions to enter into internal negotiations within the context of external negotiations in order to decide who is competent regarding the issue under discussion. On occasion, that has even barred the EU from participating altogether. This has sometimes placed the EU at a disadvantage compared with other actors (Jupille and Caporaso 1998).

Because the clear determination of respective competences is virtually impossible, the Presidency of the Council speaks on behalf of the Community and its member states in GCC negotiations, conveying common positions agreed during Council sessions. Since such common positions in part affect areas of member states' competence, they must be approved by consensus. As a result, the process of reaching agreement among all EU members and the European Commission can be cumbersome and can delay necessary action and inhibit the EU's capacity to demonstrate leadership in international negotiations. In addition, the outcome of these internal negotiations is likely to reflect lowest-common-denominator positions even before the EU enters into bargaining with third parties (Gupta and Ringius 2001; Metz et al. 2001).

NEGOTIATING CLIMATE CHANGE: INTERNAL AND EXTERNAL ORGANIZATION

As the previous section has suggested, the nature of global climate change is such that negotiation of related foreign policy draws on both the supranational and inter-governmental characteristics of the EU. Barkdull and Harris (2002) identify three main theoretical standpoints in environmental foreign policy and international relations. *State-centric theory* currently has limited applications in the EU case due to the diversity of actors and the absence of a hierarchical structure headed by a single leader. Elements of *societal theory*, emphasizing the brokering of interest group politics, have clear resonance with the current structure of the EU (see, for example, Michaelowa 1998). *Systemic theory*, assuming that foreign policy is directed by systemic factors rather than domestic politics, is also appropriate as the diversity of state interests often means that systemic concerns help to form a common ground. However, with the number and variety of EU actors involved in different elements of negotiation of the GCC regime over time, it is not surprising that an analysis highlights elements of all these theories. This section outlines the various actors and illuminates ways in which they can affect foreign policy formation.

Reaching a Common Position: The European Council

Although the European Parliament's role has to some extent been strengthened by recent treaty reforms, the Council of the European Union is still the most powerful EU body. The Council is the EU's primary decision-making, legislative, and co-ordinating authority. It consists of a representative at ministerial level of each member state. In the case of climate policy, the Council usually consists of the environment ministers. The members of the Council are bound by instructions from their respective governments, but they are also obligated – as parts of a Community institution – to act for the EU's common good (Kapteyn and van Themaat 1998). Decision-making procedures and Council voting rules vary in different areas of EU competence. Most matters, including in the environmental field, are now decided by qualified majority voting. However, Article 175, paragraph 2, provides that the Council must act unanimously on matters primarily of a fiscal nature and 'measures significantly affecting a member state's choice between energy sources and the general structure of its energy supply'.

In the context of the GCC negotiations, the Council plays a key role both in the run-up to negotiations and in signing and ratifying the relevant agreements. Before a negotiating session, the EU and its members meet in Council formation to discuss and agree on a common position. The fact that the Council was able at various stages of the GCC negotiations to reach agreement on specific internal climate policies, such as burden sharing among member states, has both ensured a common EU line and strengthened the EU's leadership role during those negotiations. However, this leadership is sometimes much less effective in the 'heat' of complex international negotiations, in which a Party is expected to think and act quickly. The EU has often shown a lack of flexibility and expeditiousness. It clearly has an effectiveness gap in international negotiations on areas of mixed competence like climate change (Gupta and Grubb 2000; Yamin 2000).

The process of reaching a common position within the Council is characterized by bargaining and uncertainty until the last moment. Success depends considerably on the leadership exerted by the country holding the Presidency at a given moment, which essentially has to broker a deal among the other member states and the Commission. Historically, smaller member states have been more prone to reaching consensus, and to a certain extent foregoing their strict national positions, in favor of an overall EU position (see Chapters 4 and 6). This was clearly the case for Sweden and Belgium in 2001. Bigger member states, in contrast, tend to try to impose their own priorities, or to broker agreement with other bigger states or influential smaller states, exhibiting elements of state-centric theory (see Chapters 2 and 3). After COP1 in 1995, an Ad Hoc Group on Climate Change was established at Council level. This co-ordinating group has been the main unit on international climate policy, and has enhanced the EU's effectiveness in reaching a common position (Oberthür and Ott 1999).

Once international agreement is reached, the Council is responsible for signing and concluding the treaty or agreement. In parallel to the rules for internal decision-making, the Council normally decides by a qualified majority, but must act unanimously in areas covered by Article 175, paragraph 2. Controversy ensued with regard to the FCCC when the Council's legal service argued for the former but the UK argued for the latter on the basis of impact on energy policy. The problem was resolved politically when the UK abstained from voting, but similar discussions arose regarding the approval of the Kyoto Protocol. In order to settle the dispute, the Council adopted a declaration, according to which further national obligations to reduce GHG emissions should be decided by consensus. It managed, however, to ratify the Kyoto Protocol by qualified majority voting under Article 175, paragraph 1.

Negotiating Internationally: The Role of the Presidency

The Council is headed by the Presidency, which is held alternately by each member state for six months. The Presidency is assisted by the Commission and the member states holding the preceding and the upcoming presidencies – the so-called 'Troika'. The Presidency co-ordinates the formation of the EU's common position and presents it at international negotiations (Oberthür and Ott 1999). Its management of the process is vital for the effectiveness of the EU negotiating stance. The Presidency determines the agenda of the Council, chairs its sessions and co-ordinates the member states at negotiations. Thus, the effectiveness of the EU in international negotiations is influenced by the Presidency's internal management and external negotiating skills and tactics. For example, in the run-up to Kyoto, EU climate policy did not progress much during the Italian and Irish presidencies in 1996. In the first half of 1997, however, the Dutch presidency drove the process forward significantly, resulting in the approval by the Council of a proposal for internal sharing of GHG emissions (the EU Burden Sharing Agreement), as well as negotiating a proposal for a 15 per cent reduction target for 2010 and an intermediate 7.5 per cent reduction target for 2005. Similarly, German leadership – and insistence – on the matter of using the Kyoto market mechanisms as 'supplemental to domestic actions' (Depledge 2000) ensured that during its presidency, in the first half of 1999, the EU concluded and presented a negotiating proposal on this matter (on Germany, see Chapters 2, 12, 14 and 15). However this focus on supplementarity arguably resulted in it falling behind in the preparation of a host of other issues pertaining to the negotiations, and thus inhibited effective leadership in the subsequent presidencies held by Finland and Portugal.

In terms of negotiating with third parties, the Presidency's leadership is also conditioned by practical and political considerations. For instance, two of the three presidencies immediately before COP7 (2001) – France and Belgium – adopted considerably different co-ordination and negotiation practices, perhaps reflecting the differences in capacities and posture between a large and a medium member state. France, which held the Presidency during the second half of 2000, at times (especially during COP6) clearly pushed for what seemed like its own agenda in internal co-ordination, and reluctantly watched the UK unilaterally enter into tentative negotiations with the US aimed at securing agreement between the EU and the US (Grubb and Yamin 2001; on the UK, see Chapter 3). In contrast to this state-centric mode of operation, Belgium, which held the Presidency during the second semester of 2001, took a more consensus-based approach internally and relied considerably on the assistance of the Commission as well as on a division of labor between member states, especially at the technical level – thus fitting the societal theory model of operation.

The rotation of the Presidency every six months presents a problem because it has not allowed the EU's GCC policy to develop in a coherent and stable way (Oberthür and Ott 1999) and has not provided negotiating partners with a steady arrangement (Oberthür 1999). The EU constitution proposed in 2005 addressed this problem of continuity by recommending a rotation system, whereby three member states co-ordinate their presidencies over a one and one-half-year period (van Schaik and Egenhofer 2005). This strengthened joint presidency model would build on the existing informal 'Troika' system. It aims to promote increased continuity of policies and approaches based on a common multiyear work program and also labor sharing among the three countries. One way in which this might be achieved is to divide labor between three member states of different sizes, such as large (Germany), medium (Portugal) and small (Slovenia).[1] Although discussions on the EU constitution have stalled since the rejection by the French and Dutch citizens, this mechanism may yet be considered to improve the co-ordination of the Presidency.

KEY ACTORS WITHIN THE EU

Member States

On 1 May 2004, 10 new countries joined the EU, bringing the total number of member states to 25. Each member state differs in their GHG emissions and energy profiles (European Environment Agency 2005). The disparity in emissions characteristics, abatement costs, possible impacts of climate change, and relative level of economic development in the different member states presents a significant obstacle to agreement on a common GCC policy (Collier 1997), and to a considerable extent acts as a microcosm of the GCC discussions at the global level. Indeed, the EU-25 now include Annex I and non-Annex I countries (Cyprus and Malta), economies in transition (e.g., Poland), and countries integrating different regional groupings such as the Eastern European Group (e.g., Slovakia) and even the Asian Group (Cyprus again). The EU therefore should, on the one hand and more than ever, be able to anticipate several of the global debates that will take place on the future of GCC policy. On the other hand, it might also become a testing ground for the policy development and negotiation challenges of increasingly complex and differentiated future global climate developments.

Analyzing approaches to the GCC debate since the 1990s shows that member states (EU-15) can switch between 'lead states', 'support states', 'swing states' and 'veto states' (a typology devised by Manners 2000).[2] The UK, for example, has changed from being a highly Eurosceptic veto state under John Major to a lead state under Tony Blair's premiership. Because the member states' positions cannot be portrayed in detail, suffice it to say that EU countries tend to pursue their national interests through the Council. However, without the EU's collective 'weight', individual member states' interests might simply not be able to prevail in the international policy arena in the face of such influential negotiating partners as the US, Japan, China, India and Brazil. This has clearly been the assumption, and indeed the operational basis, upon which European climate policy has been pursued from the outset, and is evidenced in the development of internal regulatory instruments in this field, such as the 2003 emissions trading directive.

The European Commission

Within the EU, the European Commission, comprising independent Commissioners from all member states, has a range of important functions, from elaborating legislative proposals to ensuring the implementation of European law. It is organized in Directorates-General, which may hold rather different views on climate change-related issues. Then EU Commission President Romano Prodi frequently had to mediate between often-opposing viewpoints proposed by the then Commissioner for Transport and Energy, Loyola de Palacio, and Commissioner for Environment, Margot Wallström. Proposals brought up by the Commission need to be approved by the simple majority of the Commissioners. This natural tension will remain a structural feature of the policy area across EU institutions.

In other environmental regimes (e.g., ozone), the Council has mandated the Commission to conduct the negotiations on behalf of the EU (Macrory and Hession 1996; Oberthür 1999). With international negotiation of GCC policy, however, member states could not agree to follow the same procedure. The Council did not grant the Commission's request to endow it with negotiating authority for the Kyoto Protocol; many members simply opposed transferring more competences to the EU level (Oberthür and Ott 1999). Instead, as described above, the Presidency fulfils the task of co-ordinating and presenting the EU position. As a result, the Commission plays a limited role. It participates as an equal partner in establishing the EU common position, on which it and all member states must agree.

Business and Industry

The business and industrial community has taken an increasing and diversified interest in the GCC negotiations (Carpenter 2001). In the early days of the FCCC, the focus was on fighting any restrictions on fossil fuel use, with an association of transnational corporations known as the Global Climate Coalition acting to discredit GCC science and contest the predicted nature of GCC (Gough and Shackley 2001). With improving science and a growing concern for environmentalism, many businesses have now accepted the need for action. Amongst the business organizations in support of the GCC regime in Europe are the World Business Council for Sustainable Development, the International Chamber of Commerce, the Business Environmental Leadership Council, the European Business Council for a Sustainable Energy Future and the Union of Industrial and Employers' Confederations of Europe. In the UK, Prime Minister Tony Blair launched the Climate Group, a charity that aims to 'catalyze a new political momentum on climate change'. Many of these organizations have accredited status with the FCCC to enable them to attend negotiating sessions and to be observer at the COPs. They lobby delegations and also provide advice from a business perspective. Within the EU, business representation was active during both the emissions-trading directive negotiations and the ensuing allocation plans. Business engagement has been stronger than ever since January 2005, when the EU Emissions Trading Scheme got underway.

Environmental Nongovernmental Organizations

Environmental nongovernmental organizations (NGOs) – most of them organized in the worldwide Climate Action Network (CAN), with CAN Europe being a particularly active regional office (Düwe 2001) – have from the outset also played an important role in European GCC policy. As in other fields, the climate NGOs reflect a heterogeneous universe of stakeholders, such as generalist-versus-specialized/expert NGOs and membership-versus-non-membership NGOs. Their roles are varied and extend beyond simply raising public awareness through the media or demonstrations. Member states and the Commission have made frequent use of the competence of NGO experts, such as by funding GCC-related research projects carried out by NGOs. Many NGOs discuss issues with and lobby negotiators outside and during negotiations (Newell 2000; McCormick 1995). Due to their capacity – they build up expert technical knowledge on the issues, often beyond what is possible for governments – some delegations include experts from NGOs, particularly specialized/research-based NGOs. Through adapting to the multi-level structure of the EU, and tapping into the processes of negotiating GCC policy at a variety of stages, many NGOs have managed to actively participate in shaping the regime, rather than just playing the role of external critics (Gough and Shackley 2001; Michaelowa 1998).

Media and the Public

Public attention plays a central role in determining the EU's political agenda and in turn in shaping foreign policy. High levels of public concern about environmental problems have repeatedly prompted governments to address them, while in the absence of popular interest those same governments would be considerably less likely to take action. Arguably, both the conclusion of the FCCC negotiations in time for the Rio Summit in 1992 and of the Kyoto Protocol in 1997 were to a large degree the result of substantial public pressure (Newell 2000).

Public opinion is largely influenced by the media. As the primary source of information for most people, media coverage highlights selected issues for public scrutiny and also shapes the way people perceive the problems and possible solutions to them. In recent years, major media sources have frequently reported on various aspects of GCC. Large numbers of journalists attend the COPs; almost 1,000 were accredited at COP6 (2000) in The Hague, in comparison to little more than 2,000 members of government delegations. Since COP3 (1997), newspaper articles have been posted and distributed at the conference centers and handed to government officials (Carpenter 2001), providing negotiators with immediate public feedback. On the other hand, media interest fades in the time periods between major conferences, and the way reporters have presented the subject has in certain respects added to confusion and lack of knowledge concerning important aspects of GCC. Factual errors, an overemphasis of the findings of skeptical GCC scientists, exaggeration of abatement costs and a focus on singular extreme weather events, rather than on structural problems, have resulted in widespread misunderstandings about the science, the causes of GCC and possible responses (Newell 2000). The fact that NGOs, business, the media and the public are in different ways involved in the GCC negotiations fits closely with the societal theory of environmental foreign policy formation suggested by Barkdull and Harris (2002).

THE EU IN THE CLIMATE CHANGE REGIME: SELECTED ISSUES

Previous sections have outlined some key aspects of EU policymaking on GCC from an institutional and actor-based perspective. To illustrate the role of the institutions and the actors in shaping climate policy, this section analyses selected aspects of the GCC negotiations from the signing of the FCCC in 1992 to COP9 in late 2003.We will draw upon the available literature and official documents, as well as on our own experience in taking part in different capacities in the GCC negotiations since the mid 1990s, including multiple interactions with many different participants.

Throughout this past decade the EU has remained committed to pushing for what has generally been perceived as stringent GCC standards, somewhat in contrast with its own domestic track record. The following analysis will focus on three issues: policies and measures, emission targets, and developing countries. If one can summarize what we say below, the EU positioning was predicated on the ability of the member states and the Community to jointly fulfill their emissions-target obligations under what became the Kyoto Protocol. This fact has consistently conditioned the EU position, leading it to often renege or soften other positions in order to secure its primary objective.

Policies and Measures: Exporting Domestic Approaches or Pretext for Domestic Co-ordination?

A key element in the EU negotiation strategy leading up to Kyoto consisted of a package of so-called 'common and coordinated policies and measures', such as adopting some kind of carbon/energy tax or energy efficiency standards in the protocol. The discussion on policies and measures effectively began at COP1 in Berlin (1995), when the EU formed an alliance with a number of developing countries (minus members of the Organization of Petroleum Exporting Countries), the so-called 'green group' against the JUSSCANNZ countries.[3] The Berlin Mandate that emanated from COP1 called on developed countries to elaborate policies and measures for achieving the objectives of the FCCC, which the EU duly considered throughout 1996 and 1997. Some internal political issues remained unresolved among the member states and the Commission. Some member states sought to transfer the unresolved considerations to the international-level negotiations in the hopes of attracting support for their views (Oberthür and Ott 1999). However, as Yamin (2000) sharply observes after analyzing the EU international GCC negotiating strategy,

> 'The fact that these same countries could not get other EU member states to agree to mandatory [policies and measures] internally should have alerted them to the difficulty of trying to persuade a more heterogeneous and larger number of states ... Rather than seeing itself as a microcosm of the larger group, the EU appeared to have thought of the Ad Hoc Group on the Berlin Mandate (AGBM) process as having the ability to solve the EU's own internal problems' (Yamin 2000).

One reason for the EU's insistence on policies and measures, proposed by commentators, was that the organization and its members were acting tactically vis-à-vis the US in particular. At the beginning of the AGBM process, the US had not yet signaled a preference for GHG targets and timetables. Some in the EU therefore viewed the proposal on binding policies and measures as a way to put pressure on the JUSSCANNZ countries in general, and on the US in particular. However, the EU's continued insistence on policies and measures, even after the US's announcement on targets, suggests that the strategy did little more than irritate other negotiating partners while proving to be ultimately unsuccessful.

The text of the Kyoto Protocol does not contain any reference to binding policies and measures. Article 2 merely lists examples of policies and measures to be taken by each party 'in accordance with its national circumstances'. Article 2.4 further opens the door for the future consideration of policies and measures 'co-ordination'. Since the adoption of the protocol in 1997, policies and measures have barely received negotiators' attention. This is mainly because parties were busy crafting the protocol's key operational rules on, for example, flexible mechanisms and the monitoring, reporting and verification of parties' emissions. As a result, parties limited their activity on policies and measures to the organization of two information-sharing workshops on 'best practices' and rather surreal discussions on if and how to set up an information-sharing website. However, with the withdrawal of the US from the Kyoto Protocol, and the preparation of the next stage of international GCC negotiations (post-2012), policies and measures may once again be brought to the fore as a means to bring together heterogeneous policy responses.

Targets and Timetables: Leading, but as a 'Bubble'

After the adoption of the FCCC, there were more frequent suggestions that its aim to stabilize GHG emissions would not be sufficient to tackle the challenge of GCC, and that the first review of the adequacy of commitments should take place at COP1 in Berlin. As a proponent of additional reduction commitments, six months prior to COP1 the Alliance of Small Island States (AOSIS) proposed a draft protocol that called for a 20 per cent reduction of industrialized countries' CO_2 emissions by 2005. Parties at COP1 agreed that the convention's aim was not adequate and called for quantified commitments with specified time limits. At that stage the EU had no clear common position with regard to the strengthening of targets. Germany, however, suggested language that called for further reductions, and in October 1995 outlined a proposal for a 10 per cent reduction of CO_2 emissions by 2005 and a 15–20 per cent reduction by 2010 (see Chapter 2). It was the German proposal, as well as the US coming forward with its position on targets at COP2 (1996), which led the EU in March 1997 to come to an agreement on a collective proposal. This proposal called for developed countries' targets and timetables based on a basket of three gases (CO_2, methane and nitrous oxide),[4] and on a flat-rate reduction of 15 per cent by 2010 in relation to the 1990 base year.

As noted above, during the international negotiations on GHG targets, the EU insisted on being permitted to fulfill its obligations jointly – or as a 'bubble'. This means that the EU and its member states have a common target, which at Kyoto was set an 8 per cent cut for the whole EU, but that they can redistribute the burden internally. JUSSCANNZ, in particular Japan and Australia, considered that the extra flexibility of such a mechanism was unfair. The solution was a compromize, with targets differentiated among developed countries (an approach favored by, among others, Australia), designation of 1990 as the year upon which reductions should be based (an EU preference) and comprehensive coverage of six gases rather than the three that Europe had first proposed (a US proposal). Furthermore, rather than being based on a single-year assessment, the targets are assessed on the basis of a 5-year 'budget' or 'commitment' period starting in 2008.

The Burden Sharing Agreement and Article 4 of the Kyoto Protocol

The issue of burden sharing has historically been at the center of the EU position on climate change (Bodansky 2001), even if the actual policy development moved slowly. In March 1997 during the run up to Kyoto, the Dutch presidency negotiated an internal burden-sharing agreement (see Table 9.1).

This accounted for a 9 per cent aggregate EU reduction, with member states' limitations or reductions ranging from 25 per cent to +40 per cent. This fell short of the 15 per cent initially pledged by the EU. Following negotiations for the Kyoto Protocol, the EU pledge was reviewed under the British presidency (see Chapter 3) and lowered an 8 per cent reduction pledge, essentially giving way to the positions of the more conservative member states. Indeed, as Oberthür and Ott (1999) summarize in a characteristic illustration of the relationship between the member states in the GCC arena:

> 'a number of governments used the new situation to achieve a general relaxation of their targets. Denmark and Germany demanded adjustments and the former 'green' countries Austria and the Netherlands admitted that they would not be able to meet their ambitious targets of the first agreement. These announcements triggered distinctive resistance from Greece, Ireland, Portugal and Spain, who were now expected to limit their emission growth to a larger extent than formerly agreed. Of the main emitters of the [EU], only the UK declared that it would take a stronger commitment than before.' (Oberthür and Ott 1999)

Table 9.1 EU 1997 and 1998 burden sharing agreements

Member State	1997 (%)	1998 (%)
Austria	–25	–13
Belgium	–10	–7.5
Denmark	–25	– 21
Finland	0	0
France	0	0
Germany	–25	–21
Greece	+30	+25
Ireland	+15	+13
Italy	–7	–6.5
Luxembourg	–30	–28
Netherlands	–10	–6.0
Portugal	+40	+27
Spain	+17	+15
Sweden	+5	+4.0
United Kingdom	–10	–12.5
EU–15 total	**–9.2**	**–8**

Source: Lacasta, Dessai and Powroslo (2002: 43).

The 1998 burden sharing agreement thus settled the scores after Kyoto, allowing the EU and its member states to focus on the development of their own implementation plans, which have been under way ever since. A policy debate took place in March 2005 at the European Council on strategies and targets for post-2012, and the issue of burden sharing featured between the lines. The new member states are likely to contribute to a somewhat different outlook in the EU as regards international GCC policy. However, due to diverse economic situations and emissions profiles of member states, any future emissions reduction policies in the EU will almost inevitably encompass burden sharing, even if this is on renewed premises. Once again, it can be instructive to view the EU acting as being a sort of microcosm of international negotiations on climate change.

Policy toward the Developing Countries

Traditionally, the EU has taken a conciliatory approach to developing country participation in the GCC regime, in stark contrast to other OECD developed countries such as the US and Australia, which have argued that developing countries should take on targets or limitations during the Kyoto

Protocol's first commitment period (see Chapter 15). In the run-up to Kyoto the EU argued for a 'graduation' – which would require developing countries joining the Organisation for Economic Cooperation and Development (OECD), like South Korea and Mexico, to take on commitments – and 'evolution' – because regulation of developing countries emissions in the long term is an environmental necessity (Grubb et al. 1999). The EU proposal for developing countries is fairly consistent with its own 'bubble' concept, encouraging global burden sharing. Whether this coalition holds in the future remains to be seen, but with the discussion on second commitment-period targets nearing, the EU will surely have to take a leadership and mediating role in order to 'keep the family together'. This will be one of the biggest challenges the EU faces in the years to come.

CONCLUSION: WHAT FUTURE FOR EU FOREIGN POLICY ON GLOBAL CLIMATE CHANGE?

For the EU to continue to assert itself internationally, and to be a leader of international GCC policy, it needs to ensure that its internal procedures and practices are robust and effective. We are currently at an important stage of determining just how capable the EU might be of taking (or keeping) the lead in international climate policy. If the decision-making processes remain the same, the accession of 10 new member states in May 2004 has made the process more complicated and cumbersome, reducing the EU's flexibility and opportunities to lead by example at future GCC negotiations. Although the debate about the EU Constitution is taking a backseat following rejections by France and the Netherlands, new practices mentioned in that document may yet bring about important changes in internal working arrangements. These might, in turn, reduce the complexity of internal foreign policy negotiation, and thus increase the scope for EU leadership in international GCC policy. As the Constitution has not been ratified (and might never be), and because the membership of the EU has substantially expanded, it remains to be seen exactly how much change might come from the Constitution and negotiations surrounding it.

This discussion of some of the history of Europe's GCC foreign-policy process does, however, highlight potential key areas for change within the existing system. The EU cannot afford paralyzing itself by regular haggling over competences. It needs to at least agree on a common mid and long-term GCC strategy, and on a *modus operandi* for international negotiations, which assigns clear negotiating and decision-making powers to the Commission and/or to the member states according to their perceived 'comparative advantages'. Negotiators need to be authorized to act with flexibility. That is, they need the freedom to enter into bargaining with third parties without being forced to assure consensus among all EU member states in the case of each modification to originally agreed EU negotiating positions. Member states must therefore give up some (paralyzing) control in order to ensure higher overall effectiveness of the EU's negotiating practice, which in turn would improve its relationship to negotiating partners.

Like in other environmental regimes, the Commission, rather than the Presidency, might take the lead on GCC negotiations. This would not be incompatible with a clearer division of labor among the Commission and the member states. The Commission could work closely with the member states in both preparing and negotiating a common position. Such a division of labor would lead to continuity, stability and consistency in EU negotiating positions on climate change. The Commission seems to be better equipped for this task because it is, by definition and in practice, the 'guardian of the treaties' (i.e., protector of EU policies and regulations). Gupta and Grubb (2000) have even suggested that to enhance its profile, GCC should become part of the EU's Common Foreign and Security Policy. Experience thus far shows that we are still short of this degree of integration, but nonetheless GCC has been brought up with some regularity at EU summits with third parties, such as the 2004 EU–Russia Summit. Foreign affairs ministries have also created a so-called Green Diplomacy Network that regularly addresses GCC issues, with the aim of increasing foreign policy co-ordination, for example in the preparation of joint diplomatic demarches.

In addition, the EU needs to enter into negotiating sessions with more elaborate positions in order to ensure a greater impact on the overall design of the GCC regime. It also very importantly needs to develop potential fall-back positions ahead of time to allow for more flexibility and quick moves in the decisive phases of international GCC bargaining processes. The internal process for reaching a common position needs to be streamlined by, for instance, having a system of 'lead countries' to prepare draft common negotiating positions to be decided by the European Council. Some member states have historically provided most of the intellectual capital for selected issues, and at times such a system of lead countries has been adopted during negotiating sessions. The EU would nonetheless benefit from having such a practice made more permanent and effective between negotiating sessions. The 2004 Irish presidency, in conjunction with the 2005 Luxembourg and UK presidencies, proposed a working arrangement that points towards such long-term planning, the identification of lead negotiators and a focus on more than just the EU itself during negotiations. These informal procedures have already been put to work and proved rather successful. However, the real test will come with the future GCC regime complex and multi-issue negotiation that will certainly test EU negotiators to their limits, perhaps even more so than before.

As the EU has developed, it has begun to play a more important role in the GCC regime. A number of key future roles can be identified. One is to bring the US back on board the Kyoto process. Free riding on climate change by the US and some other developed countries raises deeper issues of equity for the international community as a whole (Soroos 2001; see Chapter 15), and may increase the reluctance of developing countries to take mitigation commitments of their own. Having the US, the world's largest emitter of GHGs, outside a key 'global' climate change instrument such as the Kyoto Protocol cannot be sustained for a long period of time. The Bush administration proposals on addressing climate change with technological solutions have been met with skepticism from most Europeans. The EU seems to be the only party in the GCC regime willing to bridge, and capable of bridging, the divide between the Umbrella Group and the developing countries. With the entry into force of the protocol, parties are already preparing for the negotiations of future GCC policy approaches. Evidence suggests that the EU plans to continue trying to lead on this issue, with a 2005 communication on the post-2012 strategy, and indeed the Spring 2005 European Heads of State and Government Council strategy for future global climate policies, emphasizing proposals for bringing developing countries on board.

This suggests that the EU remains committed to exerting international leadership on environmental policy in general (as is seen in several other fora) and GCC policy in particular. Indeed, arguably without the means to exert international 'hard' power on such matters as global security, the EU has for a number of years been exerting 'soft' power in the areas of environmental policy and development (the EC and its member states combined are by far the biggest overseas development aid donors). The European Council has reiterated that this trend is likely to continue, although the effectiveness of this leadership will once again be determined by the underlying political will and the EU's internal operating procedures.

In conclusion, the EU's main priorities in the mid to long-term future relate to its ability to implement effective domestic GCC policies (leading by example); to reform internal processes so as to ensure it is more capable of leading in the international negotiations; and to prepare itself adequately for 'second generation' GCC policies by, in particular, bridging the gap between the US and the developing world. If the EU manages to devise more efficient ways to co-ordinate its many voices – perhaps at times without reaching unanimity – it will be able to lead the international GCC process into the future. An analysis of the roughly two decades of EU GCC policy reveals a mixed record that, in spite of all its shortcomings, gives rise to guarded optimism for continued international co-ordination and cooperation on the management of the global commons.

NOTES

* Suraje Dessai received support from Fundação para a Ciência e a Tecnologia, Portugal (grant SFRH/BD/4901/2001). Portions of this chapter first appeared in Lacasta, Dessai and Powroslo (2002). The authors would like to thank Louise van Schaik and Ambassador Bo Kjellen for useful comments. Any errors or omissions are, however, the authors' full responsibility.

1. As of this writing in early 2006, the prospects for adoption of the proposed EU constitution look slim.
2. See also Sprinz and Vaahtorama (1994) on the perceived vulnerability to environmental damage and abatement costs and their relation to states' interests in international environmental policy, as suggested in Barkdull and Harris (2002).
3. The JUSSCANNZ countries were Japan, the US, Switzerland, Canada, Australia, Norway and New Zealand. This coalition subsequently became known as the Umbrella Group after Switzerland left and Russia, Iceland and the Ukraine joined.
4. The UK and the Netherlands had wanted a more comprehensive strategy that included three additional gases that were, in the end, included in the Kyoto Protocol (see Chapters 3 and 4.)

REFERENCES

Barkdull, John and Paul G Harris (2002), 'Environmental change and foreign policy: a survey of theory', *Global Environmental Politics* **2**(2): 63–91.

Bodansky, Daniel (2001), 'The history of the global climate change regime', in Urs Luterbacher and Detlef. F. Sprinz, *International Relations and Global Climate Change*, Cambridge: MIT Press, 23–40.

Carpenter, Chad (2001), 'Business, green groups and the media: the role of non-governmental organizations in the climate change debate', *International Affairs* **7**: 313–28.

Collier, Ute (1997), 'The EU and Climate Change Policy: The Struggle Over Policy competences', in Ute Collier and Ragnar Löfstedt, *Cases In Climate Change Policy: Political Reality in the European Union*, London: Earthscan.

Collier, Ute and Ragnar Löfstedt (eds) (1997), *Cases in Climate Change Policy: Political Reality in the European Union*, London: Earthscan.

Craig, Paul and Grainne De Burca (1998), *EU Law*, Oxford: Oxford University Press.

Dahl, Agnethe (2000), 'Competence and subsidiarity', in Joyeeta Gupta and Michael Grubb (eds), *Climate Change and European Leadership: A Sustainable Role for Europe?*, Dordrecht: Kluwer Academic Publishers.

Depledge, Joanna (2000), *Tracing the Origins of the Kyoto Protocol: An Article-By-Article Contextual History*, UN Doc FCCC/TP/2000/2, available at: www.unfccc.int.

Dessai, Suraje (2001), 'Why did the Hague climate conference fail?' *Environmental Politics* **10**: 139–44.

Dessai, Suraje, W Neil Adger, Mike Hulme, John Turnpenny, Jonathan Köhler and Rachel Warren (2004), 'Defining and experiencing dangerous climate change', *Climatic Change* **64**(1–2): 11–25.

Dessai, Suraje, Nuno S Lacasta and Vincent Katharine (2003), 'International political history of the Kyoto Protocol: from the Hague to Marrakech and beyond', *International Review of Environmental Strategies* **4**(2): 183–205.

Dessai, Suraje and E Lisa Schipper (2003), 'The Marrakech Accords to the Kyoto Protocol: analysis and future prospects', *Global Environmental Change* **13**(2): 149–53.

Düwe, Matthias (2001), 'The climate action network: global civil society at work?', *Review of European Community and International Environmental Law* **10**(2): 177–89.

European Environment Agency (EnEA) (2005), *Annual European Community Greenhouse Gas Inventory 1990–2003 and Inventory Report 2005*, Technical report No. 4/2005, available at http://reports.eea.eu.int/technical_report_2005_4.

Gough, Clair and Simon Shackley (2001), 'The respectable politics of climate change: the epistemic communities and NGOs', *International Affairs* **77**(2): 329–45.

Grubb, Michael, Christiaan Vrolijk, and Brack Duncan (1999), *The Kyoto Protocol: A Guide And Assessment*, London: Earthscan.

Gupta, Joyeeta and Michael Grubb (eds) (2000), *Climate Change and European Leadership: A Sustainable Role for Europe?*, Dordrecht: Kluwer.

Gupta, Joyeeta and Lasse Ringius (2001), 'The EU's climate leadership: reconciling ambition and reality', *International Environmental Agreements* **1**(2): 281–99.

Grubb, Michael and Farhana Yamin (2001), 'Climatic collapse at the Hague: what happened, why, and where do we go from here?', *International Affairs* **77**(2): 261–76.

Hession, Martin (1995), 'External competence and the European Union', *Global Environmental Change* **5**(2): 155–6.

Jupille, Joseph and James A Caporaso (1998), 'States, agency and rules: the European Union in global environmental politics', in Carolyn Rhodes (ed.), *The European Union in the World Community*, Boulder, CO: Lynne Rienner.

Kapteyn, PGJ and Pieter VerLoren van Themaat (1998), *Introduction to the Law of the European Communities*, Dordrecht: Kluwer Law International.

Lacasta, Nuno S, Suraje Dessai and Eva Powroslo (2002), 'Consensus among many voices: articulating the European Union position on climate change', *Golden Gate University Law Review* **32**(4): 351–414.

Luterbacher, Urs and Detlef F Sprinz (eds) (2001), *International Relations and Global Climate Change*, Cambridge: MIT Press.

Macrory, Richard and Martin Hession (1996), 'The European Community and climate change', in Tim O'Riordan and Jill Jäger (eds), *Politics of Climate Change: A European Perspective*, London: Routledge.

Manners, Ian (2000), *Substance and Symbolism: An Anatomy of Cooperation in the New Europe*, Aldershot: Ashgate.

McCormick, John (1995), 'Environmental policy and the European Union', in Robert V. Bartlett, Priya A. Kurian and Madhu Malik (eds), *International Organizations and Environmental Policy*, Westport, CT: Greenwood Press.

Metz, Bert, Marcel Berk, Marcel TJ Kok, Jelle G van Minnen, Adre de Moor and Albert Faber (2001), 'How can the European Union contribute to a COP-6 agreement?: an overview for policy makers', *International Environmental Agreements* **1**(2): 167–85.

Michaelowa, Axel (1998), 'Impact of interest groups on EU climate policy', *European Environment* **8**(5): 152–60.
Newell, Peter (2000), *Climate For Change: Non-State Actors and the Global Politics of the Greenhouse*, Cambridge: Cambridge University Press.
O'Riordan, Tim and Jill Jäger (1996), *Politics of Climate Change: A European Perspective*, London: Routledge.
Oberthür, Sebastien (1999), 'The EU as an international actor: the protection of the ozone layer', *Journal of Common Market Studies* **37**(4): 641–59.
Oberthür, Sebastien and Hermann E Ott (1999), *The Kyoto Protocol: International Climate Policy For the 21st Century*, Berlin: Springer Verlag.
Soroos, Marvin S (2001), 'Global climate change and the futility of the Kyoto process', *Global Environmental Politics* **1**(2): 1–9.
Sprinz, Detlef and Tapani Voahtoranta (1994), 'The interest-based explanation of international environmental policy', *International Organization*, **48** 1: 77–105.
van Schaik, Louise and Christian Egenhofer (2005), Improving the climate: will the new constitution strengthen the EU's performance in international climate negotiations?', *CEPS Policy Brief No. 63*, available at www.ceps.be.
Yamin, Farhana (2000), 'The role of the EU in climate negotiations', in Joyeeta Gupta and Michael Grubb (eds), *Climate Change and European Leadership*, Dordrecht: Kluwer Academic.

10. Common policy on climate change: land use, domestic stakeholders, and EU foreign policy

Martina Jung, Axel Michaelowa, Ingrid Nestle, Sandra Greiner and Michael Dutschke*

INTRODUCTION

This chapter analyzes the foreign policy of the European Union (EU) on one of the most contentious issues in international negotiations on global climate change (GCC): the treatment of greenhouse gas (GHG) emissions by sources and removals by sinks from land use and forestry. In the international climate negotiations, this issue is formally referred to as Land use, Land-use Change and Forestry (LULUCF). It was previously lumped under the label of 'sinks', with this term being defined by the United Nations (UN) Framework Convention on Climate Change (FCCC) as 'any process, activity or mechanism which removes a greenhouse gas, an aerosol or a precursor of a greenhouse gas from the atmosphere', eg when CO_2 is sequestered from the air by biomass (e.g., trees and other plants) during photosynthesis.[1] While nobody denies the importance of forest protection and afforestation of degraded land for global climate protection, there was an intense argument among states involved in the GCC negotiations as to whether GHG removals (sequestration) from the atmosphere by sinks should be used to offset GHG emissions from fossil fuels. Proponents have praised the positive effects of forestry projects on biodiversity and watershed protection, and the low costs of sequestration compared to mitigation in the energy sector. Opponents have argued that sequestration is difficult to measure and monitor and that it is reversible.

These views have fed into the environmental foreign policies of the EU and its member states in this issue area. In analyzing their policies on LULUCF we do not rely on a single theoretical approach to environmental foreign policy (see Barkdull and Harris 2002). We instead integrate aspects of different approaches in the process, showing how power, interests and ideas at the domestic, European and international levels are important in shaping European policies and actions in this area of the GCC regime.

CARBON EMISSIONS AND REMOVALS FROM LULUCF: THE KYOTO PROTOCOL PROCESS

Forestry has played an important role in the GCC debate right from its beginning. In 1977, United States (US) physicist Freeman Dyson suggested an afforestation program on a planetary scale to sequester GHG emissions (Dyson 1977). The 1989 Nordwijk conference, one of the first global policy meetings on climate change, likewise called for a global effort on afforestation of 12 million hectares per year in the early twenty-first century (Nordwijk Conference 1989). However, at the 1992 UN Conference on Environment and Development, forestry suffered a setback. Originally, there were plans to negotiate a forest convention. The negotiations ran into an impasse and the convention never materialized. However, emissions and removal of GHGs from that atmosphere by LULUCF are an obligatory element of national emissions inventories under the FCCC. They are mentioned remarkably often in the text, although the non-binding targets of the FCCC refer only to emissions.

Until the first conference of the parties (COP) to the FCCC in 1995, the LULUCF issue came up in the context of 'Activities Implemented Jointly' (AIJ), a pilot phase for projects to reduce or sequester GHGs abroad. In the context of AIJ, project developers forged ahead and initially focused strongly on forestry projects, although these were still controversial. The first GHG-offset projects, started in the late 1980s, were all forest conservation or afforestation projects. Costa Rica, which was the pioneer in hosting AIJ projects, developed a large-scale forestry program.

In the run-up to the late-1997 Kyoto conference, the question of LULUCF was sidelined. In April 1997 there was only a vague statement about sinks in the negotiating text, but no substantial discussion had started. The issue's late introduction – serious negotiations only started in October 1997 (Depledge 2000: 48) – and a lack of background material made it the most opaque topic in the Kyoto negotiations. The FCCC Secretariat provided documentation on LULUCF only in late October, and observers thought that only a minuscule number of negotiators really had an overview of the issue (Depledge 2001: 171–3). New Zealand was the only country that made an early, fairly comprehensive proposal on the treatment of removal and emissions in the area of LULUCF.

Thus, the rules on carbon sinks that came out of the Kyoto negotiations were produced in a rash manner, and the consequences were heavy. The Kyoto Protocol defined legally binding targets for industrialized countries and some countries in transition. Sinks were integrated in different ways, some of which were sketchily defined and later became contentious.

Forests: Afforestation, Reforestation and Deforestation

Due to the differentiated treatment of forests and other land use, the need emerged to tell one from the other. For the layperson, a forest is an area with trees. However, for legal purposes more detailed criteria have to be developed. While many negotiators at Kyoto thought that the UN Food and Agriculture Organization's (FAO) definition of forests would be used, later events proved them wrong. The Intergovernmental Panel on Climate Change (IPCC) proposed a different definition. In principle, LULUCF is not considered in the calculation of the base year emissions inventory. So, all parties to the Kyoto Protocol that had a net sink in 1990 would profit if the sink were to persist during the commitment period.[2]

Annex I countries have to account for afforestation, reforestation and deforestation in their emissions inventories under the Kyoto Protocol (Article 3.3). It was decided that only land that was deforested before 1990 is eligible for reforestation. The 1990 threshold was introduced in order to avoid the perverse incentive that forests would be cut for making land available for generating credits from reforestation.

Had only afforestation, reforestation and deforestation been allowed as eligible activities in Kyoto, a lot of the later controversy would have been avoided. However, Article 3.4 of the protocol opened the floodgates for a whole series of new modalities by stating that 'additional human-induced activities related to changes in greenhouse gas emissions by sources and removals by sinks in the agricultural soils and the land-use change and forestry categories' could be adopted by the first meeting of the parties, and that these modalities could be used only from the second commitment period onwards. However, at the insistence of Japan (Depledge 2000: 49), the text of the protocol makes this limitation void by stating that 'a Party may choose to apply such a decision on these additional human-induced activities for its first commitment period'.

LULUCF in the Clean Development Mechanism

The Clean Development Mechanism (CDM) – a mechanism allowing GCC projects in countries without emissions targets (non-Annex B countries) to be credited towards targets of Annex B countries – was negotiated by a completely different group than the issue of LULUCF. There was no co-ordination of language between the groups, leading to grave problems of interpretation at later stages. The CDM text only refers to 'emission reduction'. A footnote to the text referring to further discussion on the inclusion of LULUCF was removed for editorial reasons. This footnote, which had not been opposed, had read: '"Mitigation of climate change" will include "through removals by sinks" in accordance with the decision of the parties regarding the treatment of sinks elsewhere in the Protocol' (Depledge 2000: 76). Non-governmental organizations (NGOs) immediately claimed that this meant that sequestration projects would be excluded, while negotiators stressed that the wording was meant to include them and that the issue had not been discussed.

At the Marrakech COP in 2001 it was decided that afforestation and reforestation would be eligible project activities under the CDM. After Marrakech, the attention turned to the more technical issues of modalities and procedures concerning LULUCF in the CDM, including non-permanence, baselines, additionality, leakage, and socio-economic and environmental impacts. Final decisions on most of these issues were taken at COP9 in December 2003.

POSITIONS OF NON-GOVERNMENTAL STAKEHOLDERS WITHIN THE EU

In this section we describe positions on LULUCF taken by stakeholders within the EU. Despite a high level of forest cover, a considerable forest industry and the important role played by forests in the cultural life of some European nations (e.g., the proverbial love of forests in Germany), stakeholders within the EU have predominantly opposed LULUCF in the GCC regime. Many stressed the fact that LULUCF could water down the Kyoto targets and deviate efforts from the main goal of the Kyoto Protocol, namely GHG emissions reductions at the source. Knowing these and other preferences of domestic actors is important for understanding the translation of their positions into policies adopted and implemented by governments. This is an issue stressed by societal theories of environmental foreign policy (Barkdull and Harris 2002: 74).

Environmental NGOs

Environmental NGOs should be favorable in principle to incentives that enhance vegetation cover, especially in marginal agricultural areas. Nevertheless, many NGOs at the country level have vigorously lobbied against LULUCF. A coalition of German NGOs argued in early 1997 that sequestration should not be used to reach targets even if 'conservation and ecologically sound management of forests' were among mandatory policies and measures (Forum Umwelt und Entwicklung 1997). The German North-South NGO initiative, Germanwatch, argued for exclusion of LULUCF in formal emissions accounting, though all 'voluntary measures to create carbon sinks should receive support' (Germanwatch 1997: 8). In 1999, Germanwatch asked to exclude sequestration projects from the CDM as long as the methodological issues were unresolved. The NGO *FERN*, a European subsidiary of the *World Rainforest Movement* contributed to a position paper called 'Tree Trouble' that consistently attacked attempts to open the LULUCF issue by EU countries (FERN 2000). As support for positions opposed to LULUCF, the Norwegian organization *Norwatch* collected case studies about 'carbon plantations' in Uganda and Tanzania that resulted in negative social impacts.

Industry

Emitters and forest industries have consistently called for a broad inclusion of LULUCF, preferably full carbon accounting (ERT 1997; UNICE 2000). The Association of German Forest Owners even demonstrated at the resumed session of the sixth Conference of the Parties (COP6) in Bonn 2001 by towing a wooden blockhouse in front of the conference venue. The Confederation of European Paper Industries argued that the full annual growth of existing forests, without the threshold date for deforestation of 1990, should be accounted for (CEPI 1999). It also called for consideration of wood products as sinks (CEPI 2000). The European Landowners' Organization called for a land-based approach with a fine spatial resolution (ELO 2000). This reflected the mainly smallholder structure of European forestry holdings.

Some industries fighting most of the GCC policy instruments, such as the Association of German Chemical Industry, argued for a limitation of the Kyoto Mechanisms and did not explicitly call for an inclusion of LULUCF (VCI 1998). Cheap carbon sequestration would have destroyed their argument that GCC policy is prohibitively expensive. Industry associations representing renewable energy and energy efficiency technologies, such as the European Business Council for Sustainable Energy (e^5), unsurprisingly have lobbied against LULUCF (see EBCSE 2005).

Research Organizations

Most research institutions within the EU have taken a cautious stance concerning LULUCF. The United Kingdom (UK) based Hadley Centre sees the risk of a reversal of sinks into sources at the end of this century due to stresses from GCC that destroy forests on a large scale.The Max Planck Institute for Biogeochemistry in Germany has repeatedly argued against a broad consideration of LULUCF under the Kyoto Protocol, although voicing that conservation of primary forests should have a high priority. Article 3.3 of the protocol was criticized as it would lead to increased timber imports and thus further deforestation abroad. The same arguments were advanced by the Potsdam Institute for Climate Impact Research, which even argued that afforestation in boreal areas would increase warming by absorption of solar radiation. Another argument was put forward by the Wuppertal Institute: that the utilization of sinks would slow down necessary change in energy-production infrastructure. In 1998, the German Advisory Council on Global Environmental Change argued that enhancement of sinks should be supported in principle but not included under the Kyoto Protocol.

Compared to these opposing voices, advocates of sinks were rarely heard. In 1999, 240 German forestry researchers signed a 'Wood and Forest Manifesto' and submitted it to the government. However, neither the Federal Research Institute for Forestry and Timber Economy nor the Institute for World Forestry published anything on this issue. The European Forest Institute (2000) argued that the role of forests as defined by Article 3.3 of the Kyoto Protocol would not be very significant in meeting the reduction target, even though the removal of carbon by trees in all forests of the EU was about the same size as the EU Kyoto target. However, the institute did not explicitly ask for consideration of these forests.

The Influence of European Stakeholders on Country Positions

In Germany, the impact of the NGO positions on governmental GCC policy is clearly visible, while the forestry lobby has hardly been heard (see Chapter 2). The Social Democrat-Green government, which came into power in 1998, was strongly opposed to LULUCF. In Finland, Sweden and France, forest owners formed a strong lobby, while environmental NGOs were not addressing the issue. Nevertheless, these countries formed the hard core of European LULUCF proponents. As Finland was in an unusual situation in which its overall forests were a sink but would be counted as a source during the first commitment period, it pressed for and received an exception at COP6.[3] In the run-up to COP6, Sweden, which was then holding the EU presidency, was seen as one of three EU member states (together with France and Finland) that wanted to see the 'opening up' of domestic LULUCF activities under the Kyoto Protocol (FERN 2000; see Chapter 6).

The Netherlands is an interesting case (see Chapter 4). It does not have a significant forest area but is among the supporters of LULUCF. The Dutch government started a multi-annual program of 'forestry certificates' in 2000. Companies with fleets of cars were to pay into a 'National Green Fund' that would invest the money in afforestation contracts with forest owners and municipalities. Moreover, in the context of AIJ, the Netherlands invested very actively in forestry projects.

THE EU POSITION ON LULUCF IN THE CLIMATE CHANGE NEGOTIATIONS

The EU as an Actor in International GCC Policy

The competencies of the EU in the field of GCC are mainly based on the EC Treaty. However, the division of competencies between the member states and the EU is not always conclusive. The EU has joined most of the international environmental agreements since the mid-1970s, additionally to its member states being part of the agreements, but without having a separate vote. Co-ordination between the member states and the EU concerning GCC policy has been led by the Council Presidency. The country holding the half-yearly rotating Council Presidency co-ordinates the positions of the member states and represents the EU at the international negotiations. Before Kyoto in 1997, the member states expressed their own positions additionally to the common EU position in the international GCC negotiations. Since Kyoto, the individual member states have increasingly refrained from voicing dissenting positions, which improved the public appearance of the EU as unitary actor (Oberthür and Ott 1999: 14).

Due to the high competency given to the respective member country, this co-ordination structure implies that the policy-making is largely dependant on the country holding the Presidency, and that smaller member states have difficulties fulfilling these duties. The EU tried to mitigate these problems by installing the possibility that a 'troika', consisting of the country holding the Presidency, its successor as well as the Commission, takes the lead in the negotiations[4] (Yamin and Depledge 2004).

European delegates to international negotiations have to face a complex process which can be referred to as a 'three -level game'. Figure 10.1 illustrates the different levels and influences to be considered. Even a higher number of levels can be found if sub-national actors are also taken into account. First, national delegations have to work out their positions and then co-ordinate them EU-internally, before the EU can act at the international negotiations. This problem is not unique to GCC negotiations but a general problem in international negotiations in which the EU and Member States share competence (Oberthür 2000: 102). Due to this time-consuming EU-internal co-ordination process, the EU became known for its 'bunker mentality' ever since Kyoto (Oberthür and Ott 1999: 86). Additionally to this complex official structure, delegates have to confront pressures from interest groups like NGOs and industry lobbies.

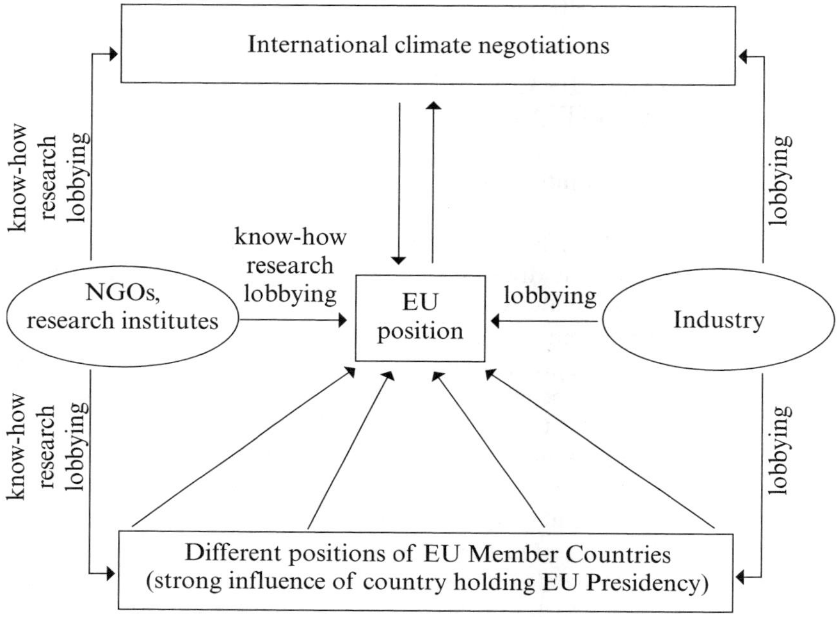

Figure 10.1 Three-level game in climate change negotiations

THE EU AND LULUCF

The EU did not have a very specific LULUCF position and surprisingly the position oscillated before the topic became relevant in October 1997. In the run-up to Kyoto, the EU initially asked for the inclusion of 'removals by sinks' in the definition of targets. However, it later demanded that sinks should be excluded from the Kyoto target, with a view to including them later, following additional research (Depledge 2000: 48).

After Kyoto, the EU asked for a balanced crediting and debiting regime under Article 3.3 and opposed the inclusion of additional activities under Article 3.4 (Zapfel and Gardiner 2002: 18). It also rejected LULUCF in the CDM, emphasizing difficulties regarding the impermanence of carbon sequestration and the possible crowding out of source-reduction projects from the CDM.

In the context of the Fifth EU Research Framework Programme, a large project called 'CarboEurope' was funded to assess European LULUCF potential. The project presented its results in Marrakech, which led some observers to the conclusion that EU opposition to LULUCF would not be continued (ENDS Environment Daily 2001). In the more technical discussions following Marrakech, the EU has actively participated in the negotiations by proposing environmental and socio-economic criteria for forestry projects in the CDM as well as a modified version of the Colombian proposal on temporary carbon credits. On the other hand, credits from forestry projects under the CDM have been excluded from trade in the first period of the European Emissions Trading Scheme. It is still unclear whether they will be included for the second period lasting from 2008–12.

POSITIONS OF INTERNATIONAL STAKEHOLDERS

Actors at the international level as well as regional and global configurations of power also had an important role in shaping European negotiating positions. Barkdull and Harris classify studies focusing on such issues as systemic theories (Barkdull and Harris, 2002: 68–9). The following paragraphs describe non-EU country positions concerning different aspects of the LULUCF discussion as well as the most important positions of relevant international NGOs and institutions, research institutes and industry outside of EU.

Positions of Non-EU States: Sinks in the GCC Negotiations

In the GCC negotiation process, several country groupings have emerged which try to strengthen their negotiation power by joining in the negotiations. The most important ones are listed in Table 10.1.

Supporters

The US, together with Canada, Japan, New Zealand, Australia and Russia were the biggest supporters of including LULUCF in general. In the negotiations in The Hague in 2000, the US, Canada and Japan presented a united position on the issue, consisting of a broad inclusion of additional activities under Article 3.4, including business-as-usual sinks, meaning the crediting of all existing natural and human sinks in the forestry and agricultural sector. This inclusion of natural and non-additional sink

Table 10.1 Important negotiating coalitions in the climate change regime

EU	European Union
G77 and China	132 developing countries and China, a very heterogeneous group of countries with sometimes conflicting interest in the GCC negotiations. Although often the chairing country speaks for the whole group, individual Parties as well as sub-groups of the G77 and China (African group, AOSIS, group of LDCs) voice their own positions in the negotiation process.
AOSIS	43 small-island states, which are threatened by sea-level rise. Most of the members are also part of the G77 and China. The states of AOSIS frequently speak with one voice in the negotiation process.
Umbrella Group	Non-EU developed countries, usually being Australia, Canada, Iceland, Japan, New Zealand, Norway, the Russian Federation, Ukraine and the US.
LDCs	The Least Developed Countries (LDCs) have become more active in defending their interest as a group.
OPEC	Oil-exporting countries.
Environmental Integrity Group	Formed prior to COP6 and includes Mexico, Republic of Korea, Switzerland, Liechtenstein and Monaco.
Central Group 11	Parties with Economies in Transition (EITs) being part of Annex I of the FCCC, except the Russia and Ukraine.
Group CAC&M	Central Asia, Caucasus and Moldova, i.e. the non-Annex I EIT Parties.

activities would have covered a great part of their reduction target, thus substantially weakening the Kyoto Protocol. The argument of the US, Canada and Australia in pushing for a broad inclusion of additional activities was that they had only agreed to the reduction commitment in Kyoto assuming that a certain amount of carbon sequestration activities would be eligible under Article 3.4. Furthermore, the US, Canada and Japan, pushed for the inclusion of forestry projects in the CDM (Anderson et al. 2001: 7). Canada was in favor of a very broad inclusion of additional LULUCF activities under Article 3.4 because its forest would have represented a source, had the eligible LULUCF activities for national GHG inventories been limited to Article 3.3 only. Thus, Canada supported an equal treatment of all kind of LULUCF activities, including plantations, natural forest growth and agricultural soil sequestration to compensate for the debits accruing from its high deforestation rate (Anderson et al. 2001: 9).

The motivation of Japan to support LULUCF might be due to relatively expensive domestic emission reductions. Thus, LULUCF may be seen as an option for Japan to reduce compliance costs. Due to its high population density, Japan will probably have to mainly rely on LULUCF project activities in other countries which may be a reason for its support of the inclusion of forestry projects in the CDM. At COP6, it was the lead proponent of LULUCF together with Russia (Spott 2001). Australia was one of the few industrialized countries whose forests still represented a net source of emissions in 1990. However, the high proportion of emissions from deforestation decreased significantly after this date. Australia voiced a strong interest for the inclusion of sink activities in the international GCC negotiations and achieved to pass the 'Australian clause' at COP3. This allowed for the inclusion of net emissions from LULUCF in the calculation of base year emissions. Therefore, Australia had already gained so much in Kyoto that it remained relatively low-key, besides arguing for inclusion of regeneration in Article 3.4.

In spite of Russia's relatively passive support of LULUCF in the negotiation process, Russia knew how to exploit the issue to its own advantage. It used the negotiation on country specific forest management caps – known as Appendix Z – to double its amount of forest management activities, which are allowed to compensate for any debits under Article 3.3, while rejecting forestry projects under the CDM. The latter can be explained by the competition of cheap LULUCF credits from developing countries with Russian hot air in the market. The described country positions show that the Umbrella group was more or less united in favor of LULUCF.

The Environmental Integrity Group supports sinks in general, but under the condition that their implementation would be bound to strict sustainability criteria. Therefore, it fought for the inclusion of natural regeneration of forest under Article 3.3, the discounting of LULUCF credits to account for impermanence of carbon sequestration in biomass, and – if this was not possible – a limitation of LULUCF activities as well as the avoidance of perverse incentives for felling primary forest.

Since the beginning, Norway has been in favor of LULUCF, under the condition that business-as-usual sinks were excluded. Latin American countries supported LULUCF, in order not to be sidelined in the CDM. Among them, Colombia and Costa Rica were the most active countries. Brazil (and in the beginning also Peru) were the only Latin American countries objecting to LULUCF. In 2000, Colombia submitted a proposal for the temporary crediting of forestry CDM projects, which became the cornerstone of the respective negotiations. Some South-East Asian countries with considerable CDM forestry potential such as Indonesia could be considered LULUCF supporters, while the countries of the Organization of Petroleum Exporting Countries (OPEC) embraced LULUCF because they support anything that reduces the impact on fossil fuel use. The African countries initially belonged to the group of the LULUCF opponents. However, the scramble for CDM projects has led to a collapse of the common position, which now tends to be in favor of forestry in the CDM with an emphasis on small-scale projects.

Countries opposing sinks

One of the most influential opposition groups against sinks outside of the EU is the Alliance of Small Island States (AOSIS). Due to their high vulnerability to sea level rise, the AOSIS are very interested in fast emissions reductions and thus, only support LULUCF activities as adaptation measures. However, they did not reject the mentioning of carbon sequestration activities as sensible policies and measures under Article 2.

Brazil, India and China are the biggest representatives of the LULUCF opponents within the group of G77 and China. Brazil was in favor of excluding avoided deforestation as eligible activity of the CDM and was instrumental in pushing this through. This is surprising because Brazil would be able to benefit considerably from avoided deforestation projects. Its efforts against the inclusion of avoided deforestation may, however, be explained by the strong position of Brazil in the G77, as well as sovereignty concerns about the amount of foreign investment in the Amazon region (Michaelowa et al. 2001). Thus, the G77 and China was divided over the question of whether to include LULUCF projects under the CDM, with most of the Latin-American countries being in favor and most of the Asian countries being against them (Anderson et al. 2001: 8). At first, the group of eastern European Annex I countries (CG11) was opposing LULUCF but some CG11 countries have started to look into forestry projects in the context of Joint Implementation (JI).

International Nongovernmental Organizations

The large international environmental NGOs like the *Climate Action Network* (CAN), *Friends of the Earth* (FoI), *Greenpeace* and the *World Wildlife Fund* (WWF), as well as smaller groups like *Cornerhouse UK* and the Japanese *CASA* rejected the idea of LULUCF activities as climate mitigation measures. Skepticism about the sustainability of carbon sequestration projects, and the fear of neo-colonialist appropriation of large land areas in developing countries, as well as of negative environmental and socio-economic impacts, were some of the reasons for this position. These NGOs were instrumental in getting avoided deforestation excluded from the CDM. The NGOs have maintained their opposition even after the decisions of Bonn (see e.g., ENDS Environment Daily 2001). Nevertheless, the position of NGOs concerning LULUCF is not such a pure opposition as often perceived. Support for LULUCF stems mainly from some NGO's goal to promote sustainable development and biodiversity conservation. Especially the big North American NGOs like *Conservation International*, *The Nature Conservancy* and *Environmental Defense* (EDF) as well as the *World Conservation Union* (IUCN), have been supporting the inclusion of LULUCF in the CDM, while putting an emphasis on conservation and the linking of the different international conventions, like the Biodiversity and Desertification Convention.

International Governmental Organizations

The positions of international organizations concerning LULUCF has been either neutral or supporting, and based on similar arguments as described for the NGOs in favor of LULUCF. The newly created BioCarbon Fund administrated by the World Bank as well as the FAO and its International Fund for Agricultural Development (IFAD) see LULUCF projects as an opportunity to attract private capital to biodiversity protection, soil conservation, sustainable agriculture, forestry and community development (BioCarbon Fund 2003). The United Nations Environment Programme (UNEP) and Organisation for Economic Cooperation and Development (OECD) have been relatively reluctant to commit themselves. The Forest Stewardship Council (FSC) voiced the opinion that afforestation projects could be implemented in a sustainable manner, thus influencing the decision to exclude avoided deforestation and focus on afforestation and reforestation.

Research Institutions

International research organizations like *Resources for the Future* (RFF, Washington DC), the *World Resources Institute* (WRI, Washington DC), the *Center for Clean Air Policy* (CCAP, Washington DC), the *International Union of Forest Research Organizations* (IUFRO) and the *Union of Concerned Scientists* tried to participate in a constructive way in the LULUCF discussion, citing both advantages and disadvantages of LULUCF projects. Also other organizations related to forestry and agriculture like the *Consultative Group of International Agricultural Research* (CGIAR) support LULUCF.

Industry

International industry umbrella organizations represent the national organizations that are overwhelmingly in favor of LULUCF. The *World Business Council for Sustainable Development* (WBCSD) that has more than 150 large companies as members saw full accounting of sinks as a precondition for a competitive and sustainable forest industry. The federations of electricity utilities argued for full inclusion of LULUCF in the Kyoto Mechanisms (Edison Electric Institute et al. 1998). The International Forest Industry Roundtable (IFIR) asked for the use of FAO forest definitions and full crediting of sequestration between 1990 and 2007, including forest management and forest products (IFIR 1998). Industry did not much lobbying for avoidance of deforestation, as it did not feel that it could profit from such projects actively apart from reducing overall compliance costs. Peugeot has financed a 12,000 ha afforestation project in Brazil aiming at sequestering 183,000 t CO_2 annually. United States utilities have been financing several small sequestration projects through the *International Utility Efficiency Partnership*. The most unusual forestry investor is the *International Automobile Federation* that wants to make Formula One car racing emissions neutral, including production and transport of racing cars.

The overwhelming majority of GCC project consultants and carbon brokers support land use and forestry in the context of climate policy. Some of them have specialized on forestry projects such as the Dutch foundation FACE (*Forests Absorbing Carbon dioxide Emission*). However, the delay of decisions on forestry projects until COP9 led to a diversification of companies that also had originally focused on LULUCF, e.g., *Trexler and Associates* in the US and *Ecosecurities* in the UK. The *US Business Council for a Sustainable Energy* has, however, been opposing LULUCF.

THE POST-KYOTO NEGOTIATING PROCESS ON LULUCF

Recognizing the lack of scientific background, the session of the Subsidiary Bodies in June 1998 asked the IPCC to write a report on LULUCF. This lead to a break in negotiations as everybody was waiting for the publication of the report eventually finalized in May 2000 (IPCC 2000). Its main message was that all relevant forms of terrestrial carbon could be measured accurately enough to implement the Kyoto Protocol. Overall, the report cautiously supported the pro-sink stakeholders. As it obviously could not solve the differences in opinion, afterwards, the debate flared up again.

LULUCF became the pivotal issue at COP6 in The Hague in October 2000. From the outset, the EU stressed its position of excluding LULUCF in the CDM and not to allow for additional LULUCF activities under Article 3.4 of the Kyoto Protocol in the first commitment period. Until late in the conference proceedings, the EU repeated its concerns regarding risk, scale and uncertainty. However, it was obvious that other Parties did not accept this position. Moreover, Finland was fighting the EU position from within by circulating a paper demanding that forest management including forest conservation should be eligible for credit. Even a proposal for compromise tabled by chairman Pronk did not lead to an agreement. This paper suggested a three-tiered system for Article 3.4, discounting forest management activities by 85 per cent but already including agricultural soil sinks in full. It also included a limit to 50 per cent of the reduction target or 2.5 per cent of base year emissions. In the last night, during negotiations among Ministers and Heads of Delegations behind closed doors, the idea to include a wide range of LULUCF activities but to limit the scale by caps for individual countries became a concrete option. These negotiations were like an oriental bazaar deal where numbers for the caps floated around. Yet, the conference ended in failure. There was a bitter exchange among EU ministers to find a scapegoat for the failure, specifically between the UK on one side and France and Germany on the other. To avoid a complete breakdown of the process, it was decided to continue the negotiations half a year later in Bonn at COP6.

In early 2001, the US decided to withdraw from the Protocol. Instead of dooming the process, the remaining Parties proceeded in finding a compromise. This situation especially strengthened Russia's negotiation power, since its ratification was needed for the entering into force of the Kyoto Protocol The EU was aware of the need for additional concessions. During the preparations for the second part of the conference the idea of setting individual caps for the different countries was taken up and further developed among EU LULUCF experts. As LULUCF activities under Article 3.4 had to be included for the first commitment period in order to reach consensus, they produced a method to calculate country-specific caps for forest management.

Concerning LULUCF in the CDM, the EU still had some hope in the beginning to exclude them. Especially countries that had been actively involved in the AIJ pilot phase, including many Latin American countries on the host-country side and Australia, Canada, and the US on the investor side, started to push for the inclusion of forestry in the CDM. The EU could not mobilize a coalition of LULUCF opponents of the same order (Scholz 2002: 57–9). Already before the start of negotiations, Michael Grubb, a reputable GCC policy researcher and advisor to the EU, was quoted as saying that the EU's position had shifted 'a long way' since The Hague (Spott 2001). Now Russia and Japan became the key players. A compromise in the LULUCF issue was crucial to involve both countries, particularly Russia.

Conference chairman Pronk produced another tentative compromise after intensive consultations with the heads of delegations from all the relevant groups which in the end was accepted. It included afforestation and reforestation measures in developing countries with a cap as well as a wide range of additional activities in Annex-I countries. The EU decided to accept this proposal without any changes if the other Parties agreed to do the same.[5] It is obvious that the final agreement differs widely from the ideas the EU had originally voiced in the negotiations. Therefore, it is of interest to analyze when and how the EU position changed and what had been the decisive factors determining this change.

Probable sources of influence are the pressure groups, notably the environmental NGOs. Their presence does not pass unnoticed in the negotiation process. When the EU was elected 'Fossil of the day' in The Hague for its LULUCF strategy, an award granted by the environmentalists to the nation that most hinders a successful and environmentally sound outcome of the negotiations, the EU sinks expert team did mind this criticism. It is, however, impossible to assess now, whether the EU's strong return to its previous position has partly been triggered by the pressure from NGOs. However, as far as the position of the EU on LULUCF is concerned, the influence of the environmentalists relies not so much on exercising pressure in the traditional NGO manner, but rather on providing expert input to the process. The NGOs wished to steer the negotiations concerning LULUCF into the same direction as the EU. In Bonn, the NGOs furthermore put a strong emphasis on the need to get an agreement. Many of the activists have closely followed the process over a number of years. They have a precise idea of what is possible in the negotiations; thus they try to cooperate with the EU rather than to criticize it severely. This is probably a quite effective way to get some influence on the EU position. The NGOs are integrated into the European policy process. Both at the conferences and between sessions the national government and EU representatives meet with important environmental NGOs and regularly exchange views on a very informal basis.

The negotiators are interested to hear the opinions of the well-informed NGO-representatives. They sometimes present new ideas to the lobbyists in order to check them for potential flaws; at other occasions, an NGO person might draw the attention of a government representative to a special detail to which he has so far not paid any special attention. This is especially true during the sessions when a lot of new input from the other Parties has to be processed rapidly and adequate reactions must be formulated at high speed. At those moments the suggestions, calculations and remarks from well-informed members of the NGO community can be very helpful and are consequently being listened to. This would of course hardly be the case if the EU felt the NGOs to be opponents that constantly try to make the EU look ridiculous.

Generally, the EU position can only be changed very slowly as a lot of member states have to be convinced. This turns out to be negative when speed is a decisive factor, but can also prevent hasty and unfounded reactions. It leads to a situation where real arguments have more weight than doubtful but well presented input or unjustified pressure. Thus, the failure of The Hague came about, as the pressure by the Umbrella Group did not make the EU yield. In the end, however, it led to a pragmatic outcome in spite of the strong EU position at the outset. Obviously there is now the risk that in future rounds of negotiations the opponents of EU positions will repeat this attrition process.

After the EU had failed in excluding LULUCF from the CDM, the EU strategy was to participate in the discussion on technical issues and influence the modalities and procedures of forestry projects towards conservative approaches. Concerning the influence of non-state actors, it is obvious that the more technical negotiations became, the less LULUCF opponents prevailed. At COP9 in Milan in December 2003, the rules and modalities for LULUCF projects in the CDM had to be finalized and agreed upon. In the sessions prior to Milan, the EU had been actively involved in the development of a proposal for a crediting regime which accounts for impermanence, as well as for binding sustainability criteria for CDM forestry projects. The modalities agreed on in Milan reflect the initial European position only to a relatively small extent. For example, concerning criteria for environmental and social impact assessment of LULUCF projects the EU only got a mention in the preamble that host countries could define such criteria if they wished to do so. The EU's acceptance of this text is the result of the general wish to not further postpone final decisions on forestry in the CDM. While especially China slowed down a fast agreement on the negotiating text by insisting on a very low threshold for the definition of small-scale projects, the EU compromised a lot for the sake of finding an agreement at COP9. Last implementation issues regarding LULUCF in the CDM were finalized at COP10 in Buenos Aires, in December 2004. With the Russian ratification of the Kyoto Protocol, it finally entered into force on February 16, 2005. In Montreal at the end of 2005, the first ever meeting of the Parties to the Protocol (MOP1) ran parallel to the Conference of the Parties to the Convention (COP11). One of the main subjects dealt with at the COP/MOP was the future of the climate regime post-2012. The most important new LULUCF-related development after COP10 was the so-called Papua New Guinea Proposal.[6] Based on a presentation at a government expert seminar in May 2005, Papua New Guinea – supported by eight other countries from Latin America and Africa – asked for the inclusion in the COP agenda of approaches to reduce emissions coming from deforestation, and it prepared a submission on the issue together with Costa Rica. As deforestation represents a large share of global CO_2 emissions not covered by the Kyoto Protocol so far, the general idea of the proposal was well received. However, technicalities of implementing incentives remain (Wittneben et al. 2006). This issue will certainly play a major role in ongoing land use-related discussions of the climate change regime.

SUMMARY AND CONCLUSION

Our analysis of stakeholder positions and processes at domestic, EU and international level applies a range of different theoretical approaches used in the literature on environmental foreign policy (see Barkdull and Harris 2002). Therefore, our three-level approach addresses influences of domestic actors on country positions (societal theory) as well as the influences of structural characteristics of the EU and the international regime on the negotiation process (systemic theory). Aspects of power, interests as well as ideas are considered at all levels.

The formulation of European GCC policy is a very complex process since it is not only subject to the pressure of various domestic interest groups but also a time-consuming EU-internal co-ordination process. Furthermore, the power constellation in the international negotiations as well as influences exerted by different international stakeholders affect the European position. For example, NGOs and research institutes participate in the negotiation process by sharing knowledge on the subject with policy-makers. Lobbying from all kinds of interest groups enters through other informal channels. Due to the necessity to first agree on a common position EU-internally, a change in the official EU negotiation position is rather slow and inflexible. This is a disadvantage when speed is decisive for reacting to new developments in the negotiation process.

The EU – an opponent of the use of LULUCF for the mitigation of climate change – could only bring through a very limited amount of its positions concerning this subject in the GCC negotiation process. One reason for this is the refusal of the US to ratify the Kyoto Protocol. This made it necessary to give concessions to Japan and Russia which were both calling for a broad use of LULUCF. Furthermore, internal differences with some EU member countries arguing in favor of LULUCF weakened the EU position. After it was clear that LULUCF would be a part of the Kyoto framework, the EU participated actively in the development of modalities and procedures for LULUCF. While the EU worked towards conservative rules and modalities regarding LULUCF, it was willing to compromise to avoid a further postponement of the decisions on LULUCF.

NOTES

* Sandra Greiner's contribution, which was made when she was with the Program on International Climate Policy, reflects her personal views.

1. Forests and soils can also be sources of GHGs (e.g., when trees are cut and burned, they release carbon). As emissions as well as removals (sequestration) of GHGs have to be considered if including the land-use sector in emissions inventories, the term 'sink' is not very precise. Nevertheless, it has been used widely in negotiations to refer to land use, land-use change and forestry.
2. This is called gross-net accounting. Australia, a country with net emissions in 1990, managed to introduce an exception allowing a net-net calculation.
3. Countries in which Article 3.3 creates a debit can be compensated by activities under Article 3.4, up to the amount of 30 megatons of CO_2 equivalent.
4. Prior to the Amsterdam Treaty in 1997, the Troika was made up of the current, past and future presidencies.
5. Personal communication with Rosemarie Benndorf, February 13, 2003.
6. At COP9, the Brazilian research institute IPAM presented a first proposal for a mechanism creating incentives for avoiding deforestation, the so called 'compensated reduction approach'.

REFERENCES

Anderson, David, Robert Grant and Chris Rolfe (2001), 'Taking credit. Canada and the role of sinks in international climate negotiations', *Climate of Change, Position Paper 2001*, Vancouver: David Suzuki Foundation.

Barkdull, John and Paul G Harris (2002), 'Environmental change and foreign policy: a survey of theory', *Global Environmental Politics* **2**(2): 63–91

BioCarbon Fund (2003), *Background Brief: Carbon Sinks, Climate Change and Sustainable Development* (May 2), available at http://carbonfinance.org/biocarbon/home.cfm.

Confederation of European Paper Industries (CEPI) (1999), *Implications of the Kyoto Protocol for the European Pulp and Paper Industry*, Brussels.

Confederation of European Paper Industries (CEPI) (2000), *CEPI's Views on the Climate Change Challenge*, Brussels.

Depledge, Joanna (2000), *Tracing the Origins of the Kyoto Protocol: An Article-by-Article Textual History*, Bonn: FCCC/TP/2000/2; UCL.

Depledge, Joanna (2001), 'The Organization of the Kyoto Protocol Negotiations: Lessons for Global Environmental Decision-making', Doctoral Dissertation, London: University College London.

Dyson, Freeman (1977), 'Can we control the carbon dioxide in the atmosphere?', *Energy* **2**: 287–91.

Edison Electric Institute (EEI), Federation of Electric Power Companies (FEPC) and International Union of Producers and Distributors of Electrical Energy (UNIPEDE) (1998), *Common EEI/FEPC/UNIPEDE Position Paper For COP-4 (Buenos Aires) Flexibility Mechanisms Under the Kyoto Protocol*. Buenos Aires.

European Business Council for Sustainable Energy (EBCSE) (2005), Website of the European Business Council for Sustainable Energy, available at www.e5.org/.

Environmental Data Services (ENDS) (2001), 'Great potential in EU carbon sinks', *Environment Daily* (19 November).

European Forest Institute (2000), *Kyoto Forests in EU: Insignificant Compared to the Carbon Dioxide Emissions and the Carbon Sink of all EU Forests*', Helsinki.

European Landowners Organisation (ELO) (2000), *ELO Statement for COP6*, Brussels.

European Round Table of Industrialists (ERT) (1997), *Climate Change. An ERT Report on Positive Action*, Brussels.

Forests and the European Union Resource Network (FERN) (2000), 'Special report climate change', *EU Forest Watch* **47**.

Forum Umwelt und Entwicklung (1997), *Zehn Kernforderungen für den Klimagipfel '97 in Kioto*, Bonn.

Germanwatch (1997), *Anforderungen an ein Klimaschutzprotokoll – worum geht's in Kyoto?*, Bonn.

Intergovernmental Panel on Climate Change (IPCC) (2000), *Land Use, Land-Use Change, and Forestry. Special Report of the Intergovernmental Panel on Climate Change*, Cambridge: Cambridge University Press.

International Forest Industry Roundtable (IFIR) (1998), *Statement of the International Forest Industry Roundtable on the Role of Forests and Forest Products in the Kyoto Protocol*, Buenos Aires.

Michaelowa, Axel, Sandra Greiner and Michael Dutschke (2001), *Positionen von Ländern und waldpolitisch engagierten Organisationen zur Einbeziehung von Aufforstung und Walderhalt in CDM und Joint Implementation sowie in die nationalen Treibhausgasinventare*, Hamburg: mimeo.

Nordwijk Conference (1989), *Nordwijk Declaration*, Nordwijk.

Oberthür, Sebastian (2000), 'The EU in International Environmental Regimes and the Energy Charter Treaty', in Joyeeta Gupta and Michael Grubb (eds), *Climate Change and European Ladership*, Dordrecht: Kluwer: 83–108.

Oberthür, Sebastian and Hermann Ott (1999), *The Kyoto Protocol*, Heidelberg: Springer.

Scholz, Sebastian (2002), *Die Rolle des Waldes als Kohlenstoffsenke im Clean Development Mechanism des Kyoto Protokolls der Klimarahmenkonvention der Vereinten Nationen*, Freising: Diplomarbeit, Fakultät für Forstwissenschaft und Ressourcenmanagement der Technischen Universität München.

Spott, Peter (2001), 'Kyoto best hope for blunting global warming: EU', *Christian Science Monitor* (July 18).

Union of Industrial and Employers′ Confederation of Europe (UNICE) (2000), *EU Strategy Responding to Climate Change*, Brussels.

Verband der Chemischen Industrie eV. (VCI) (1998), *Position of the Association of the German Chemical Industry For Concrete Policies and Measures and the Flexible Instruments Under the Kyoto Protocol to the UNFCCC*, Frankfurt.

Wittneben, Bettina, Wolfgang Sterk, Hermann Ott and Bernd Brouns (2006), In From the Cold: The Climate Conference in Montreal Breathes New Life into the Kyoto Protocol', Report of the Wuppertal Institute for Climate, Environment and Energy (30 January 2006), available at www.wupperinst.org/cop11-mop1/index.html.

Yamin, Farhana and Joanna Depledge (2004), *The International Climate Change Regime: A Guide to Rules, Institutions and Procedure*, Cambridge: Cambridge University Press.

Zapfel, Peter and David Gardiner (2002), 'Climate policy instruments and strategies: EU and US perspectives', in Johan Albrecht (ed.), *Instruments For Climate Policy: Limited Versus Unlimited Flexibility*, Cheltenham: Edward Elgar.

11. Climate policy in the EU: international regimes and policy diffusion

Sebastian Oberthür and Dennis Tänzler

INTRODUCTION

Climate policies in the European Union (EU) have evolved considerably since the 1990s. Significant progress has been made both in individual EU member states and at the EU level. EU member states have introduced or strengthened a wide variety of policy instruments on global climate change (GCC), including energy taxes, insulation standards for new and existing buildings, emissions trading schemes, support for an enhanced use of renewable sources of energy and combined heat and power production, voluntary agreements with industry, and more. The policy mix is constantly evolving.[1] At the EU level, some of the core GCC policy instruments include voluntary agreements starting in the late 1990s with European, Korean and Japanese car manufacturers on reducing carbon dioxide (CO_2) emissions from new cars, the EU Landfill Directive of 1999, the EU Renewable Energy Directive of 2001 and the EU Emissions Trading Directive of 2003. Further measures, such as proposals for regulating certain fluorinated gases or for promoting bio-fuels, are under development within the framework of the European Climate Change Program, which was established by the European Commission in 2000 (see European Commission 2003, 2004; EEA 2005: 32–41).

In this chapter, we examine the forces driving the diffusion of three particular GCC policy instruments in the EU that, together, provide an important underpinning for the EU's environmental foreign policy (EFP), namely GCC policy plans and strategies, emissions trading, and CO_2/energy taxes up to February 2005, when the 1997 Kyoto Protocol to the 1992 United Nations (UN) Framework Convention on Climate Change (FCCC) entered into force. The protocol does not require its parties to implement any specific policies or measures, although the EU adamantly but unsuccessfully pursued binding policies and measures in the Kyoto negotiations. Article 2 of the protocol only provides a basis for a 'soft' learning approach within the framework of the regime, which has been pursued by means of workshops on 'good practice' (see Oberthür and Ott 1999: 103–14), as well as a database for policies and measures of Annex I (developed country) parties reported in their third national communications (UNFCCC 2003). Both the FCCC and the Kyoto Protocol oblige their parties in virtually the same words to 'formulate, implement, publish and regularly update national and, where appropriate, regional programs containing measures to mitigate climate change' (Article 10[b] of the Kyoto Protocol and Art. 4.1[b] of the FCCC). However, this obligation remains unspecific and vague (e.g., no deadline is specified) and therefore hardly constitutes a 'hard' legal obligation.

Research on policy diffusion rooted in comparative politics (e.g., Walker 1969; Gray 1973; Kern 2000) that has recently gained prominence especially in the area of environmental policy (e.g., Kern et al. 2001; Busch and Jörgens 2005) might suggest policy learning through effective communication as the central causal mechanism explaining the spread of the particular policy instruments (see Rogers 1995: 5; Kern et al. 2001: 10–11). In this framework, diffusion of GCC policy instruments in the EU might be explained with reference to the existence of pioneering states, the crossing of a threshold of diffusion, the self-dynamics of diffusion/learning and certain conditions of the EU and its member states that may be more or less conducive to adopting new policies (e.g., Kern et al. 2001). In any event, the major causal pathway accounting for the policy diffusion is thought to lead from one state to another.

In contrast, we argue that the international regime on global climate change, and more specifically the Kyoto Protocol of 1997, has been an important force driving the diffusion of GCC policy instruments *within* the EU in recent years. International institutions are either not considered at all in analyses of policy diffusion or are thought to fulfill only a facilitative function by helping to spread innovative ideas from one country to another (Kern 2000: 275; Kern et al. 2001: 9–10; Busch and Jörgens 2005: 91–4). In contrast, we assign major explanatory value to the influence of international institutions by starting from the hypothesis that they can play a substantial role as drivers and triggers of processes of policy diffusion in at least two ways. Short of directly requiring states to take specific action, which would represent 'central regulation' (Kern 2000: 31–3), international institutions can provide special incentives to implement particular innovative policy instruments and/or put diffuse pressure on state governments to take action in order to be able to comply with their international commitments. They may also serve to induce learning processes at the national level. Thus, the unspecific 'compliance pull' of the Kyoto Protocol or particular incentives provided by it may increase the willingness of EU governments to implement innovative policy instruments, as may redefinitions of their interests induced by the Kyoto process.

In terms of the typology of theoretical approaches underlying the contributions to this volume (see Barkdull and Harris 2002: 69), our analysis focuses on the systemic level and employs interests and, to a lesser extent, ideas as prime variables. By investigating the interface between the international regime on climate change and domestic GCC policies within the EU, we primarily treat the EU and its member states as unitary actors, without going into much detail of the underlying societal foundations or state factions. In so doing, we do not want to deny the relevance of the societal and state-centric levels of analysis. Given the greater research effort required, complementing our approach with these analytical approaches is simply beyond the scope of this chapter (see the chapters in Part I of this volume). In line with a number of moderate approaches that attempt to combine rationalist and constructivist approaches to the analysis of international relations and foreign policy (e.g., Fearon and Wendt 2002), we refer to both national interests and ideas/learning, paying particular attention to states' interests as influenced by their perceptions and ideas.

While we focus on the influence of the international GCC regime on the implementation of domestic policies, this analysis is immediately relevant for understanding the EU's foreign policy on global climate change. There can be no doubt that the state of development of national GCC policies within the EU is an important factor influencing EU EFP on this issue. The importance of domestic factors such as existing policies is well established in foreign policy analysis (see Barkdull and Harris 2002: 76–7, 79–80). Similarly, the literature on leadership in international environmental affairs has highlighted the importance of the domestic underpinning of international 'leadership by example' (Grubb and Gupta 2000; Young 1991). Investigating the forces driving the development and spread of GCC policies within the EU therefore also promises to hold relevant lessons regarding the EU's past and future environmental foreign policy on GCC and to allow for insights into the conditions of EFP in general (cf Barkdull and Harris 2002).

In this perspective, international institutions and domestic policies can thus be conceptualized as feeding back on each other over time. First, the EFPs of various states result in a particular design of an international institution. Subsequently, the international institution influences the conditions of the EFPs of these states. Finally, states' EFPs in turn provide the basis for further institutional development at the international level.

In the remainder of this chapter, we attempt to substantiate our claim by investigating the development and spread of GCC policy plans and strategies (section 2), emissions trading (section 3), and CO_2/energy taxes (section 4) in the EU up to February 2005. In doing so, we investigate whether, to what extent and how the diffusion of GCC policy has been influenced by the international GCC regime and the Kyoto process in particular. In each case, we correlate the diffusion process with the international political process, in particular the beginning of the negotiations on the Kyoto Protocol in 1995 and its adoption in 1997. We subsequently examine the evidence and arguments supporting and potentially invalidating our hypothesis. In this respect, particular attention is paid to analyzing to what extent the diffusion can be understood and explained without reference to the international regime but instead to potential alternative explanations. As mentioned above, a fuller analysis would ideally investigate the respective domestic political processes in order to identify how and why the three policy instruments analyzed were introduced. Such an analysis is, however, beyond the scope of this chapter.

We do not examine the 10 new member states that acceded to the EU on 1 May 2004, instead focusing our analysis on the 'EU-15' member states existing before the EU's most recent enlargement. This focus on the EU-15 is justified by the fact that most of our analysis addresses the time before enlargement. Furthermore, conditions of GCC policies differ significantly between the 'old' and the 'new' member states. In particular, GCC policies in the 10 new member states have been driven to a large extent by the accession process and the need to implement the EU's *acquis communautaire* (see Chapters 5 and 13). Moreover, due to the greenhouse gas (GHG) emissions reductions that occurred in these countries in the 1990s (see Chapter 5), the Kyoto Protocol cannot be expected to provide a significant driving force for their domestic GCC policies.

Our analysis provides evidence that the international climate regime has indeed had an important influence on the spread of certain GCC policy instruments within the EU. The role of international institutions in processes of policy diffusion can go well beyond a purely facilitative and informative function by exerting pressure and providing incentives for governments to adopt innovative policies. The GCC regime thus already had significant positive effects prior to the entry into force of the Kyoto Protocol – a finding that may be surprising for those doubting the viability of the international regime (e.g., Victor 2000). As a result, the foundations for EU leadership in international GCC policy have been strengthened.

THE DIFFUSION OF CLIMATE PROTECTION PLANS AND STRATEGIES

While other kinds of plans and strategies in environmental policy have already been the subject of research on policy diffusion, the spreading of climate protection plans or strategies has received little attention to date. A sizeable literature has emerged examining the diffusion of 'national environmental plans' following the adoption of Agenda 21 at the Rio Earth Summit in 1992, which calls for the elaboration of such plans (Jänicke and Jörgens 1998: 28, 36). In contrast, the increasing elaboration and adoption of plans or strategies specifically aiming at climate protection has received scant attention.

Figure 11.1 illustrates the diffusion of climate protection plans in the EU and its member states up to 2005. Plans (whether called plans, programs, strategies or guidelines) sharing the following characteristics were taken into account. First, they had to provide for an integrated approach towards achieving a national GHG emissions target by including a selection of policies and measures to be implemented to achieve that target. Secondly, these policies and measures had to cover a variety of sectors or take into account developments in a variety of relevant sectors in an integrated way. Finally, the plan had to have become official government policy by being formally adopted or approved (in contrast to a plan that may have been approved only by an environment ministry).[2]

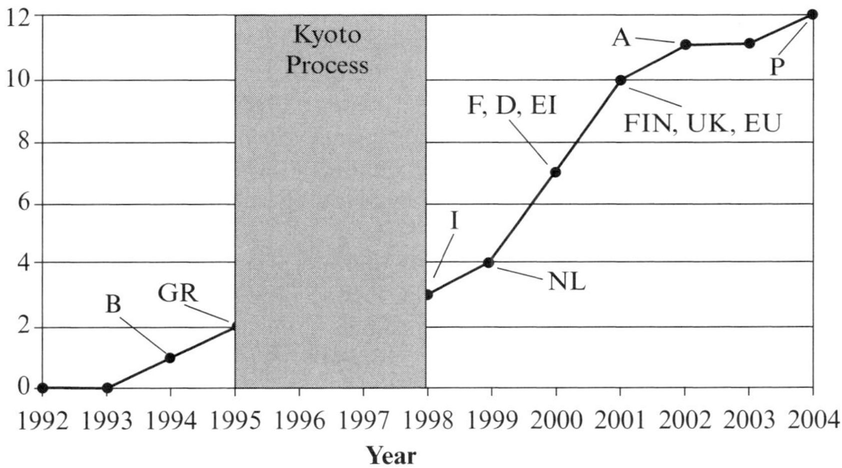

Notes: Austria (A); Belgium (B); European Union (EU); France (F); Finland (FIN); Germany (D); Greece (GR); Ireland (EI); Italy (I); the Netherlands (NL); Portugal (P); United Kingdom (UK).

Source: EU member states' national communications to the FCCC Secretariat (available at <www.unfccc.int>). See also Table 11.1 and the references list.

Figure 11.1 The adoption of climate protection plans in the EU and its member states

Eleven of the then 15 EU member states and the EU itself had adopted such plans by February 2005. The last of them (Portugal) adopted a climate protection plan in June 2004. In Spain, discussions on taking a similar step have started (see Chapter 7). Nine of the 11 existing plans were adopted between 1998 and 2002, with the remaining two dating from 1994 (Belgium) and 1995 (Greece). In 2000, the EU itself adopted the European Climate Change Program for the elaboration of suitable policies and measures to mitigate GCC, which resulted in a plan for the adoption of first measures in 2001 (European Commission 2000, 2001, 2003).

Some of these plans followed up on earlier related activities. For example, the German climate protection program of 2000 built upon the work of an inter-ministerial working group, 'CO_2 Reduction', which was established in 1990 to identify policies and measures for reducing energy-related CO_2 emissions by 25 to 30 per cent from their 1987 levels by 2005 (Loske 1997: 283–90; see Chapter 2). Similarly, the Dutch climate protection plan was developed on the basis of previous work (see Chapter 4), such as the National Environmental Policy Plan 2 of 1993 and the CO_2 Reduction Plan of 1996 – the so-called '750 million letter' from the Ministry of Economic Affairs to Parliament (see Netherlands 1997: 45–64). In the case of Italy, the Inter-ministerial Committee for Economic Planning adopted a first planning instrument in 1994 that was confined to a number of energy efficiency measures. Only the National Plan of 1998 elaborated on inter-sectoral policies and measures needed to achieve the Italian Kyoto target with a view to integrating them into a common strategy (see Montini 2000: 10–13). Neither of the preceding policy planning instruments constituted a full-blown plan as defined above.

The chronology of events supports the interpretation that the Kyoto process, and thus the international GCC regime, provided a major rationale for the elaboration and adoption of climate protection plans within the EU. The majority of the plans were adopted after the conclusion of the Kyoto Protocol in late 1997. The EC and its member states have a reduction commitment of –8 per cent under the protocol. Political agreement on sharing the burden of this commitment, based on Article 4 of the protocol, was finalized in 1998 and resulted in commitments ranging from +27 per cent (Portugal) to –28 per cent (Luxembourg) (Oberthür and Ott 1999: 147–9). This burden sharing was formalized as part of the decision of the EU and its member states to ratify the Kyoto Protocol in spring 2002 (EU 2002). The EU and all 15 of the member states officially ratified the protocol on 31 May 2002. The implementation of their Kyoto targets required the elaboration of strategies, including concrete policies and measures, because a concerted and co-ordinated effort pursued over a longer period of time is essential

Table 11.1 Climate protection plans, emissions trading and CO_2/energy taxes: status within the EU (2005)

Country	Climate Protection Plans	Emission Trading	CO_2/Energy Taxes
Austria	'Strategie Österreichs zur Erreichung des Kyoto-Ziels' 2002 (Austria 2002)	Study published in 2000, waited for the EU trading scheme. (Austria 2001: 81)	Introduced in 1996 (Austria 2001: 80)
Belgium	'The National CO_2 Programme' 1994 (Belgium 1994)	–	Introduced in 1993 (OECD 2001)
Denmark	–	Adopted in 1999, introduced in 2001 (DEA 2001)	Introduced in 1993 (Denmark 2000: 51)
Finland	'National Climate Change Strategy' 2001 (Finland 2001: 13)	–	Introduced in 1990 (Finland 2001: 100)
France	'Programme national de lutte contre le changement climatique' 2000 (France 2000)	–	–
Germany	'National Climate Change Programme' 2000 (Germany 2000)	Working group set up in 2000 (Germany 2000: 141)	Introduced in 1999 (Germany 2000: 36–37)
Greece	'Climate Change – The Greek Action Plan for the Abatement of CO_2 and Other Greenhouse Gas Emissions' 1995 (Greece 1997:1)	–	–
Ireland	'National Climate Change Strategy' 2000 (Ireland 2000)	Advisory group set up in 1998: participation in the EU and international trading scheme intended (Ireland 2000: 4)	–
Italy	'National Plan for the Reduction of GHG emissions' 1998, updated in 2002 (Montini 2000)	–	Introduced in 1999 (Speck/ Ekins 2000: 11)
Luxembourg	–	–	–
Netherlands	'National Climate Policy Implementation Plan' 1999 (Netherlands 2001: 52)	Commission appointed in 2000 (Netherlands 2001: 63–64), trading system announced for 2005 (ENDS daily, 16 April 2002)	Introduced in 1988 as a fuel tax (OECD 2001: 51), 1990, 1992 & 1996 expanded (Netherlands 2001: 50)
Portugal	'Plano nacional para as alteracoes climaticas' (Portugal 2004)	–	–
Spain	–	–	–
Sweden	–	–	Introduced in 1991, reforms in 1993, and 1996 (Sweden 2001: 13)
UK	'UK Climate Change Programme' 2001 (UK 2001a)	Announced 2001; entry into force in 2002 (UK 2000)	Introduced in 2001 (UK 2001a: 72)
EU	'European Climate Change Programme' 2001 (European Commission 2001a)	Adopted 2002, entry into force 2003, operational 2005 (European Commission 2001, European Union 2003)	Adopted in 2003, entry into force 2004 (European Union 2003a)

Source: Author's compilation. See also, references.

for achieving the GHG emissions reduction commitments. Formally, the EU and its member states were not yet required to implement the protocol, since its entry into force was still uncertain. However, as the leader in the international efforts the EU was committed to bringing the protocol into effect. Preparing for implementation by elaborating a climate protection plan thus constituted a rational response to international developments.

The time lag between the adoption of the Kyoto Protocol and adoption of the climate protection plans may be easily explained by the time needed to elaborate such a plan, which can be estimated at one to two years, on average. In that respect, the Italian plan of 1998 can even be assessed as being rather early – also taking into account that the EU member states agreed on the final internal distribution of their reduction commitment under the Kyoto Protocol only in June 1998. Nevertheless, such an early elaboration of a plan is compatible with the hypothesis that it was driven by the Kyoto process for two reasons. First, the writing on the wall that emissions would have to be reduced was evident since 1995, when negotiations on the Kyoto Protocol started. Secondly, the EU member states had, in March 1997 during the international negotiations on the protocol, agreed on preliminary burden sharing (amounting to an overall reduction of 9.2 per cent). The Italian commitment under both burden-sharing agreements changed only slightly, from –7 per cent to –6.5 per cent (Ringius 1999; Oberthür and Ott 1999: 147–9).

A simple model of horizontal policy diffusion is insufficient for fully explaining and understanding the diffusion process. Disregarding the international process, the spread of climate protection programs within the EU, as depicted in Figure 11.1, might be interpreted as a process of 'horizontal diffusion'. In such a conceptual framework, building upon the experiences gained in some pioneer countries (e.g., Belgium, Greece), others followed suit until a critical mass was achieved. However, such an interpretation would have to come to grips with several anomalies. First, after the first movers, the second wave of plans occurred a few years following the Kyoto conference. Secondly, the plans of this later wave generally refer to the emissions targets under the Kyoto Protocol as modified in the EU burden-sharing agreement of 1998. Finally, the climate protection programs principally give Kyoto as a major rationale.

It is also difficult to argue that the adoption of climate protection plans followed a model of 'forced diffusion' because of central regulation (Kern 2000: 33) or that the international GCC regime induced a learning process similar to the Agenda 21 process with respect to the diffusion of national environmental plans (Kern et al. 2001: 18). Neither the FCCC nor the Kyoto Protocol establishes a hard, legally binding commitment to elaborate and adopt such a plan. The language that they do contain on the elaboration of policy plans, however, is nearly identical in both legal instruments (see above). Any attempt to explain the diffusion of climate protection plans as being forced by international commitment or as the result of a learning process induced by the international regime would therefore have to answer the question of why such plans were elaborated following Kyoto but not following the adoption of the FCCC in 1992. The preparation of GCC policy plans can hardly be assumed to require 6 years and more. If the Kyoto Protocol had directly forced the elaboration of climate protection plans or induced a related learning process, then the FCCC of 1992 should have already done so because the relevant provisions in both treaties are virtually identical. Only by taking into account other commitments contained in the Kyoto Protocol, in particular the emissions limitation and reduction commitments, can we explain the diffusion of GCC policy plans in the EU following the adoption of the protocol.

Trying to explain the spread of climate protection plans throughout the EU with reference to a potential self-dynamic and learning process at the domestic level also runs into major difficulties. Such an explanation may be put forward in particular in the cases where full-blown climate protection plans built upon earlier planning efforts (see above). Under these circumstances, governments may have 'learned' from experiences with earlier attempts that a more encompassing planning instrument was warranted. Any such interpretation would have to explain why these countries felt inclined to further develop their plans and how they all may have come to accept that further action was warranted at approximately the same time irrespective of the Kyoto Protocol. Given that Kyoto provides the major reference point in the existing plans, it is difficult and even implausible to design such an explanation without accepting a major influence of the Kyoto process. Besides, such an alternative explanation is hardly suitable for explaining why the other countries that had not gone through similar preparatory stages nevertheless elaborated similar plans.

Overall, the balance of evidence strongly suggests that the Kyoto Protocol, and more specifically its emissions targets, was a critical force driving the diffusion of climate protection plans throughout the EU. Even though the entry into force of the protocol was not ensured yet, governments could anticipate and expect that they would be obliged internationally to curb their GHG emissions; they were acting on the assumption that the Kyoto Protocol would eventually need to be implemented. To a large extent, this unspecific 'compliance pull' of the protocol thus triggered the spread of these policy instruments within the EU.

EMISSIONS TRADING

Emissions trading as a policy instrument to combat GCC has its main roots in the United States (US). The US had gained relevant experience in particular with its national sulfur dioxide emissions-trading program that was implemented in 1992 as a major component of its strategy to combat 'classical' air pollution. By and large, this experience has been evaluated to have been positive: emissions were reduced in excess of the actual requirements and the overall costs turned out to be much lower than anticipated (see Ellermann et al. 1999; Aulisi et al. 2000). Consequently, it was the US that introduced the concept of emissions trading into the international GCC negotiations and was the major driving force advocating the establishment of an international emissions trading scheme as part of the Kyoto Protocol (Grubb et al. 1999: 128–31; Oberthür and Ott 1999: 187–205).

In contrast, the EU traditionally had far more limited experience with the concept of emissions trading (for some examples see Klaasen 1999). It generally relied more heavily on command and control regulation, while market instruments made only little and late headway in EU environmental policies (Clinch 1999: 6). Consequently, the EU and its member states remained reserved, skeptical or even hostile vis-à-vis the concept of emissions trading in the Kyoto process (Victor 2000: 5). As a result, the EU only agreed to the introduction of emissions trading into the Kyoto Protocol in exchange for a stronger outcome on the literal emissions targets (Grubb et al. 1999: 94; see Chapters 9 and 12).

The EU thereby exploited the major rationale of the concept of emissions trading. Such trading is meant to provide geographical flexibility. Emissions reductions do not necessarily need to occur by taking action in a certain country (or company), but can be implemented wherever they can be achieved at lowest cost. In permit-trading systems, the overall amount of available permits is generally capped. Assuming a certain limit of emissions, the costs of achieving the target should be minimized by granting geographical flexibility. At the same time, assuming that a certain amount of resources is generally available for emissions abatement, higher emissions reductions should be achievable with emissions trading than without it.

The Kyoto Protocol permits parties to use emissions trading without requiring them to do so. Due to the political stand-off on the issue between industrialized and developing countries, Articles 3 and 17 of the protocol provided only a skeleton for an emissions trading system, only 'the bare minimum of enabling language' (Grubb et al. 1999: 129). Subsequently, more detailed rules were agreed at the seventh conference of the parties to the FCCC in 2001. Thus emissions trading from the beginning was a voluntary instrument. No country is legally required to use this mechanism, even less so to establish a *domestic* emissions trading system. However, early movers could be expected to gain additional benefits by acquiring expertise in running and administrating emissions-trading systems that were expected to provide new business opportunities (e.g., brokers, exchange markets, other services) (UNCTAD 1995; OECD 1998).

Driven by these forces, the concept of GHG emissions trading gained support within the EU following the adoption of the Kyoto Protocol, despite the organization's original skepticism and opposition. Serious discussions on and preparations for establishing domestic emissions trading systems started in several member states in the second half of the 1990s (Hasselknippe and Høibye 2001). Denmark passed a plan for national emissions trading for the electric power industry in 1999 (Danish Parliament 1999), which entered into force in 2001 (Hasselknippe and Høibye 2001: 20).

The leader among EU member states can be said to be the United Kingdom (UK) (see Chapter 3), where the design of a domestic emissions trading system had been discussed intensely since the late 1990s (Lord Marshall 1998; UK 2000, 2001a). The domestic UK debate provides clear anecdotal evidence for the impact of the Kyoto Protocol. A task force established in March 1998 under the lead of Lord Marshall concluded in its final report of November 1998: 'The issue over tradable permits is not so much *whether* they have a role in helping the UK to meet its targets as *how much* and *when*. Trading is already on its way. A system of international GHG emissions trading is provided for in the Kyoto Protocol' (Lord Marshall 1998: 11). A business-led initiative was recommended to design a 'dry run pilot' domestic trading scheme for GHG emissions. Consequently, the UK Emissions Trading Group was set up in June 1999 to take forward the design of a UK scheme. An emissions trading system then became part of the UK's climate change program of 2001 (UK 2000: 8; UK 2001, 2001b). Emissions trading is complementary to the UK's climate change levy, with participating companies receiving an 80-per cent discount from the levy (UK 2001a).

In addition, concrete preparatory work for establishing domestic emissions trading systems was undertaken in, *inter alia*, the Netherlands, Sweden and Ireland until the end of 2002 (Netherlands 2001; Sweden 2001; Ireland 2000). In spring 2002, the Netherlands (see Chapter 4) announced the launch of an emissions trading scheme in 2005 (ENDS Daily, 16 April 2002). In other member states, discussions on different design options and approaches had started.[3] That these preparatory activities did not result in the establishment of further domestic emissions trading schemes within the EU is also due to the fact that member states, such as Ireland (2000: 27) wanted to ensure the compatibility of their domestic systems with relevant international and EU rules.

This decision was more than appropriate because the discussions on setting up an EU-wide system of GHG emissions trading advanced at high speed after the European Commission first declared in 1998 that 'the Community could set up its own internal trading scheme by 2005' (European Commission 1998: 20). After initial work had been carried out in 1999 (CCAP 1999; FIELD 2000), the Commission presented a 'Green Paper' on a harmonized EU-wide emissions trading scheme in order to initiate European discussion, including multi-stakeholder participation (European Commission 2000a: 87). In October 2001, the Commission presented a proposal for an EC Directive on an EU-wide emissions trading scheme (European Commission 2001a: 581). EU environment ministers agreed on the rules for the EU-wide trading scheme in December 2002 (European Council 2002) and the resulting EU directive entered into force in October 2003 (European Council 2003a). The EU started a pilot phase of the scheme in 2005, which is to be followed by a full-blown system that will become operational in 2008 (from the beginning of the 2008–12 Kyoto Protocol commitment period).

Kyoto tended to move all EU member states towards accepting and implementing emissions trading and thus facilitated the adoption of the EU Emissions Trading Directive. The EU and its member states undertook a reassessment of their position on GHG emissions trading in response to Kyoto. Observers of, and participants in, this reassessment process generally agree that the Kyoto Protocol process was crucial in bringing this change about. Due to the Kyoto Protocol, governments accepted that emissions trading had become a reality. If every member state was able to engage in international emissions trading under the Kyoto Protocol, even original opponents of the concept had a strong incentive to do so in order to reduce the costs of implementing the Kyoto targets. Furthermore, governments got familiar with and attached to the general concept through their involvement in the detailed international discussions. As one element of the international process, a large number of side-events examining various aspects of emissions trading were held at the international negotiating sessions, and an uncounted number of studies analyzing various design options and their implications were undertaken and published (e.g., OECD 1998). The considerable effort spent in designing the international system was not to be invested for nothing. It led to an internalization and increasing 'ownership' of the concept of emissions trading in domestic political systems.[4]

Overall, the Kyoto process thus triggered a learning process about the concept of emissions trading, enhanced its general acceptance and provided special incentives for its implementation in the EU. The importance of other factors such as 'instrumental leadership' by the European Commission notwithstanding (Wettestad 2005), the Kyoto process has been a major force driving the diffusion of the concept of emissions trading and its implementation in the EU and its member states. Given the history of the European discussion on this concept and previous widespread opposition to it in most of Europe, there is hardly any plausible alternative explanation available.

Interestingly, the diffusion mechanism in this case originally involved a transfer from the US, as the major proponent of the emissions trading concept, to other parts of the world, in particular Europe, through an international institution. On this basis, the spread of the emissions trading concept throughout Europe may have been interpreted as a simple expression of US hegemony or as a horizontal diffusion process, employing an international institution as the vehicle and as a facilitator. However, such an interpretation fails to explain the continued drive towards emissions trading in Europe even after US President George W Bush rejected the Kyoto Protocol and the US largely withdrew from Kyoto-related international discussions in early 2001 – even strengthening the EU's efforts (Wettestad 2005). Therefore, the international regime in which the emissions trading concept is enshrined has driven the diffusion process. As a result, the EU has, ironically, become the front-runner in developing the concept (gaining various first-mover advantages), from which the US and others may learn and to which they may have to adapt in the future (Legge and Egenhofer 2001).

CARBON DIOXIDE AND ENERGY TAXATION

Carbon dioxide/energy taxes are particular types of environmental or 'eco' taxes. In order to internalize the otherwise external costs associated with the use of natural resources, CO_2/energy taxes aim at reducing the emission of GHGs through a reduction in energy consumption. In contrast to a system of tradable emission permits, eco-taxes or charges do not require setting an absolute cap on emissions. They also do not ensure achieving a certain environmental standard (e.g., certain emissions standards) or objective (eg stabilization of GHG concentrations in the atmosphere at a certain level). Thus, the policy instrument of taxation differs from command and control instruments (e.g., regulations). The Kyoto Protocol does not call for the implementation of a CO_2/energy tax or any other eco-tax. Despite attempts to push for agreement on legally binding policies and measures in the negotiations, the protocol eventually left the choice of the means of implementing its emissions targets up to each country. The protocol's mandatory, binding emissions targets do, however, create unspecific pressure to implement effective policies and measures to mitigate GCC and abide by these targets ('compliance pull').

Taxation of CO_2 and energy diffused throughout the EU at a moderate pace from the beginning of the 1990s (see Figure 11.2).[5] This policy instrument was first introduced between 1988 and 1991 by the Nordic member states. The diffusion process gained momentum in the second half of the 1990s. Between 1996 and 2001, five member states introduced CO_2/energy taxes. Whereas small member states were the first movers and innovators, Germany and the UK followed in 1999 and 2001 respectively. The introduction of CO_2/energy taxation in France in 2001 was prevented only when the tax plan was ruled unconstitutional by the French constitutional court (ENDS Daily, 3 January 2001). In addition, Ireland had originally announced its intention to introduce a GHG tax in 2002, but after postponing it in late 2002 the plan was eventually abandoned in 2004 because CO_2 emissions have been decreasing (Ireland 2000: 4; ENDS Daily, 6 December 2002; ENDS Daily, 13 September 2004). Overall, nine of the EU-15 states had introduced CO_2/energy taxes by February 2005.

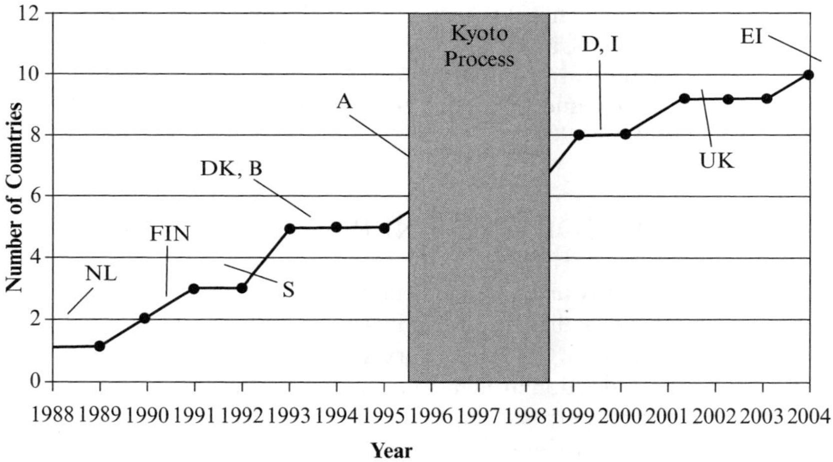

Notes: Austria (A); Denmark (DK); European Union (EU); Finland (FIN); Germany (D); Italy (I); the Netherlands (NL); Sweden (S); United Kingdom (UK).

Sources: See national communications to the FCCC Secretariat, available at www.unfccc.int, and references cited in Table 11.1.

Figure 11.2 Introduction of CO_2/energy taxes in the EU and its member states

In addition, EU-wide harmonized energy and/or CO_2 taxation has been under discussion for more than 10 years. In the second half of the 1990s the debate shifted, from an early proposal by the European Commission to introduce a combined CO_2/energy tax to consideration of a broadened framework for harmonizing tax levels for energy products.[6] For a long time, however, none of these proposals received the unanimous support by EU member states required for adoption of any regulation regarding energy taxation. The harmonization of minimum tax levels in particular met with resistance by some southern EU member states (especially Spain; see Chapter 7). It gained renewed attention in the wake of the discussion on the implementation of the Kyoto Protocol. The European Council, for example, stressed in its June 1999 conclusions regarding climate change and the Kyoto Protocol that there was a need for 'an appropriate framework for energy taxation' (European Council 1999).

After 6 years of difficult negotiations, EU finance ministers reached agreement in March 2003, and the directive was finally adopted in October 2003 (European Council 2003b). The directive sets minimum tax levels for virtually all energy products including oil, gas and coal. As a result of numerous compromises required to get unanimous backing, it will have few implications for current tax levels on most energy products in the EU-15 states that adopted the directive. However, the directive, which entered into force on 1 January 2004, might serve as a starting point for upward harmonization of EU energy tax rates in the longer term. In addition, the directive will require the 10 new EU member states to raise their energy tax rates substantially, which was also a motivation for finalizing the deal in fall 2003 (ie before the EU's enlargement) (ENDS Daily, 28 October 2003).

The diffusion of CO_2/energy taxes within the EU can be interpreted as being at least partially driven by the international GCC regime. Even the first movers on the issue in the EU may be said to have introduced such a tax partially motivated by the rise of the GCC issue on the international political agenda at the end of the 1980s and the beginning of the 1990s. With the negotiation and the adoption of the Kyoto Protocol, the incentive and pressure for EU member states to introduce CO_2/energy taxes as a means of limiting GHG emissions increased in the mid-1990s. Consequently, the number of EU states having implemented such a tax has steadily risen since that period (see Figure 11.2). That this increase was far less steep than in the case of climate protection plans (see above) may be explained by the fact that taxes are a far more consequential policy measure which and are thus more likely to faces serious opposition.

However, a number of factors other than the international GCC regime can plausibly be referred to for explaining the diffusion of this policy instrument. On various occasions, CO_2/energy taxes were implemented by adding an increment to already existing taxes on fossil fuels/mineral oils. They may thus be considered instruments for raising tax revenues while sailing under a 'green climate' flag, when in fact they could be motivated by general fiscal objectives. In some cases, such as the German eco-tax introduced in 1998–99, an important motivation was to lower the fiscal burden on labor by recycling the income from the indirect taxation of the use of natural resources and the environment. Finally, in cases such as Germany, the introduction of CO_2/energy taxation may more easily be explained by reference to a change in government, and in particular the coming into power of the green party (which occurred in Germany in 1998; see Chapter 2).

Notwithstanding these limitations, the international GCC regime and the Kyoto Protocol in particular provided a major argument in the domestic debates about the introduction of CO_2/energy taxes in the respective EU member states. For example, the GCC plans of both Ireland and Germany give the GHG emissions reduction commitments under the Kyoto Protocol as a major rationale for introducing eco-taxes. Overall, however, claiming that the GCC regime was crucial for the spread of CO_2/energy taxation in the EU after 1990 would overstate the case. The Kyoto Protocol represented only one of several factors driving the diffusion of CO_2/energy taxes within the EU. Other factors appear to have been more influential on average and in several, if not most, cases might have led to the introduction of such taxes even in the absence of the Kyoto Protocol.

CONCLUSION

The Kyoto Protocol has influenced the adoption and diffusion of policy instruments in the EU and its member states, thus changing the conditions for the EU's EFP on climate change. Analyzing the spread of three instruments of GCC policy (climate change plans, emissions trading, CO_2/energy taxes) within the EU and its member states against the backdrop of the evolution of the international GCC regime provides evidence international institutions driving policy diffusion. Hence, international institutions have a crucial role in shaping the overall EFP approach of the EU. Ignoring the Kyoto Protocol, the cases presented might easily be inadequately interpreted as processes of 'simple' horizontal policy diffusion. However, an adequate understanding of these processes requires taking account of the influential role of the Kyoto Protocol in shaping the EU's interests.

The Kyoto Protocol drove the diffusion of GCC plans, emissions trading schemes and eco-taxes in at least two different ways. First, the protocol exerted pressure on national actors, and was used by them as powerful support, to take effective 'nonspecific' action to fulfill international commitments (nonspecific 'compliance pull'). While not requiring any specific policy instrument to be implemented, such pressure can drive the spread of policy instruments (especially of 'soft' instruments such as GCC plans). Secondly, the protocol created significant incentives to implement the innovative policy instrument of emissions trading (albeit without requiring its implementation). As a result of such incentives and a learning process advanced by related international discussions on innovative concepts, international institutions can contribute to actors' reassessing and changing their preferences in this respect (e.g., emissions trading). Thus, in both cases the international regime can be regarded as the driving force shaping both states' interests and their specific policy orientation. In contrast, CO_2/energy taxes may have diffused across the EU even in the absence of the international regime on GCC. The contribution of the international institution in this case was moderate and probably not crucial.

The Kyoto Protocol drove the spread of emissions trading schemes and the elaboration of GCC plans in the EU before its provisions acquired legally binding force – on the basis of the expectation that entry into force was likely to occur eventually. Such a 'hurrying obedience' (or, in our context, a hurrying response) of governments vis-à-vis international regulation is in line with previous findings of research into the effects of international institutions (see, e.g., Oberthür 1997). Despite various predictions of doom for the Kyoto Protocol (e.g., Victor 2000), it has therefore already had noticeable positive effects on the development of GCC policies in the EU and its member states – that is, in a significant part of the industrialized world.

This early effectiveness of the Kyoto Protocol has fed back into the leadership role of the EU in the international GCC negotiations. This feedback effect is most obvious in the case of emissions trading. While the EU had been a skeptic and even an opponent of international emissions trading in the negotiations on the Kyoto Protocol, in response to the protocol several of its member states and the EU as a whole have initiated implementation of domestic emissions trading schemes. As a result, the EU has become a frontrunner in developing and supporting the international emissions trading system. Together with other implementing activities, the Kyoto Protocol has triggered in the EU and its member states this feedback process, thus reinforcing the EU's leadership role in international GCC politics. It also demonstrates how domestic policies, which themselves are inspired and driven by international institutions, shape the 'national interest' and become driving forces of EFP. It illustrates the importance of the societal underpinnings of EFP in terms of interests and ideas.

Our analysis provides evidence for the fruitfulness of employing both 'interests' and 'cognition/ideas' as factors explaining policy outcomes (cf Barkdull and Harris 2002). In particular, the emissions trading case demonstrates the potential complementary character of the underlying analytical approaches and the fruitfulness of combining them. While we have pointed to the influence of the Kyoto Protocol on the incentive structure of the EU and its member states regarding the implementation of all policy instruments investigated, the acceptance of emissions trading within the EU can hardly be understood without the change of perceptions and ideas induced by the GCC regime. The regime's influence on both material interests and perceptions were decisive in bringing about EU emissions trading.

Whereas the Kyoto Protocol has thus strengthened its own support, this may not suffice to sustain progress in international GCC policy. The Protocol appears to have seen its support grow primarily in the EU, which has been among its supporters from its inception. While EU leadership on GCC may thus be reinforced, thereby also shaping a specific European approach to EFP, the implementation gap between Europe and other Kyoto Protocol parties, on one hand, and the US on the other, may actually be widened. To resolve the challenge of narrowing this gap and promoting climate protection in the US as well, forces other than the diffusion of policies driven by the Kyoto Protocol may be required.

NOTES

1. For an overview of developments in the power sector see Vrolijk (2003). For a general overview, see European Commission (2004).
2. The empirical analysis in this section is based on Oberthür/Tänzler 2006. It takes into account various national climate protection plans as of February 2005. See Table 11.1 for an overview and a list of sources. Additional information was gathered from member states' most recent national communications to the FCCC Secretariat (available at the FCCC's website, http://www.unfccc.int).
3. For example, in Germany a Working Group on Emissions Trading for Combating the Greenhouse Effect was initiated by the Ministry for the Environment (Germany 2000: 141). (For a detailed discussion of emissions-trading policy development in Germany and Norway, see Chapter 12.)
4. Based on one author's (Oberthür) observations while watching or participating in various EU internal discussions. See also Christiansen and Wettestad (2003) and Wettestad (2005).
5. See also Kern et al. (2001: 23). For a comprehensive analysis of eco-taxes in several European states, see Speck and Ekins (2000) and Ekins and Speck (2002).
6. On the so-called 'Monti proposal', see European Commission (1997). For accounts of the debate on CO_2/energy taxation within the EU, see Zito (2002).

REFERENCES

Aulisi, Andrew, Daniel J Dudek, Joseph Goffman, Michael Oppenheimer, Annie Petsonkand Sarah Wade (2000), *From Obstacle to Opportunity: How Acid Rain Emissions Trading is Delivering Cleaner Air*, New York: Environmental Defense.

Austria (2001), *Third National Climate Report of the Austrian Federal Government. The Third National Communication in Compliance with the Obligations under the FCCC (Federal Law Gazette No. 414/1994)*, Vienna: Austrian Federal Ministry of Agriculture, Forestry, Environment and Water Management.

Austria (2002), *Strategie Österreichs zur Erreichung des Kyoto-Ziels. Klimastrategie 2008/2012 (18 Juli 2002)*, Wien: Bundesministerium für Land und Forstwirtschaft, Umwelt und Wasserwirtschaft.

Barkdull, John and Paul G Harris (2002), 'Environmental change and foreign policy: a survey of theory', *Global Environmental Politics* **2**(2): 63–89.

Belgium Federal Department of the Environment (1994), *The National CO_2 Programme*, available at www.environment.fgov.be/Root/tasks/atmosphere/klim/set_en.htm (April).

Busch, Per-Olof and Helge Jörgens (2005), 'International patterns of environmental policy change and convergence', *European Environment* **15**(2): 80–101.

Center for Clean Air Policy (CCAP) (1999), *Design of a Practical Approach to Greenhouse Gas Emissions Trading Combined with Policies and Measures in the EC*, Washington DC: Center for Clean Air Policy.

Clinch, J Peter (1999), *Environmental Policy Reform in the EU*, Environmental Studies Research Series Working Paper 99/06, Dublin: University College Dublin, Department of Environmental Studies.

Christiansen, Atle C and Jørgen Wettestad (2003), 'The EU as a frontrunner on greenhouse gas emissions trading: how did it happen and will the EU succeed?' *Climate Policy* **3**(1): 3–18.

Danish Energy Agency (DEA) (2001), *Danish CO_2 Cap and Trade Update (as of 18 December 2001)*, Copenhagen: Danish Energy Agency/Ministry of Economic and Business Affairs.

Denmark (2000), *Climate 2012: Status and Perspective for Denmark's Climate Policy*, Copenhagen: Danish Ministry of the Environment and Energy.

Danish Parliament (1999), *Act No. 376: Bill on CO_2 Quotas for Electricity Production* (2 June), available at www.ens.dk/graphics/publikationer/laws/actonco2quotafor electricityproduction.pdf.

ENDS Daily (2001), 'Court sinks French energy tax plan', Issue 863, 3 January.

ENDS Daily (2002), 'Netherlands backs CO_2 emissions trading', Issue 1197, April 16.

ENDS Daily (2002), 'Ireland confirms plan for national carbon tax', Issue 1345, December 6.

ENDS Daily (2003), 'EU energy products tax directive adopted', Issue 1540, October 28.

ENDS Daily (2004), 'Ireland drops plan to introduce carbon tax', Issue 1724, September 13.

Ekins, Paul and Stefan Speck (2002), 'Evaluating environmental taxes: recent experiences and proposals for the future', in Peter Clinch, Kai Schlegelmilch, Rolf-Ulrich Sprengerand and Ursula Triebswetter (eds), *Greening the Budget: Budgetary Policies for Environmental Improvement*, Cheltenham: Edward Elgar .

Ellerman, Denny, Richard Schmalensee, Paul L Joskow, Juan Pablo Montero and Elizabeth M Bailey (1999), 'Summary evaluation of the US SO_2 emissions trading program as implemented in 1995', in Steve Sorell and Jim Skea (eds), *Pollution for Sale: Emissions Trading and Joint Implementation*, Northampton, MA: Edward Elgar.

European Commission (1997), *Proposal for a Council Directive Restructuring the Community Framework for the Taxation of Energy Products (COM(97)30 final)*, Brussels: European Commission.

European Commission (1998), *Climate Change: Towards an EU Post-Kyoto Strategy COM(1998) 353*, Brussels: European Commission.

European Commission (2000), *Communication from the Commission on EU policies and measures to reduce greenhouse gas emissions: towards a European climate change programme (ECCP) (COM (2000)88)*, Brussels: European Commission.

European Commission (2000a), Green Paper on Greenhouse Gas Emissions Trading within the European Union (COM (2000)87). Brussels: European Commission

European Commission (2001), *Communication From the Commission on the Implementation of the First Phase of the European Climate Change Programme (ECCP) (COM(2001) 580 final)*, Brussels: European Commission.

European Commission (2001a), *Proposal for a Council Decision Concerning the Approval, on Behalf of the European Community, of the Kyoto Protocol to the United Nations Framework Convention on Climate Change and the Joint Fulfillment of Commitments there under (COM(2001) 579)*, Brussels: European Commission.

European Commission (2003), *Second ECCP Progress Report: Can We Meet Our Kyoto Targets?*, Brussels: European Commission.

European Commission (2004), *Report 'Catching up With the Community's Kyoto Target' under Decision 280/2004/EC of the European Parliament and of the Council Concerning a Mechanism for Monitoring Community Greenhouse Gas Emissions and for Implementing the Kyoto Protocol COM(2004) 818 final*, Brussels: European Commission.

European Council (1999), *Presidency Conclusion, Cologne European Council, 3 and 4 June 1999*, available at http://europa.eu.int/council/off/conclu/june99/june99_en.htm (February).

European Council (2002), *2473rd Council meeting. Environment* (9 December), Brussels: European Commission.

European Council (2003a), 'Directive 2003/87/EC of the European Parliament and of the Council of 13 October 2003 Establishing a Scheme for Greenhouse Gas Emission Allowance Trading Within the Community and Amending Council Directive 96/61/EC', *Official Journal of the European Union* **275** (25 October): 32–48.

European Council (2003b), 'Council Directive 2003/96/EC of 27 October 2003 Restructuring the Community Framework for the Taxation of Energy Products and Electricity', *Official Journal of the European Union* **283** (31 October): 51–70.

European Environment Agency (EnEA) (2005), *Analysis of Greenhouse Gas Emission Trends and Projections in Europe 2004. Technical Report 7/2004*, Copenhagen: European Environment Agency.

European Union (2002), 'Council Decision 2002/358/CE of 25 April 2002 Concerning the Approval, on behalf of the European Community, of the Kyoto Protocol to the United Nations Framework Convention on Climate Change and the Joint Fulfilment of Commitments thereunder', *Official Journal of the European Communities* **130** (15 May): 1–20.

Fearon, James and Alexander Wendt (2002), 'Rationalism vs constructivism: a skeptical view', in Walter Carlsnaes, Thomas Risseand Beth Simmons (eds), *The Handbook of International Relations*, London: Sage.

Foundation for International Environmental Law and Development (FIELD) (2000), *Designing Options for Implementing an Emissions Trading Regime for Greenhouse Gases in the EC. Final Report to the European Commission, DG Environment*, London: Foundation for International Environmental Law and Development.

Finland Ministry of the Environment (2001), *Finland's Third National Communication under the United Nations Framework Convention on Climate Change*, Helsinki: Ministry of the Environment.

France (2000), *National Programme for Tackling Climate Change*, available at www.effet-de-serre.gouv.fr/maine.cfm?page=en/solutions/solutions.htm.

Germany (2000), *Nationales Klimaschutzprogramm. Beschluss der Bundesregierung vom 18 Oktober 2000 (Fünfter Bericht der Interministeriellen Arbeitsgruppe 'CO2-Reduktion')*, available at www.bmu.de.

Gray, Virginia (1973), 'Innovation in the States: a diffusion study', *American Political Science Review* **68**(4): 1174–93.

Greece Ministry for the Environment, Physical Planning and Public Works (1997), *Second National Communication to the UNFCCC*, Athens: Ministry for the Environment.

Grubb, Michael and Joyeeta Gupta (2000), 'Leadership: theory and methodology', in Joyeeta Gupta and Michael Grubb (eds), *Climate Change and European Leadership: A Sustainable Role for Europe?*, Dordrecht: Kluwer.

Grubb, Michael, Christiaan Vrolijk and Duncan Brack (1999), *The Kyoto Protocol: A Guide and Assessment*, London: The Royal Institute of International Affairs and Earthscan.

Hasselknippe, Henrik and Geir Høibye (2001), *Meeting the Kyoto Protocol Commitments: Domestic Emissions Trading Schemes*, Norway: Energy and Environment Confederation of Norwegian Business and Industry.

Ireland (2000), *National Climate Change Strategy*, available at www.environ.ie/ (February).

Jänicke, Martin and Helge Jörgens (1998), 'National environmental policy planning in OECD countries: preliminary lessons from cross-national comparisons', *Environmental Politics* **7**(2): 27–54.

Kern, Kristine (2000), *Die Diffusion von Politikinnovationen. Umweltpolitische Innovationen im Mehrebenensystem der USA*, Opladen: Leske and Budrich.

Kern, Kristine, Jörgens Helge and Martin Jänicke (2001), *The Diffusion of Environmental Policy Innovations: A Contribution to the Globalisation of Environmental Policy*, Discussion Paper FS II 01–302, Berlin: Social Science Research Center.

Klaassen, Ger (1999), 'Emissions trading in the European Union: practice and prospect', in Steve Sorell and Jim Skea (eds), *Pollution for Sale: Emissions Trading and Joint Implementation*, Northampton, MA: Edward Elgar.

Legge, Thomas and Christian Egenhofer (2001), *After Marrakech: The Regionalisation of the Kyoto Protocol (CEPS Commentary November 2001)*. Centre for European Policy Studies, available at www.ceps.be/Commentary/Nov01/Marrakech.htm.

Loske, Reinhard (1997), *Klimapolitik. Im Spannungsfeld von Kurzzeitinteressen und Langzeiterfordernissen*, Marburg: Metropolis.

Marshall, Lord (1998), *Economic Instruments and the Business Use of Energy*, available at www.hm-treasury.gov.uk/mediastore/otherfiles/marshall.pdf.

Montini, Massimiliano (2000), *Italian Policies and Measures to Respond to Climate Change*, Milano: Fondazione Eni Enrico Mattei.

Netherlands Ministry of Housing, Spatial Planning and the Environment (1997), *Second Netherlands National Communication on Climate Change Policies*, The Hague: Ministry of Housing, Spatial Planning and the Environment.

Netherlands Ministry of Housing, Spatial Planning and the Environment (2001), *Third Netherland's National Communication on Climate Change Policies. Prepared for the Conference of the Parties under the Framework Convention on Climate Change*, The Hague: Ministry of Housing, Spatial Planning and the Environment.

OECD (1998), *International Emissions Trading under the Kyoto Protocol*, Paris: OECD.

OECD (2001), *Environmentally Related Taxes in OECD. Issues and Strategies*, Paris: OECD.

Oberthür, Sebastian (1997), *Umweltschutz durch internationale Regime. Interessen, Verhandlungsprozesse, Wirkungen*, Opladen: Leskeand Budrich.

Oberthür, Sebastian and Hermann E Ott (1999), *The Kyoto Protocol: International Climate Policy for the 21st century*, Berlin: Springer.

Oberthür, Sebastian and Dennis Tänzler (2006), 'Internationale Institutionen und die Diffusion von Klimapolitiken innerhalb der EU', *Zeitschrift für Umweltpolitik & Umweltrecht*, **29**(3).

Portugal (2004), *Plano nacional para as alteracoes climaticas. Sintese. Approved through Resolução do Conselho de Ministros*, Lisbon: Instituto do ambiento.

Portugal Ministry of the Environment (1997), *Portugal's Second Report*, Lisbon: Ministry of the Environment.

Ringius, Lasse (1999), 'Differentiation, leaders, and fairness: negotiating climate commitments in the European Community', *International Negotiation* **4**(2): 133–66.

Risse, Thomas (2002), 'Constructivisim and international institutions: toward conversations across paradigms', in Ira Katznelson and Helen V Milner (eds), *Political Science: The State of the Discipline*, New York: WW Norton.

Rogers, Everett M (1995), *Diffusion of Innovations*, New York: Free Press.

Spain (1997), *Segunda Communicación Nacional de Espana*, Madrid: Ministerio de Medio Ambiento.

Speck, Stefan and Paul Ekins (2000), *Recent Trends in the Application of Economic Instruments in EU Member States Plus Norway and Switzerland and an Overview of Economic Instruments in Central and Eastern Europe: Update of Database of Environmental Taxes and Charges. Report to DG Environment*, London: Forum for the Future.

Sweden (2001), *Sweden's Third National Communication under the United Nations Framework Convention on Climate Change*, Stockholm: Ministry of the Environment.

UNCTAD (1995), *Controlling Carbon Dioxide Emissions: The Tradable Permit System*, Geneva: United Nations.

UK Department for Environment, Food and Rural Affairs (2001), *The UK's Third National Communication under the United Nations Framework Convention on Climate Change*, London: Department for Environment, Food and Rural Affairs.

UK Department for Environment, Food and Rural Affairs (2001a), *The UK's Climate Change Programme*, London: Department for Environment, Food and Rural Affairs.

UK Department for the Environment, Transport and Regions (2000), *A Greenhouse Gas Emissions Trading Scheme for the United Kingdom: Consultation Document*, London: Department for the Environment, Transport and Regions.

UNFCCC (2003), *'Good Practices' in Policies and Measures among Parties included in Annex I to the Convention (FCCC/WEB/2003/1 30 May 2003)*, Bonn: Secretariat of the Framework Convention on Climate Change.

Victor, David (2000), *The Collapse of the Kyoto Protocol and the Struggle to Slow Global Warming*, Princeton: Princeton University Press.

Vrolijk, Christiaan (2003), *Climate Change and Power Economic Instruments for European Electricity*, London: Earthscan.

Walker, Jack L (1969), 'The diffusion of innovations among the American states', *American Political Science Review* **63**(3): 880–99.

Wettestad, Jørgen (2005), 'The making of the 2003 EU emissions trading directive: an ultra-quick process due to entrepreneurial proficiency?', *Global Environmental Politics* **5**(1): 1–23.

Young, Oran R (1991), 'Political leadership and regime formation. On the development of international institutions in international society', *International Organization* **45**(3) (Summer): 281–308.

Zito, Anthony (2002), 'Integrating the Environment into the European Union: The History of the Controversial Carbon Tax', in Andrew Jordan (ed.), *Environmental Policy in the European Union: Actors, Institutions and Processes*, Sterling, VA: Earthscan.

12. Meeting Kyoto commitments: EU influence on Norway and Germany

Guri Bang, Jonas Vevatne and Michelle Twena

INTRODUCTION

This chapter explores how the environmental policies of the European Union (EU) influence domestic climate policy in individual European countries. Environmental policy has been part of joint EU policy for years, and now a common policy on global climate change (GCC) is developing. At the same time, in 2004 the EU expanded its membership from 15 to 25 states. Foreign and environmental policies are increasingly co-ordinated for a broad range of European countries. The question is whether the influence of the EU on national policies is growing as a result of increasing integration. We look specifically at two countries: Germany, an EU member state; and Norway, a non-member.[1] Both of these countries have expressed strong prior national preferences for policy choice. Thus the question is to what extent the policy decisions of the EU have been able to affect the ultimate choice of policy instruments in these countries.

To address this question empirically, we focus on the policy formulation of a particularly relevant and recently debated GCC policy issue, namely, emissions trading (ET). At the Kyoto Conference in 1997, parties to the United Nations (UN) Framework Convention on Climate Change (FCCC) acknowledged that trading in emissions allowances was a cost-effective instrument to achieve greenhouse gas (GHG) emissions reductions. The specific details of how the instrument could be implemented were not decided at the time, but were left to subsequent international GCC negotiations. It was not until the seventh conference of the parties in Marrakech in 2001 that agreement was finally reached on the issue. Since the third conference of the parties in Kyoto (1997), ET schemes have been central to foreign policy discourses in the EU, Germany and Norway, and hence provide a good basis for a case study into how the EU may have influenced the formation of national GCC policy in these two European countries. We focus on the objectives that national governments have sought to meet, examine the values and principles underlying those objectives, identify the means through which each country has attempted to achieve their desired goals, and go on to investigate how this process may have been influenced by the EU.

The chapter is organized as follows: Section two lays out the analytical framework underpinning our analysis. We deploy interest-based theory and examine the role of ideas in international GCC policy, following the typology of approaches to environmental foreign policy (EFP) in Barkdull and Harris (2002). We note how their typology of explanatory approaches can be used to explain policy instrument choice. As a background to understanding EU influence, the third section of this chapter systematically lays out the policymaking process for an emissions trading scheme (ETS) in the EU. The fourth section presents country profiles for Norway and Germany. We then go on to discuss in more detail the processes through which ET has been introduced as a policy instrument alternative in the two countries. In the fifth section, we examine the extent to which the EU appears to have influenced policy instrument choice in Norway and Germany and suggest reasons for this. Finally, we draw some conclusions, and discuss problems and implications of using this analytical framework.

FRAMEWORK FOR POLICY ANALYSIS

Our investigation focuses on how interest group bargaining, power structures and ideas interacted at national and international levels to determine GCC policy instrument choice in Norway and Germany between 1997 and 2004. We apply multiple levels of analysis – systemic, societal and state-centric – to explain and understand EU influence on policy instrument choice. Following Jupille et al. (2003), we attempt to integrate rationalism and constructivism in studying EU influence in both case studies. First, we use a social constructivist approach to explain how norms and values shape EFP preferences (Hasenclever et al. 1997; Wendt 1999). According to our interpretation, the two countries developed key objectives – environmental leadership, respect for the precautionary and polluter pays principles and a determination to protect their competitive position within the internal market – on the basis of essentially socially constructed norms. These normatively oriented objectives cannot be explained and understood purely on rational grounds since there is little economic gain from adopting the precautionary principle or being an environmental leader. Rather, these objectives are perceived to be important as they give 'soft power', prestige and respect among European and international peers, and reinforce domestic values. Social constructivist approaches at the systemic and societal levels are therefore used to explain how these key norm-based objectives helped guide perceptions and evaluations of which policy instrument would best meet these goals.

Secondly, theories focusing on power relationships at the societal and state levels are essential for demonstrating how ideas are introduced and incorporated into the national policy program. At the societal level, elite groups have a central role in agenda setting. Politicians, bureaucrats, stakeholders and researchers exercise (varying degrees of) power to suppress unwanted issues, and to keep them from the policy agenda (Bachrach and Baratz 1962; Lukes 1974). Power in decision-making situations can be expressed as the ability of elites to shape preferences and suppress alternative conceptions, thereby protecting their own positions and world-views (Lukes 1974). In environmental policymaking, this is reflected in the established conception that governments are unlikely to achieve environmental agreements against unified industry opposition; large firms must participate in the implementation of environmental policy (see Barkdull and Harris 2002). At the state level, executive-legislative relations are central for defining the policy agenda. In parliamentary systems it matters for policy outcome what the constellation of the parliamentary majority looks like, and whether it shares the same policy objectives and conceptions as the executive/prime minister. For instance, in Norway the result of minority coalition cabinets has been that the parliamentary majority can override cabinet decisions and the GCC policy agenda of the cabinet.

Thirdly, interest-based theory offers the best explanation of how interests are pursued once they have been defined. At the systemic level, this approach assumes that rational actors in Europe cooperate in order to achieve joint gains. In a tradeoff between calculating environmental damage versus the economic costs of abatement policies, states are assumed to pursue their national interests by employing policy that achieves economic growth, or at least no loss (Underdal 1998; Barkdull and Harris 2002). In the case of Norway and Germany, the process of European integration led to the development of a common GCC policy, which limited national autonomy and restricted policy instrument choice. At the societal level, interest-based approaches assume GCC policy instrument choice to be the result of bargains struck among different constituencies with a stake in GCC policy (Sprintz and Weiss 2001; Underdal 1998). In both Norway and Germany, stakeholder groups were actively involved in bargaining over whether and how to implement the EU's ETS. To explain policy instrument choice, therefore, we focus on identifying the groups that participated in this process, their relative influence, and the strategies and tactics they employed (Barkdull and Harris 2002: 74–5).

We focus in particular on these three aspects of GCC policymaking, and discuss how EU influence was decisive for GCC policy instrument choice in Norway and Germany.

The timeframe of our analysis begins in December 1997 at the Kyoto Summit, and progresses through July 2003, when an ETS was established for all EU members. It ended in July 2004, when the Norwegian government announced plans to adopt a domestic ETS, duplicating the EU system. Our time-limited case study approach allows a structured, focused comparison and points out actors, discourse, power relationships and interests involved in GCC policy instrument transfer that took place in Norway and Germany.

EMISSIONS TRADING AS A NEW POLICY INSTRUMENT IN EUROPE

Before we go on to discuss the potential impact of EU membership on policymaking at the domestic level in Germany and Norway, we take a brief look at how ET emerged as a policy instrument at the EU level.

The EU: from Skeptic to Policy Innovator

Today it is hard to imagine a more enthusiastic proponent of ET as a GCC policy instrument than the EU. In July 2003, after a lengthy consultation process, the EU passed a directive establishing a ground-breaking internal ETS for carbon dioxide (CO_2), which went into operation in 2005. Yet an observer at the Kyoto summit of 1997 could be forgiven for expressing surprise at this somewhat unexpected turn of events. The EU had, after all, put up fierce resistance to the flexibility mechanisms during conference negotiations. The EU delegation at Kyoto only agreed to the inclusion of Article 17 of the Kyoto Protocol (paving the way for an ETS) under severe pressure from the United States (US), which made it clear that agreement on targets and timetables – a cornerstone of EU policy – was contingent on the inclusion of market-based instruments.

How did this significant u-turn come about? We begin with a survey of the key events leading up to the establishment of an internal EU ETS, which is summarized in Table 12.1. We then investigate the features of the new scheme before going on to explore events at the domestic level.

At Kyoto, the EU position was one of 'risk-prevention leadership'. This refers to its commitment to the precautionary principle (see European Commission 2000b), and to an environmental policy historically based

Table 12.1 The EU and emissions trading

Year	Event	Significance
Dec 1997	Kyoto Protocol (Article 17)	– Commits the EU to reducing GHG emissions by 8 per cent from 1990 levels during the period of 2008–12 – Introduces ET as one of three flexible mechanisms to be operationalized internationally
Jun 1998	Commission Communication to the Council and Parliament: 'Climate Change: Towards an EU Post-Kyoto Strategy' (European Commission1998: 353)	– Launches target date for ETS as 2005: 'the Community could set up its own internal trading regime by 2005 as an expression of its determination to promote the achievement of [Kyoto] targets in a cost-effective way' – Advocates a step-by-step approach to the development of ETS
May 1999	Commission Communication to the Council and Parliament: 'Preparing for Implementation of the Kyoto Protocol' (European Commission1999: 230)	– States urgent need for an informed debate on the instruments of ETS; proposes Green Paper – Suggests that ETS should initially be limited to CO_2 in 'key sectors', but should gradually be extended to other sectors and gases
Mar 2000	Green Paper on GHG ET within the EU (European Commission 2000a: 87)	– Establishes consultation process, 'which will allow all stakeholders, both governmental and non-governmental, to give their opinions on how the EU should strike the right balance in the use of ET'
Jul 2000	European Climate Policy Program established	– Working group investigates climate policy alternatives; in May 2001 recommends ETS be established 'as soon as practicable'
Oct 2001	Proposal for a Directive for GHG ET within the EU (European Commission 2001: 581)	– Begins legislative process at EU and member-states levels
Dec 2002	Council presents Amended Proposal for a Directive of the European Parliament (EP) and Council	– Establishes scheme for GHG ET within the EU – Amended by EP in Spring of 2003
Jul 2003	EP and Council approve Proposal for Directive establishing internal ETS (European Commission 2003: 403)	– ETS to be operative in January 2005 and to be extended to all member states of the enlarged EU
Oct 2003	Directive of EP and Council published in *Official Journal* of EU	– Directive establishing EU ETS officially enters into force – Directive implemented with start of ET in January 2005

on 'stringent CO_2 reductions, fierce protection of the internal market's international competitiveness and a desire for leadership in global climate change policy' (Damro and Méndez 2003: 75). From its conception in the 1980s, EU GCC policy had centered on regulation and harmonization, and targets and timetables. Flexibility mechanisms, such as ET, had been viewed with caution and even suspicion, as it was feared that such policy instruments would only serve to create loopholes allowing some countries (the US in particular) to buy their way out of making domestic emissions reductions. More to the point, it was thought that such instruments would ultimately pose a threat to EU competitiveness by hampering economic interests and undermining EU attempts to establish itself as a global environmental leader.

Article 17 of the Kyoto Protocol introduced ET as one of three flexibility mechanisms available to assist industrialized countries to meet their emissions reduction targets. The subsequent EU dialogue, initiated by the European Commission during the 2 years proceeding Kyoto (see European Commission 1998: 353, 1999: 230), could thus be seen as a reaction to the EU's international commitment to the establishment of a global GHG trading system. The Commission expressed repeated support for a public dialogue on an internal ETS, suggesting that 'the best preparation for the Community and its member states might be to develop their own ET experience' (see European Commission 1999: 230). In keeping with this sentiment, a Green Paper launching an EU consultation process on ET and advocating a bottom-up pathway was published in the spring of 2000. Soon after, a working group comprising representatives from member states, industry and environmental pressure groups was established under the auspices of the European Climate Policy Program. In May 2001 the group gave its support for the launch of an EU ETS. This somewhat informal process was formalized in the autumn of 2001 when the Commission put forward their 'Proposal for a Directive for Greenhouse Gas Emissions Trading within the European Community', thus beginning the legislative process at the member-state and EU levels (European Commission 2001: 581). The European Council and Parliament tabled amendments to the Directive in late 2002 and early 2003 respectively (European Commission 2002, 2003). The Proposal was finally approved in July 2003, and the Directive entered into force three months later, establishing an internal ETS that came into operation in January 2005.

Features of the EU Emissions Trading Scheme

The EU ETS gives member states flexibility in meeting the EU Burden Sharing Agreement targets by establishing a multinational trading system in

GHG emissions allowances.[2] The underlying motive behind the scheme is to encourage installations (individual firms), to make emissions reductions at the lowest cost to the economy, driving efficiency and fostering innovation. Central features of the mandatory system are limited coverage, permit allocation by member states, non-compliance penalties, an opt-out for individual installations, and the inclusion of project-based mechanisms under the Kyoto Protocol's Clean Development Mechanism (CDM). These elements are discussed further below.

Although the Directive covers all Kyoto GHGs in principle, it states that for the first phase, ET will be confined to CO_2. Carbon dioxide accounts for over 80 per cent of EU GHG emissions, and it is estimated that units participating in the scheme will be responsible for 46 per cent of CO_2 emissions in 2010. The scheme applies to energy-intensive companies such as power plants, steel factories, oil refineries, paper mills and glass and cement installations. The mandatory system, with unspecified caps, gives member states the responsibility to set reduction targets and allocate emissions allowances. Ninety-five per cent of permits must be distributed free of charge in 2005–2007 (falling to 90 per cent from 2008), but governments may choose to auction the remainder. Allocation must be based on an emissions path compatible with national targets, and must also comply with the criteria laid out by the Commission in Annex III of the Directive. Member states were required to submit National Allocation Plans to the Commission by 31 March 2004, detailing the quantity and allocation of allowances in order to minimize market distortions. Only five members met this deadline, but by the end of 2004 all EU Members had submitted their final plans.[3]

Strict non-compliance penalties of €40 per ton of CO_2 already apply in 2005–2007, rising to €100 per ton of GHG emissions from 2008. A strong financial deterrent was implemented in the interest of efficiency and political acceptability, and drew from the experience of the US sulfur dioxide trading scheme, which achieved an 'excellent compliance record' as a result of the imposition of monetary penalties (Christiansen and Wettestad 2003: 5). Individual firms (but not sectors) can opt out of the trading system if they can demonstrate a commitment to making equivalent emissions reductions. The opt-out is restricted to the first phase of the scheme, and is most likely to be used by Britain (where an ETS has already been established) and possibly also Germany (where there is a history of voluntary environmental agreements). The ETS accept credits from CDM projects from 2005, and from joint implementation schemes from 2008 (in accordance with the amended Linking Directive[4]).

INTRODUCING ET IN NORWAY AND GERMANY

Based on the framework described in section two, we now embark on a comparative analysis of GCC policy in Norway and Germany. First, we examine how key socially constructed objectives remained more or less constant for both countries throughout the ETS process, while their understanding of which instruments best met these objectives changed during our period of analysis. Secondly, although key objectives were socially constructed, once new means of achieving these national goals were available (through the EU ETS), domestic bargaining took place in order to maximize the outcome for the state and stakeholder groups. Before examining the ET programs of Norway and Germany, we present short country profiles to highlight the special circumstances that have been important in GCC policymaking. We then move on to point out how the transition to ETS occurred in each of the countries.

Norway and Germany: Country Profiles

In the early 1990s, German GHG emissions were significantly reduced. The trend continued in the late 1990s, but at a slower pace. Emissions are projected to fall by 33.6 per cent by 2010 (see Table 12.2). In contrast, Norwegian emissions are growing steadily and projections highlight an increasing gap between likely emissions levels and Kyoto commitment targets (see Figure 12.1).

Table 12.2 Emissions commitments and trends: Germany and Norway

Country	Kyoto commitment %	Actual emissions (2000) %	Projected emissions (2010) %	'Gap' between projections and commitments in 2008–12 %
Germany	–21[5]	–19.1	–33.6	–12.6
Norway	+1	+6	+22	+21

Sources: National Communications to the FCCC Secretariat (GFME 2001; NME 2002b).

Norway

Norway's distinctive energy and industry profile is characterized by a heavy reliance on hydroelectric power, oil and natural gas extraction, and by high demand for transport. Hydropower fuels domestic energy consumption (99 per cent of electricity production), and forms the basis for the energy-

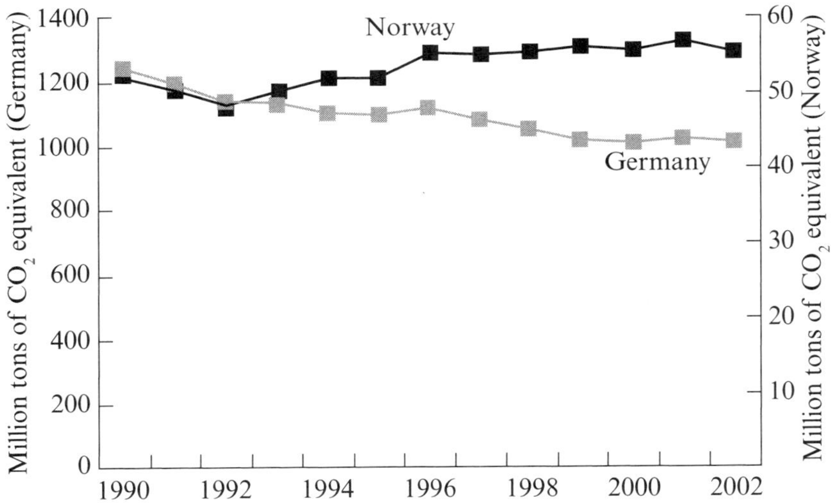

Source: UNFCCC 2004.

Note: Million tons of CO_2 equivalent. The left axis shows German absolute emissions, while the right axis shows those of Norway.

Figure 12.1 Emissions of GHGs in Germany and Norway, 1990–2002

intensive process industry, which currently accounts for a fifth of the country's GHG emissions (NME 2002b: 61). A decentralized settlement pattern makes transport the largest emitter, producing 31 per cent of GHGs in 1999, while the petroleum industry was responsible for a further quarter in 2000 (Statistics Norway 2003). Oil and gas are key sectors, contributing 30 to 47 per cent to export revenue, and between 11 and 25 per cent to GDP in 1998–2000 (NME 2002a: 17). Taken together, these factors lead to high abatement costs for Norway. The introduction of a CO_2 tax and the onset of an economic recession resulted in falling GHG emissions in the early 1990s (see Figure 12.2). However, levels soon began to rise, and are projected to increase by 22 per cent between 1990 and 2010 (see Table 12.2). Norway ratified the Kyoto Protocol in May 2002, but faces considerable challenges in meeting its commitment target of a 1 per cent increase relative to 1990 levels.

Germany

In contrast to Norway, Germany has managed to implement GHG emissions cuts of 19 per cent since 1990. Largely owing to the structural changes resulting from reunification, it is well on track to meeting its Kyoto target of a 21 per cent reduction from 1990 levels by 2005 (see Table 12.2). With the exception of the transport sector, it has managed to reduce emissions in all source groups through measures including reduced coal and lignite consumption, economic and industrial restructuring, and improved energy efficiency and management practices (IEA 2002). Limited indigenous energy resources make Germany increasingly reliant on imports of natural gas, oil and even coal – which still fuels 50 per cent of electricity production. Although hard coal remains heavily subsidized (IEA 2002), the state has also given financial support to 'green' electricity. Renewables currently account for only 2.6 per cent of total energy use (IEA 2002), but Germany has a quarter of the world's installed wind power capacity. There is no doubt that Germany achieved significant emissions reductions during the 1990s. However, the opportunity awarded by reunification was unique, and it will be much more politically and economically costly to reduce emissions further and reach the government's ambitious national target of a 25 per cent reduction in CO_2 emissions from 1990 levels by 2005.

Figure 12.2 illustrates the structural differences between Norway, Germany and the EU. In the next sections we discuss how their contrasting economic profiles caused our case study countries to develop very different national positions on the idea of ET as a GCC policy instrument.

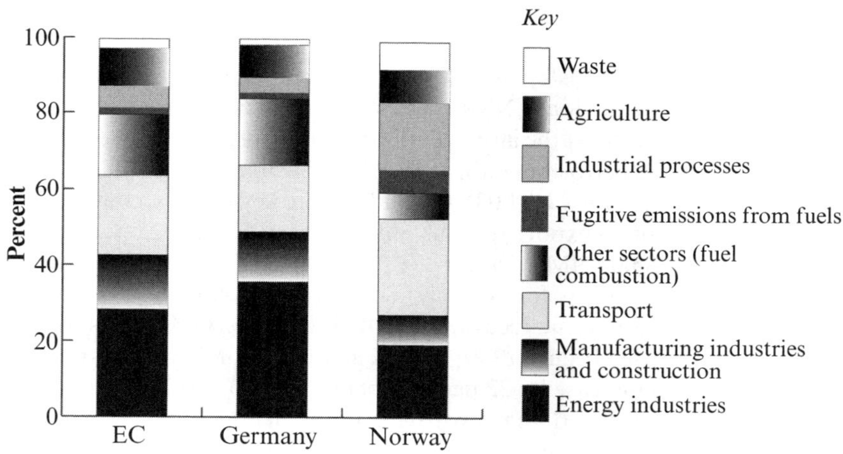

Source: UNFCCC 2004.

Figure 12.2 Relative GHG emissions in the EU, Norway and Germany, by source (2002)

Norway and Emissions Trading

Objectives and underlying values and principles

We begin our analysis by drawing on constructivist claims that domestically and internationally developed understandings and norms shape national foreign policy, in this instance in the GCC issue area (Finnemore 1996; Wendt 1999; Betsill 2000). This section outlines the way in which internal and external socialization processes produced a set of Norwegian environmental goals, including respect for the precautionary and polluter pays principles, a desire to demonstrate environmental leadership, and a willingness to protect the international competitiveness of Norwegian industry.

In many ways, Norway was ahead of the EU in terms of its willingness to accept the flexibility mechanisms that were agreed upon at the Kyoto Conference. The shift from command and control instruments to a more market-oriented approach took place in the early 1990s. Underlying these approaches were the polluter pays principle and the precautionary principle, which were incorporated into Norwegian environmental policies in the 1970s and 1980s respectively (Jansen and Osland 1996: 188–9). By the close of the 1980s, a desire to be an instigator in international environmental politics began to gather momentum (Andresen and Butenschøn 2001: 338). This was partly driven by Prime Minister Gro Harlem Brundtland's position as Chairperson of the UN World Commission on Environment and Development. The resulting report, *Our Common Future,* was published in 1987. It singled out GCC as a key issue. This spilled over into the national political arena (Bang 2004: 268–9), fostering strong public awareness of environmental concerns in the run-up to the 1989 parliamentary election. It also contributed to the country's willingness to be among the first to adopt a unilateral policy target of stabilizing CO_2 emissions at 1989 levels by 2000.

However, it soon became apparent that the stabilization target conflicted with vital economic interests, such as the expansion of the oil industry and plans to export natural gas to Europe. This led to a distinct change in policy objectives, based on the principles of international cost-effectiveness, flexibility and ET. Norway supported US initiatives to introduce the idea of flexibility mechanisms at the GCC negotiations as early as 1991. It abandoned the stabilization target a few years later (Hovden and Lindseth 2002: 149; see Chapter 8).

Hence, the interplay of international and domestic forces over time produced a set of objectives that formed the basis of Norwegian EFP. The country's reliance on energy exports made it impossible to ignore the implications of its unilateral domestic actions on the international competitiveness of its industries. At the societal level, popular support for green politics was in step with a state-led desire for environmental leadership, which the Brundtland process had set in motion (Bang 2004: 268–9). We believe that these goals are hard to explain solely within a rationalistic framework because Norway's high abatement costs, combined with its low vulnerability to climate change, suggest that if hard national interest alone had been the priority Norway would have been a 'dragger' rather than a 'pusher' (Sprinz and Vaahtoranta 1994: 81).

Bearing in mind how ideas and norms shaped Norwegian GCC policy, in the two sections that follow we turn to rational choice-based theories that emphasize elite power and interest-based bargaining. We illustrate that once social and economic forces had defined *perceptions* of the national interest, state and non-state actors adopted rationalist, goal-maximizing methods to defend this interest. Elite theory explains how dominant state and societal groups, inspired by the international climate process, brought ET onto the domestic policy agenda. Interest group theory demonstrates how different groups formed new coalitions to argue their case at the national and European level.

Policy instruments

Norway's first policy response to the GCC challenge was to implement a set of environmental taxes and regulations. First and foremost, a CO_2 tax was adopted in 1991, which became – and still is – the country's chief GCC policy instrument. Industries exposed to international competition, such as the processing industry, were exempt from the tax, which was levied on almost two-thirds of CO_2 emissions and covered just under half of total GHG output (NME 2001; Vevatne et al. 2004). As a follow-up to Norway's Kyoto commitments and to a state-appointed Green Tax Commission Report, in 1998 the center-coalition government proposed broad environmental tax reforms to incorporate the sectors previously enjoying exemptions from the tax (NME 1998). While environmental NGOs supported the government's recommendations (see Chapter 8), business associations and energy intensive industries were highly critical. Between 1995 and 1997, they pressed for voluntary agreements, voicing skepticism that taxation would deliver the desired environmental benefits and arguing that the measure was just a ploy to broaden the government's tax base at their expense.

After the Kyoto Protocol was signed in 1997, however, their attention quickly turned to ET. Business and industry began to lobby for the introduction of an international ETS in place of the CO_2 tax because it allowed them to substitute expensive domestic mitigation measures with cheaper emissions reductions abroad (Bang 2004: 210). Opposition parties, which had close links to the energy-intensive industries, claimed that ET would improve efficiency, and suggested that industry's competitiveness problem could be solved by the allocation of free quotas. Under duress, the minority government agreed to examine the potential benefits of a national ETS. Given the highly institutionalized links between industry elites, the major political parties holding the majority in parliament (in particular the Labor and Conservative Parties) and interest groups (Kasa 2000), it is perhaps unsurprising that when an economically and environmentally efficient solution to the climate problem presented itself in the form of ET, it very quickly made it onto the agenda.

From green taxes to domestic emissions trading

Our claim is that motivation to introduce ET was independent of the EU, and inspired by pre-Kyoto proposals originating in the US. Instead, European influence was confined to shaping the *design* of the domestic ETS that Norway now has implemented.[6] Support for an ETS gathered momentum within Norway during 1998, when industry elites realized that they stood to suffer from broad environmental tax reform. At the international level, they seized the opportunity to respond to the Kyoto Protocol's opening for enterprizes to participate in ET, and on the domestic front they drew upon the loyalty of their political allies to gain parliamentary leverage.

Domestic academics were also warming to the prospect of ET. Important academic contributions emerged when, in December 1999, a government-appointed Commission presented its evaluation of ET as a policy instrument for Norway. To a large extent, its recommendations were adopted in June 2001 when the new Labor minority government proposed that the CO_2 tax be replaced with an ETS including all Kyoto gases from 2008 (NOU 2000: 1).[7] The plan was the result of fierce lobbying on the part of industry. It was also staunchly backed by the Ministry of the Environment, which supported ET on ideological grounds and stood to gain responsibility for administering the new system.[8]

Following the General Election of 2001, the new center-right government (Liberals, Christian Democrats and Conservatives) signaled a change in environment policy when, in March 2002, it presented a supplementary report to the Storting (NME 2002a). Among other measures, it proposed a mandatory ETS for industries *not* covered by the CO_2 tax, designed to be a supplement to the tax in the pre-Kyoto period (2005–2007). The shift to a broad ETS from 2008 had wide parliamentary support, and was backed by powerful industry groups, such as the petroleum sector, which stood to gain financially from an ETS, as the price of emissions permits was expected to be significantly lower than the carbon tax rate. On the other hand, the proposed pre-Kyoto ETS was controversial, and was heavily criticized by industry, and the Labor and Progressive Parties, who preferred voluntary agreements.

Key features of the government's proposed 2002 scheme were as follows:

- *Coverage*: participation limited to sectors exempt from the CO_2 tax. Some GHGs to be included from 2005 (e.g., CO_2, methane, nitrous oxide, perfluorocarbons and sulfur hexafluoride), but all to be incorporated in 2008, thereby covering approximately 80 per cent of emissions.
- *Allocation*: free allocation of emissions allowances (2005–2007). Opportunity to auction and discriminate between sectors from 2008.
- *Compliance*: standardized, binding and transparent monitoring; verification and reporting procedure; sanctions for non-compliance.
- *Clean Development Mechanism*: credits from CDM projects to be included from 2005.

EU influence on the Norwegian ETS

Norway has long stated its preference to implement a system that is compatible with an EU ETS, appreciating that its European Economic Area (EEA) agreement might otherwise be jeopardized (EEA members must comply with EU directives in several areas, including this one). The government and industry have actively lobbied the EU Commission to influence the design of the EU ETS, the former calling for 'sufficient flexibility to reflect national circumstances' (Brende 2002). However, their efforts have met with little success. Norwegian objections to the EU system have been threefold. First, Norway wanted to 'opt-in' the major polluting domestic sectors that are excluded from the European scheme in order to make the system as broad as possible. Secondly, it would have liked to include more GHGs. The EU Directive is restricted to CO_2, which covers 46 per cent of EU GHG emissions, but only 27 per cent of Norway's. A third bone of contention has been the allocation of emissions rights. Norway preferred free allocation in 2005–7, followed by auctioning during the Kyoto period (2008–12), but the EU system allows 5 per cent of allowances to be auctioned from 2005, and only 10 per cent from 2008.

In spite of these objections, in March 2004, the Norwegian Minister for the Environment announced his country's intention to implement a domestic ETS in 2005 that would be compatible with the EU ETS (Brende 2004). Perhaps of greatest surprise is that Norway was not granted any special treatment on the basis of its unusual economic structure, which many had expected would be a necessary condition for Norwegian entry into the EU system. As a result, the processing industry, representing Norway's main potential source of emissions reductions, will not be covered at all, but will instead take on responsibilities in the form of a voluntary agreement with the government to reduce GHG emissions. Meanwhile, Norway's road transport, oil and gas sectors will continue to be covered by the CO_2 tax until 2008. If a broader ETS is introduced, it could imply a huge tax relief for these sectors because permit prices will be lower than the tax level.

Despite Norway's previous reservations, it is now prepared to compromise on its more ambitious system – in terms of gases and sectors – in order to adapt to the EU scheme. While it has a legal right to reject the Directive, it would be difficult to justify on political and economic grounds. Such a move could endanger its EEA membership and the tremendous benefits that come from it. Norway's willingness to join an ETS, which covers as little as 11 per cent of its GHG emissions (Torvanger 2004), gives an indication of the strength of EU influence on its GCC policies.

Germany and Emissions Trading

Objectives and underlying values and principles

While the Kyoto Protocol put ET on the agenda for the EU (see Chapter 11), it was EU interest that sparked the domestic debate in Germany. The objectives that the German federal government has sought to achieve through its GCC policy have been constant since the issue reached the political agenda in the late 1980s. With considerable focus on making environmentally efficient emissions reductions domestically, Germany has also been keen to play a leading role internationally without risking the competitiveness of its large manufacturing industry. There are several values and principles underlying these goals, which can best be understood and explained along constructivist-theoretical lines. First, the precautionary principle has long held a central position in German environmental policy. Secondly, green values have run strong in German public opinion, which has given all political parties good reason to promote proactive environmental policies. Thirdly, as a major economic force in Europe, Germany has wanted to show leadership. Climate change has provided the country with a real opportunity to do this, in contrast to other issue areas, such as international peacekeeping. Finally, it has been important to ensure that Germany remains an attractive location for industry. Promoting equal opportunities for firms, especially in the common market, has therefore been an important aspect of GCC policy (Bang 2004: 291–8; Sprinz 2001: 13–15; see Chapters 2 and 14).

As with the Norwegian case, a norm-based explanation accounts for Germany's willingness to commit itself to the GCC issue. Systemic and societal pressures have caused international competitiveness to pervade the national interest, but also ensured that environmental concerns remain important. Modernizing outdated factories in eastern Germany in the early

1990s resulted in impressive windfall cuts in national GHG emissions without exposing the manufacturing and process industries to negative economic consequences, but it is beyond doubt that economic sacrifices will need to follow if the government is to realize its self-imposed medium-term target of reducing CO_2 emissions by 40 per cent from 1990 levels by 2020, provided the rest of the EU achieves a reduction of 30 per cent during the same period. These factors confirm that rationalism alone cannot explain Germany's GCC policy choices, and that a norm-based explanation contributes a fuller understanding.

In the following two sections, however, we apply interest and power-based approaches for analyzing how state and non-state actors responded to the idea of ET introduced from outside the EU. German elites established common interpretations of the ET issue that became dominant in the policy debate and shaped the policy agenda, while decisive interest group bargaining took place once the idea had entered the political arena and was seen as a novel means of pursuing the national interest.

Policy instruments

Germany approached the GCC challenge with a set of command and control policy instruments that had been tried and tested in other policy fields. Three types of regulations were preferred: environmental laws, ordinances and technical specifications. With a strong legalistic-bureaucratic tradition in German politics, these policy instruments were preferred by both the Christian-Conservative-Liberal and the Social Democrat-Green coalitions. In particular, a voluntary agreement between industry associations and government, directed at CO_2 emissions reductions, had broad political support. A 1996 agreement committed industrial associations – representing four-fifths of industrial energy consumers – to reducing CO_2 emissions by 20 per cent from 1990 levels by 2005 (Bang 2004: 229). With this policy measure Germany achieved both emissions reductions *and* consensus in policymaking, as well as avoiding legally mandatory regulation of the industrial sector. In other words, the decision to implement a voluntary agreement instead of carbon taxes was made in 'the shadow of the law' (Würzel et al. 2003: 120).

When the Red-Green coalition took office in 1998 it became evident that the benefits of GHG-reductions due to reunification was wearing off, and that intensified policy measures were needed to continue the trend of national emissions cuts. The government initiated a five-step ecological tax reform to reinforce GCC policy. In addition, it presented a new and intensified national GCC protection program in October 2000, where voluntary agreement and ecological tax reform were presented as the most vital ingredients. The scheme did not mention ET as a serious alternative for German GCC policy. Although researchers began discussing the topic academically in 1997 when it became part of the Kyoto Protocol, policymakers did not approach the issue until 2000–2001.[9] Our claim is that ET entered the German policy agenda as a direct result of the internal discussions in the EU, and the policymaking process leading to the proposed Directive in 2001.

Emissions trading as a new policy instrument

Germany managed to cut its GHG emissions by 18.5 per cent from 1990 levels by 2000. The federal government pointed to its command and control approach – combining mandatory and voluntary policy measures – as the main reason for this reduction (GFME 2001a, 2001b), although reunification certainly had an important part to play. Furthermore, the government specifically stated that a new policy instrument must not reduce the effectiveness of its current climate protection program. As late as 2000, Chancellor Gerhard Schröder publicly opposed an ETS, while Environment Minister Jürgen Trittin supported it (Würzel et al. 2003: 124). Moreover, strong industrial pressure groups in Germany were reluctant to accept the introduction of an ETS. In particular, the German Industry Association, the Chemical Industry Association, and the energy intensive sector voiced strong opposition. A 2002 survey pointed out that general knowledge about ET was quite low in German industry, especially among small and medium-sized firms (Santarius and Ott 2002: 5). Many companies would have liked to preserve the voluntary agreement, and expressed concern that an ETS might not take into account the 'early action' reductions implemented through the voluntary agreement since 1995.[10] Meanwhile, other business sectors, such as banking, strongly supported the implementation of an ETS (Würzel et al. 2003: 127–8), which is perhaps of no great surprise given the likely benefits that the creation of a new financial market were expected to bring them.

The discourse on ET in Germany was primarily a response to activities at the EU level, but also a reaction to developments in other member states (i.e., Denmark and the UK). Germany found itself in jeopardy of losing its leadership role in this issue area. It appeared that if the country did not form a clear position on the issue, an EU-wide ETS could be imposed from above, with Germany having little or no influence over it. Hence, in realizing that the most effective means of pursuing the national interest had changed, German industrial and political elites put the ETS on the domestic policy agenda.

In the tradition of consensus policymaking, the Federal government's reaction to a changed policy agenda was to establish the German Emissions Trading Group (AGE) in October 2000. The Working Group brought together representatives from the federal and länder governments, the

German Bundestag, trade and industry (30 companies and nine industrial associations), and environmental groups and agencies. The level of disagreement within the group was high. The chemical and aluminum industries were so opposed to ET that they ran an ad-campaign in 2001 warning of the adverse economic consequences of ET for German industry (Butzengeiger et al. 2003: 221). The Federal government, meanwhile, was starting to warm to the principle of an EU-wide ETS. It saw the scheme as a means through which domestic GCC policy objectives could be met cost effectively. Furthermore, ETS could potentially lift Germany to the forefront of an innovative environmental trading scheme, and in doing so revive its role as an environmental pioneer.

In September 2001, AGE presented a report that was widely supportive of an EU-wide ETS. It contained the following recommendations:

- *Coverage*: launch of EU ETS with a voluntary 3-year pilot scheme including economic incentives for participation. Participation limited to firms with proven ambitious reduction commitments; scheme initially limited to CO_2, but to be extended to all six Kyoto GHGs.
- *Allocation*: absolute emissions caps, allocation of permits through hybrid system combining grandfathering and auctions.
- *Compliance*: standardized, binding and transparent monitoring, verification and reporting procedure; penalties for non-compliance.
- *JI and CDM*: emissions credits to be included.

These recommendations were accepted by the Federal government and presented in a position paper at an EU Commission discussion later that month. In addition, German industry pressured the government to propose an 'opt-out' clause, giving member states the right to withdraw sectors from the ETS pilot phase. Furthermore, the Federal government was pushed into proposing a 'pooling-model' that would allow industry sectors to pool emissions allowances in order to maximize cost efficiency. Finally, Germany proposed that 'early action' be acknowledged (Butzengeiger et al. 2003: 221).

Since 2002, ET has been accepted as an unavoidable development in Germany. Rather than *opposing* the implementation of the policy instrument outright, the federal government and German interest groups have sought to *influence* the proposed EU scheme. Yet, to what extent have they been successful? Germany's interests were perhaps better defended in the EU Directive proposed in October 2001 than in the final form of the Directive passed in July 2003, but the ultimate outcome was largely what the Germans had campaigned for. In line with German requests, permits are allocated free of charge (with a small proportion available for auction), the system covers CO_2 (to be extended to other GHGs in the future), monitoring and reporting is standardized and binding, penalties are imposed for non-compliance, and credits from CDM and JI are incorporated in the first and second phase, respectively.

Calls for 'early action' to be taken into account have been met insofar as member states are free to decide the distribution of permits and can therefore choose to allocate allowances on the basis of historical emissions and to select an early base year if they wish to do so. Furthermore, fierce lobbying from the German Chemical Industry Association at home and in Brussels appears to have paid off, as chemical installations are excluded from the first phase of trading. On the other hand, three elements that Germany had pressed for were *not* included in the final scheme: the voluntary, 3-year pilot phase; absolute rather than flexible emissions caps; and an 'opt-out' clause giving member states the right to withdraw sectors from the scheme. Since the approved opt-out applies to installations rather than to sectors, it will be much harder – if not impossible – for an entire sector to opt out of the EU ETS. It is also improbable that pooling will materialize in the absence of a mandatory obligation, as companies achieving emissions below their allocations are unlikely to want to give away surplus permits to laggards within their potential pool.

The EU-wide system came into effect in January 2005, with the German government as a staunch supporter. The sharp turnaround in attitudes towards this GCC policy instrument, which occurred between 2000 and 2003, gives a strong indication of the country's dependence on EU policy processes. While it could lobby for changes in the *design* of the ETS, as an EU member it would have been unthinkable for it to have rejected the policy instrument *per se*.

DID THE EU INFLUENCE DOMESTIC CLIMATE POLICY?

Common Objectives Frame Climate Policy, but Elites Set the Agenda

Although Norway and Germany have both arrived at the conclusion that ETS represents the best means of reaching their policy objectives, it is interesting to note that their policy goals have remained remarkably stable. We have already observed that ET was a novel policy concept that originated in the US, but was reluctantly embraced by the EU when it became apparent that the instrument represented a vehicle through which its objectives could be met. However, the EU's underlying objectives remained unchanged. Instead, it was gradually persuaded that ET represented a window of opportunity for the re-launch of its GCC policy, a means through which it could meet Kyoto targets cost-effectively, and a chance to reassert its global leadership credentials as an environmental frontrunner. According to Christiansen and Wettestad: 'A synergistic and multilevel mix of explanatory factors, including developments at the international, EU, Members State, sub-national, and even down to the personal level' heralded a U-turn in EU attitude towards ET as a GCC policy instrument, resulting in the establishment of the first regional GHG ETS of its kind in Europe (Christiansen and Wettestad 2003: 3).

Norway and Germany have had similar GCC policy objectives. In both countries, several elements have been critical: the precautionary principle, the polluter pays principle, the desire for leadership, the promotion of environmentally efficient commitments, and the protection of the industrial sector's competitiveness. The precautionary principle and the polluter pays principle are ideas that have been accepted in EFP since the first UN Conference on Environment and Development in Stockholm in 1972. To a large extent these ideas and objectives have been kept in the policy loop by environmental pressure groups, and a consistently 'green' public opinion in both Norway and Germany.

Both countries have expressed a desire for leadership on this particular issue. For Norway, a leadership role has been credible due to the country's established 'green' reputation. The environment represented a 'soft' policy issue that even a small country could try to influence (see also Chapter 4). For Germany, climate change has been one of a limited number of international areas where it could demonstrate leadership, in spite of its history. While it was important for both countries that environmental effectiveness was taken into account, there was also considerable concern that this should not be at the expense of international competitiveness. Hence, social elites in both countries were actively involved in establishing a broad consensus on policy direction that ensured that established power structures remained unchanged. The underlying principles and objectives therefore had to be balanced against economic and industrial interests. Similar policy goals in the end led to acceptance of the ETS as a preferred policy instrument.

National Interests Shape Policy Instrument Choice

In the ET discourse, significant shifts have occurred in the relative positions of Norway and Germany since the early 1990s. Norway has supported the idea of an *international* ETS from a very early stage. Although this stance conflicted with the position of environmental groups, the principle of cost-effectiveness ran deep in the thinking of government officials, stakeholder groups and other key policymakers. It provided the opportunity for a broad range of sectors and GHG gases to be incorporated in a single trading system that promoted the most cost-effective emissions reductions. In this setting, the Norwegian industrial lobby was able to push international ET onto the policy agenda, and played a major role in Norway's decision to opt for an ETS from 2008 to 2012. As for the pre-Kyoto ETS from 2005–2007, Norwegian industry was successful in protecting its interests at the EU-level through its European branch associations when it became clear that the Norwegian government preferred an early ETS to voluntary agreements. The European business lobby had a major influence on the design of the EU ETS (Markussen and Svendsen 2005). In particular, the aluminum industry was successful in securing exemptions from the European system. In Norway, this resulted in the aluminum industry taking on voluntary agreements instead of participating in the pre-Kyoto ETS.

The proposed Norwegian pre-Kyoto ETS would not have been compatible with the EU ETS, therefore Norway has been forced to take into consideration the EU approach due to the great importance of the common market for its domestic economy. Given Norway's longstanding enthusiasm for an environmentally effective international ETS, it is significant that as an EEA member it has been forced to implement an ETS that, in effect, covers only 10 per cent of domestic CO_2 emissions. The government's failure to negotiate any special considerations based on its unusual economic and energy structure points to the constraints that international factors – in particular the EU – have placed on the country's domestic policy choice. External and domestic factors interacted to redefine the parameters within which Norway's national interests could be pursued, and the EU demonstrated its power as a supranational force by influencing domestic policy.

In contrast, Germany originally preferred to make emissions reductions domestically. Its industrial structure, the process of reunification and a high dependency on fossil fuels (coal in particular) made it possible to achieve substantial reductions during the 1990s using a command and control approach. The industrial sector agreed to shoulder emissions reductions, but it preferred voluntary agreements that allowed cutbacks to be distributed within sectors. Industry associations initially resisted replacing these voluntary agreements with ET. As had been the case in Norway, the German industry lobby strongly shaped the government's positions on the issue. However, in contrast, it opposed ET because the voluntary agreement was more economically attractive. Germany, along with Norway, has accepted the EU ETS as the new preferred policy instrument as a result of changed perceptions about the most effective means of pursuing national interests. As an EU member, Germany is bound by the economic regulations of the common market. Having first resisted the introduction of the ETS, Germany has since 2001 been active in trying to influence the system.

In other words, Norway and Germany's very different economic and energy structures to some extent explain why early positions on emissions trading were so polarized, despite their similar GCC policy objectives. The EU's sudden interest in ET, however, acted as an external force pressing both domestic arenas for change, and serving to redefine perceptions of the most appropriate means of protecting national interests among domestic elites in both countries. In Germany, ET not only entered the agenda, but also became the centerpiece of the country's GCC policy program. On the other hand, in Norway, where ET had been championed for over a decade, the comprehensive domestic system that the country had proposed was watered down so that it could be harmonized with the EU ETS. Thus, from very different starting points in the early 1990s, German and Norwegian attitudes to ET as a GCC policy instrument have converged to imitate the EU directive.

CONCLUSION

The EU-level GCC policy discourse and decision-making processes strongly influence German and Norwegian environmental foreign policies. When the EU initiated the Proposal for a Directive on ET in 2001, it helped chart the course for domestic GCC policy debate in Germany and Norway. Developments inspired by the international and European climate regimes reshaped national political and economic landscapes, and redefined the boundaries within which domestic and international interests could be pursued. Now that the scheme has been implemented, the established trading system represents a vehicle for the EU, Germany and Norway to reach their broadly common GCC policy goals. The main conclusion that can be drawn from our comparative case study is that the EU has successfully influenced the GCC-related policy contexts within EU member *and* non-member states alike. As highlighted by our analytical framework, the dynamic interaction of national factors and international forces shaped the national interest, which in turn determined policy instrument choice in Norway and Germany. Not only were international factors like the Kyoto Protocol and global competitiveness significant, but the development of the EU as an increasingly integrated political and economic union has also been an international force that neither European country could afford to ignore.

By applying an analytical framework that adopts constructivist, power-based and interest-based explanatory approaches, we have highlighted a broad range of factors affecting GCC policy instrument choice in Norway and Germany. Furthermore, our study demonstrates how Barkdull and Harris's typology of approaches to environmental foreign policy can be applied empirically, and illustrates how 'the line between one approach and another can blur' (Barkdull and Harris 2002: 68). The cognitive approach was found to be important in explaining the early phases of the development of the idea of ET, while interest group bargaining and the role of elites became more useful in understanding the later stages of policymaking. Overall, we find that a social constructivist approach provides the most convincing explanation of how norms and values shape EFP objectives, at least in the cases examined here, while rational choice-based theories give a better understanding of how interests are pursued once they have been defined. We propose that rather than seeing different theoretical approaches to EFP as competing alternatives, they are best viewed as complementary tools that can help us to paint a more comprehensive picture and allow us to simultaneously utilize concepts such as norms, power and interests within a multi-level context.

NOTES

1. Norway is a member of the European Economic Area (EEA), but it rejected EU membership twice after referenda in 1972 and 1994. (On Germany, see also Chapters 2 and 14, and on Norway, see also Chapter 8.)
2. The EU Burden Sharing Agreement is commonly referred to as the 'EU-bubble'. Within the bubble, the EU is allowed to distribute emissions reductions among member countries, thus providing a flexible and cost-effective way of reaching the EU target of cutting GHG emissions by 8 per cent.
3. See European Commission (2006) for more details about the NAP process.
4. On 20 April 2004 the European Parliament adopted an amendment bringing forward the start date of the 'Linking Directive', which will allowed credits from CDM projects to be used from 2005, rather than from 2008 (as proposed by the European Commission).
5. Strictly speaking, Germany's Kyoto commitment is –8 per cent, as this was the figure negotiated for the EU block as a whole. However, its national target according to the 'EU bubble' system is a 21 per cent reduction in GHG emissions.
6. In line with the EU, Norway implemented its domestic ETS from January 2005.
7. Although the recommendations were adopted, agreement was not unanimous.
8. The Ministry of Finance is responsible for levying the carbon tax.
9. There were earlier academic discussions on ET in the 1980s and 1990s, but it became revitalized as a result of the Kyoto Protocol negotiations, with their focus on ET as a preferred policy instrument.
10. Taking 'early action' into account refers to selecting an early base year so that companies making emissions reductions early on are not penalized.

REFERENCES

Andresen, Steinar and Siri Hals Butenschøn (2001), 'Norwegian climate policy: from pusher to laggard?', *International Environmental Agreements: Politics, Law and Economics* **1**(3): 337–56.

Bachrach, Peter and Morton S Baratz (1962), 'The two faces of power', *American Political Science Review* **56**: 941–52.

Bang, Guri (2004), 'Sources of influence in climate change policymaking: a comparative analysis of Norway, Germany, and the United States', PhD Thesis 43/04, Oslo: Department of Political Science, University of Oslo.

Barkdull, John and Paul G Harris (2002), 'Environmental change and foreign policy: a survey of theory', *Global Environmental Politics* **2**(2) (May): 63–91.

Betsill, Michele (2000), 'The United States and the evolution of international climate change norms', in Paul G Harris (ed.), *Climate Change and American Foreign Policy*, New York: St Martin's Press.

Brende, Børge (2002), *Letter of 4th December 2002, from the Norwegian Minister of Environment to the EU Commissioner for the Environment, Mrs. Margot Wallström*, available at http://odin.dep.no/md/norsk/aktuelt/pressesenter/korrespondanse/022001–110005/index-dok000-b-n-a.html.

Brende, Børge (2004), 'Norge innfører nasjonalt kvotesystem for klimagassutslipp' ['Norway introduces an emissions trading system for greenhouse gases'], *Press Release*, Ministry of the Environment (26 March), available at http://odin.dep.no/md/norsk/tema/forurensning/ arkiv/022051–070113/index-dok000-b-n-a.html.

Butzengeiger, Sonja, Axel Michaelowa and Sven Bode (2003), 'Europe: a pioneer in greenhouse gas emissions trading: history of rule development and major design elements', *Intereconomics* **38**(4): 219–28.

Christiansen, Atle C and Jørgen Wettestad (2003), 'The EU as a frontrunner on greenhouse gas emissions trading: how did it happen and will the EU succeed?', *Climate Policy* **3**(1): 1–106.

Damro, Chad and Pilar Luaces Méndez (2003), 'Emissions trading at Kyoto: from EU resistance to union innovation', *Environmental Politics* **12**(2): 71–94.

European Commission (1998), 'Climate change: towards a post-Kyoto strategy', *COM (1998) 353*.

European Commission (1999), 'Preparing for the implementation of the Kyoto protocol', *COM (1999) 230 final.*

European Commission (2000a), 'Green paper on greenhouse gas emission trading within the EU', *COM (2000) 87*, available at http://europa.eu.int/eur-lex/lex/LexUriServ/site/en/ com/2000/com2000_0087en01.pdf.

European Commission (2000b), 'Communication from the commission on the precautionary principle', *COM (2000) 1*, available at http://europa.eu.int/eur-lex/lex/LexUriServ/site/en/ com/2000/com2000_0001en01.pdf.

European Commission (2001), 'Proposal for a Framework Directive for Greenhouse Gas Emissions Trading Within the European Community.' COM (2001) 581, *Official Journal of the European Communities* **45**(33): C75E.

European Commission (2002), 'Amended Proposal for a Directive of the European Parliament and of the Council Establishing a Scheme for Greenhouse Gas Emission Allowance Trading Within the Community and Amending Council Directive 96/61/EC', *COM (2002) 680*, available at http://europa.eu.int/eur-lex/lex/LexUriServ/site/en/com/2002/com2002_ 0680en01.pdf.

European Commission (2003), 'Proposal for a Directive Amending the Directive Establishing a Scheme for Greenhouse Gas Emission Allowance Trading Within the Community, in Respect of the Kyoto Protocol's Project Mechanisms', COM (2003) 403. *Official Journal of the European Union* C96, **47**: 6.

European Commission (2006), 'Emissions Trading: National Allocation Plans', *Climate Change Homepage*, available at http://europa.eu.int/comm/environment/climat/emission_plans.htm.

Finnemore, Martha (1996), *National Interests in International Society*, Ithaca: Cornell University Press.

German Federal Ministry for the Environment, Natural Conservation and Nuclear Safety (GFME) (2001a), *Position of the Federal Government Regarding the Introduction of EU-wide Trading in Greenhouse Gases*, Berlin: GFME.

German Federal Ministry for the Environment, Natural Conservation and Nuclear Safety (GFME) (2001b), *Third Report by the Government of the Federal Republic of Germany in Accordance with the Framework Convention of the United Nations*, Berlin: GFME.

Hasenclever, Andreas, Peter Mayer and Volker Rittberger (1997), *Theories of International Regimes*. Cambridge, UK: Cambridge University Press.

Hovden, Eivind and Gard Lindseth (2002), 'Norwegian climate policy 1989–2002', in William M. Lafferty, Morten Nordskag and Hilde Annette Aakre, *Realizing Rio in Norway: Evaluative Studies of Sustainable Development*, Oslo: Prosus.

International Energy Agency (IEA) (2002), 'Germany 2002 Review', *Energy Policies of IEA Countries*, Paris: OECD/IEA.

Jansen, Alf-Inge and Oddgeir Osland (1996), 'Norway', in Peter Munk Christiansen, *Governing the Environment: Politics, Policy and Organization in the Nordic Countries*, Årahus: Nordic Council of Ministers.

Jupille, Joseph, James Caporaso and Jeffrey T Checkel (2003), 'Integrating institutions: rationalism, constructivism and the study of the European Union', *Comparative Political Studies* **36**(1/2): 7–40.

Kasa, Sjur (2000), 'Policy networks as barriers to green tax reform: the case of CO_2-taxes in Norway', *Environmental Politics* **9**(4): 104–22.

Lukes, Steven (1974), *Power: A Radical View*, London: Macmillan.

Markussen, Peter and Gert Tinggaard Svendsen (2005), 'Industry lobbying and the political economy of GHG trade in the European Union', *Energy Policy* **33**(2) (January): 245–55.

Norwegian Ministry of the Environment (NME) (1998), 'Norway's follow up of the Kyoto Protocol', *Report No. 29 to the Storting (1997–98)*, Oslo: NME.

Norwegian Ministry of the Environment (NME) (2001), 'Norwegian climate policy', *Report No. 54 to the Storting (2000–2001)*, Oslo: NME.

Norwegian Ministry of the Environment (NME) (2002a), 'Amendment to Report No. 54 to the Storting (2000–2001) Norwegian climate policy', *Report No. 15 to the Storting (2001–2002)*, Oslo: NME.

Norwegian Ministry of the Environment (NME) (2002b), *Third National Communication under the Framework Convention on Climate Change (NC3)*. Oslo: NME.

Norwegian Ministry of Finance (1998), 'Green taxes', *Proposition No. 54 to the Storting (1997–98)*, Oslo: Ministry of Finance.

Norwegian Official Report (NOU) (2000), 'Et kvotesystem for klimagasser: Virkemiddel for å mote Norges utslippsforpliktelse under Kyotoprotokollen' ['A quota system for greenhouse gases: a policy instrument for fulfilling Norway's emission reduction commitments under the Kyoto Protocol'], Oslo: Statens forvaltningstjeneste.

Santarius, Tilman and Hermann Ott (2002), 'Attitudes of German companies regarding the implementation of an emissions trading scheme', *Wuppertal Papers* **122e** (August).

Sprinz, Detlef (2001), 'Germany: European leadership, active climate policy and wall-fall profits', *German Foreign Policy in Dialogue* **2**(6) (December): 13–15.

Sprinz, Detlef and Tapani Vaahtoranta (1994), 'The interest-based explanation of international environmental policy', *International Organization* **48**(1): 77–105.

Sprinz, Detlef F and Martin Weiss (2001), 'Domestic politics and global climate policy', in Urs Luterbacher and Detlef F Sprinz (eds), *International Relations and Global Climate Change*, Cambridge, MA: MIT Press.

Statistics Norway (SSB) (2003), *Naturressurser og Miljø 2002* [*Natural Resources and the Environment 2002*]. Oslo: Statistics Norway.

Torvanger, Asbjørn (2004), 'Brende kopierer EUs kvotesystem' ['Brende Copies the EU Quota System'], *Center for International Climate and Environmental Research – Oslo* (*CICERO*) (29 March), available at www.cicero.uio.no/kvoter/brende.html.

Underdal, Arild (1998), 'Explaining compliance and defection: three models', *European Journal of International Relations* **4**(1): 5–30.

United Nations Framework Convention on Climate Change (UNFCCC) (2004), *Greenhouse Gases Database*, available at http://ghg.unfccc.int/default.htf, Bonn, Germany.

Vevatne, Jonas, Henrik Linhjem, Gunnar S Eskeland, Torleif Haugland and Anne Therese Gullberg (2004), 'Klimapolitiske virkemidler og norsk industris konkurranseevne' ['Climate policy instruments and the competitiveness of Norwegian industry'], *CICERO Report 2004:08*, Oslo: CICERO and ECON. 71pp, available at www.cicero.uio.no/media/ 3278.pdf

Wendt, Alexander (1999), *Social Theory of International Politics*, Cambridge, UK: Cambridge University Press.

Würzel, Rüdiger, Andrew Jordan, Anthony R Zito and Lars Brückner (2003), 'From high regulatory state to social and ecological market economy? "New" environmental policy instruments in Germany', *Environmental Politics* **12**(1) (Spring): 115–36.

13. Climate change policy and the enlargement of the EU

Leonardo Massai

INTRODUCTION

Since the early 1990s, the European Union (EU) has assumed a leading role in the international community's efforts to mitigate global climate change (GCC) by curbing greenhouse gas (GHG) emissions. The most important enlargement in the history of the EU occurred on 1 May 2004 when 10 Central and Eastern European Countries (CEECs) joined the organization. These new member states are the Czech Republic, Cyprus, Estonia, Hungary, Latvia, Lithuania, Malta, Poland (see Chapter 5), Romania, Slovakia and Slovenia.[1] The EU's environmental foreign policy (EFP), along with many other EU policies, has been considerably affected by this development. The consequences of the enlargement for European EFP on global climate change are investigated in this chapter. The chapter begins with an analysis of the EU's institutional structure for enlargement and their implications for European GCC policy. This is followed by a case study of the implications of enlargement for the EU's Emission Allowance Trading Scheme (EATD). The chapter also shows how environmental and, above all, economic interests play a fundamental role in the definition of the EU's GCC policy.[2] The chapter builds on the systemic approach to EFP (see Barkdull and Harris 2002) to the extent that the international climate regime and requirements in the United Nations (UN) Framework Convention on Climate Change (FCCC) and in the Kyoto Protocol (which entered into force on 16 February 2005) have contributed to the development of an advanced European GCC strategy and policy.

The procedure for the accession of the new EU states required that they fulfill many conditions for membership, notably a wide range of reforms through implementation of the so-called *Acquis Communautaire*, which consists of the EC legislation together with principles, policies and objectives at the EU level. Within this process, the EU environmental *Acquis* has played an important role, with the eastern enlargement described as a major challenge for EU environmental policy. The environmental protection standards are not uniform within the new countries; they fluctuate according to different national conditions and structures. However, the EU 'green' requirements have substantially contributed to environmental improvement throughout the region. The new member states, which have agreed to adopt national policies and to take corresponding measures in order to combat GCC within the framework of the FCCC and the Kyoto Protocol, are in a different situation compared to the old member states. This is mainly due to past environmental liabilities and to the different level of environmental protection in Central and Eastern Europe prior to the EU accession.

EU ENLARGEMENT

The legal basis of EU enlargement lies in Article 49 of the EU Treaty, which states that each applicant must respect the principles of 'liberty, democracy, respect for human rights and fundamental freedoms, and the rule of law' (wording from Article 6 of the Treaty of Amsterdam). The process behind the 2004 enlargement started in 1993 when the European Council of Copenhagen acknowledged the possibility of expanding the EU eastward. The political criteria for accession were designed in Copenhagen, and they still have a fundamental role. The criteria are the respect of democracy, human rights and the rule of law, and the capacity to collaborate with EU institutions for the achievement of community objectives. From the economic point of view, acceding countries were required to establish a market economy in order to comply with EC competition law. All candidate countries signed association agreements to establish closer collaboration and free trade with the EU, and to prepare the ground for full membership (see Steiner and Woods 2001: 8).

The accession procedure requires acceding states to integrate Community law and competences into their national legal systems and institutions. This approximation of legislation is based on the *Acquis*. The implementation of the *Acquis* had two different aspects: first, the incorporation of all the Community provisions in the various national legal systems; secondly, the full introduction of these provisions into practice. Regarding the second aspect, the European Commission has seemed very keen to make sure that regulations are not only incorporated into the national juridical framework, but also, and above all, that they can be implemented correctly and effectively.

Each acceding country discussed individually with the European Commission 30 negotiating chapters, which set the timeframe for the implementation of Community legislation. In a few cases, concessions on transitional periods were given to countries whose particular circumstances did not allow compliance with all EU requirements by the accession date. The negotiations were concluded in Copenhagen in December 2002 (EU Council 2002a: 1–3), and the European Commission officially certified the successful conclusion of the EU enlargement process. After the approval of the Council and the Parliament, taking into account the final opinion on the outcome of the negotiating process presented by the Commission, the Treaties of Accession were signed by existing EU member states and the acceding states in Athens on 16 April 2003, and the official date for the enlargement was fixed at the 1 May 2004. The treaties were then submitted for ratification by the acceding states according to their different constitutional requirements. Ratification of the existing member states' parliaments was also required. The Treaty of Accession is divided into two parts: the treaty itself and the Act of Accession, which defines the conditions of accession and the adjustments to the treaties upon which the EU is founded. These include technical adaptations of secondary legislation necessary for accession, and a clause that would avoid rejection of the accession in case the candidate country failed to ratify the treaty.[3]

After the signature of the Treaties of Accession, the new member states were granted EU 'observer status', which lasted until they became full members in 2004. The Interim Period was the time between the so-called 'cut-off date' (1 November 2002) and the date of accession. The EU legislation adopted during this phase that was neither covered in the accession negotiations nor contained in the Treaties of Accession has turned out to be a delicate issue. From the EC law perspective, CEECs were obliged to implement such provisions by 1 May 2004. More practically, in some cases and according to particular circumstances, these countries could have negotiated transitional arrangements for singular provisions. This is particularly true for the EU's GCC policy and associated provisions adopted during the interim period (see below).

THE ENVIRONMENTAL *ACQUIS* AND CLIMATE CHANGE

Negotiations and discussions on the environmental chapter of the *Acquis Communautaire* were among the most complicated. The process of compliance with the *Acquis* is very long and it requires an adequate legal and administrative system in the acceding countries that can insure the effective implementation and enforcement of the *Acquis*' provisions. National Programs for the Adoption of the *Acquis* (NPAA) were established by the new member states in order to comply with European environmental laws. The process included three phases: transposition of European law into national systems, implementation, and enforcement. Consequently, acceding countries were met with legislative, institutional and financial challenges.

European environmental provisions are a very important part of the *Acquis*, with about 300 environmental laws that represent one of the strictest schemes of environmental legislation in the world. Environmental rules, as well as any other rule within the *Acquis*, apply to new member states in the same way as to the existing members (European Commission 2001: 8–16). The result of this harmonization is a new enlarged environmental policy with stronger jurisdiction and standards, aiming at the reduction of air and water pollution, the establishment of a more efficient waste management, and improved public health (Ecotec 2001: 6–17). However, the *Acquis* chapter on the environment turned out to be one of the most controversial, and negotiations were very long and complicated. This is mainly due to the complexity of European environmental law, as well as to past environmental liabilities and different degrees of environmental protection in CEECs compared to the EU. The CEECs were required to pass new legislation that had high compliance costs: large investments and changes of institutional structure were indeed required, as were capacity building and additional staff resources.[4] With the aim to provide financial and technical aid to these countries, in 1990 the EC initiated the Phare (*Pologne, Hongrie Assistance à la Reconstruction Economique*) program (see Chapter 5). In 2000, assistance from the program was reformed under the rubric of 'pre-accession funds' and was directed at helping new members to implement EU regional policy (i.e., structural and cohesion funds) (McGuinn 2003: 18).

Within the environmental chapter, and for each environmental sector, the NPAA provided an analysis of the current situation concerning transposition, as well as implementation and enforcement, and it indicated short-term and medium-term priorities in line with the Accession Partnership, an act including the country's NPAA and a plan of the EU funds available for the accession. In several communications and reports on accession strategies, the European Commission recognized that full compliance with the environmental *Acquis* would probably require long-term perspectives, and it pointed to the importance of the integration of the environmental approximation activities in other policy areas. A set of environmental priorities to be considered within the process of approximation was also specified by the Commission, including detailed assessment of the environmental situation in each country, legislative and administrative gaps to be filled based on a detailed comparison of existing national and European environmental legislation, and investments for their implementation (e.g., large combustion plants). Thus, it is important to remember that before accession, the CEECs were free to design, together with the

EU, their own strategy to implement the environmental *Acquis* through the establishment of the NPAA. As for the GCC issue in particular, this could have represented a good chance to meet the EU accession environmental criteria through the integration of different 'climate friendly' measures included in different chapters of the *Acquis* itself. Although other priorities were at the forefront, the enlargement could have had a more positive impact on GCC policy in the region.

As pointed out above, in some cases candidate countries requested and received permission from the Commission to apply 'transitional' measures, that is, to delay the complete transferal of EU provisions into their national systems. These permissions were granted in those cases where a substantial adaptation of infrastructure in the acceding state was required. The number of transitional periods negotiated within the environmental chapter goes from two for the Czech Republic to nine for Poland. They covered important provisions, such as Directive 94/62/EC on recovery and recycling of packaging waste, Directive 91/271/EEC on urban waste water treatment, and Directive 96/62/EC on integrated pollution and prevention control (IPPC). Difficulties in meeting the EU air quality requirements for candidate countries were related to problems of emissions from stationary sources, such as industrial plants and local district heating installations, as well as emissions from large combustion plants. Fuel quality standards, as well as monitoring and data collection, also represented a tough challenge. Transitional arrangements for measures and sectors related to GCC have been granted with regard to emissions of volatile organic compounds from storage and distribution of petrol (Estonia, Latvia, Lithuania, Poland and Slovakia), air pollution from large combustion plants (Hungary and Slovakia), integrated pollution prevention control (Poland, Slovakia and Slovenia) and sulfur content of liquid fuels (Poland).

A separate chapter on GCC policy was not established during the accession negotiations because the only two measures of European law directly aimed at combating climate change were included in the air quality section. However, the two pieces of legislation officially included in the *Acquis* that explicitly covered climate change were Council Decision 99/296/EC on a monitoring mechanism for CO_2 and other GHGs, establishing a reporting and monitoring mechanism for fulfillment of EC obligations under the FCCC, and Council Decision 2002/358/EC on EC ratification of the Kyoto Protocol (European Commission 1997: 129 and 2001: 69). The monitoring decision requires member states to submit an annual assessment of anthropogenic CO_2 emissions and national inventory data on the other GHG emissions (Article 3.2 of Council Decision 99/296/EC), as well as to establish and implement a national GCC plan that includes envisaged policies and measures as well as estimated emissions. In early 2004, the

European Parliament and the Council adopted Decision 280/2004/EC concerning a new EU GHG monitoring mechanism that would enhance transparency and provide member states with a single and clearer legislative act. The document included new monitoring and reporting guidelines, which covered flexible mechanisms (i.e., International Emissions Trading (IET), Joint Implementation (JI) and the Clean Development Mechanism (CDM)), and the establishment of registries as required under the Kyoto Protocol and the Marrakech Accords.

As noted above, the implementation of measures on air, water, industrial pollution, waste and energy are going to contribute to the reduction of GHG emissions and consequently the CEECs must take into account such legislation for the preparation of national GCC plans. This is the case, for instance, of measures on the liberalization of electricity and gas markets. Additionally, there are pieces of horizontal legislation, such as the environmental impact assessment and the environmental information directives, which have an influence on climate policy and especially on JI projects when considering project cycles, public participation and the process of selecting projects.[5]

As indicated above, the *Acquis* included in the negotiation chapters covered only legislation adopted up to November 2002, thus leaving some *legal uncertainty* regarding the implementation of provisions adopted during the Interim Period. Climate policy is still a relatively new policy area and legislation on this field continues to evolve, with new provisions always in the pipeline. Legislation at the European level adopted after November 2002, which had to be further approximated by the CEECs, include, among others, the previously mentioned new GHG monitoring system, Directive 2003/87/EC establishing the Emission Allowance Trading Scheme, Directive 2004/101/EC linking emission units of JI and CDM project to the EATD, and measures related to the European Climate Change Program (ECCP).[6]

With these considerations in mind, the next section focuses on innovative GCC policy provisions in Europe, focusing on the case of the EATD and correlations among legislative processes, as well as the actors involved and EU enlargement.

THE CEECS' INVOLVEMENT IN THE EU CLIMATE POLICY PROCESS

The participation of CEECs' officials in the preparation of the EU's GCC policies has been a crucial issue also because, once these countries joined the EU, they have had a fundamental role in designing the organization's new commitments within the international GCC regime. While most of the EU legislation on GCC policy was not covered by the accession negotiations and agreements (see above), the CEECs were involved in negotiations concerning the establishment of the EU instruments on GCC. They participated in discussions on establishing the GHG trading scheme, and they were also active in some of the working groups set up within the framework of the ECCP. The ECCP identifies and develops elements of an EU strategy to implement the Kyoto Protocol and includes a range of additional EU-level policies and measures to curb GHG emissions. Climate change measures often involve different areas and sectors, and the way the ECCP actually works is an excellent example of the integration of environmental protection – mitigation of GCC – with other areas of legislation. The provisions included in the ECCP ranged from cross-cutting issues, such as a directive linking JI and CDM projects to the EATD and the review of the monitoring mechanisms, to energy issues like a directive for the promotion of combined heat and power (CHP) and issues related to transport and industry. Working groups were established in order to define a common approach and to co-ordinate different policies and measures at the EU level through the involvement of different European Commission DGs and sectors, as well as existing and new member states' representatives, stakeholders, business sectors, as well as non-governmental organizations. Measures in the ECCP considering bio-fuels, CHP, energy efficiency and renewable energies were often of particular interest to the acceding states.

The EU created an 'Interim Committee' in December 2002. The committee, which was composed of representatives of the EU, the European Commission and all 10 new member states, allowed the acceding states to comment on EU draft legislation concerning the signature of the Treaties of Accession.[7] The level of involvement of CEECs representatives in the preparation of EU legislation increased after the signature of the Treaties of Accession in April 2003, after which the CEECs could make observations and participate in the meetings (without the right to vote) until actual accession a year later. (The 10 incoming member states participated in the 2004 European Parliament elections as members.) Commissioners from the CEECs joined the Commission from May 2004, and they participated in the Intergovernmental Conference, which designed the EU constitutional reforms. After the signing of the Treaties of Accession, acceding states were automatically informed of proposals, recommendations, communications and other initiatives within the Council of Ministers or its working parties, and ministers from the new member states took part as observers in their first formal council meeting on 1 May 2003.

The CEECs' representatives discussed the structure of the EATD system with the EU Council of Ministers. The EATD was on the agenda at the second meeting of the Interim Committee (March 2003). Some of the CEECs tried unsuccessfully to reopen discussions on the system. In addition, the European Commission collaborated with the Center for Clean Air Policy (CCAP) to provide CEECs a series of ministerial-level workshops on emissions trading. This initiative was designed for helping the EU acceding countries to prepare the implementation of the EATD, through trading simulations and capacity needs assessment reports.

THE EU AND CEECS IN THE INTERNATIONAL CLIMATE CHANGE REGIME

With regard to the Kyoto Protocol first commitment period (2008–12), the EC and its member states are jointly committed to fulfilling their objectives under the so-called Burden Sharing Agreement (BSA),[8] meaning that the EU's overall cut in GHGs has been distributed among the 15 member states at the time of the agreement. This joint commitment is regulated by Article 4 of the Kyoto Protocol, according to which the EC and its member states can fulfill their obligations under Article 3, paragraph 1, in a mutual way. The EC's legal position within the FCCC and the Kyoto Protocol will be affected by the inclusion of the 10 new member states, but only in regard to future commitment periods (Michaelowa and Betz 2001: 270). Discussions on the second commitment period began in May 2005 with a seminar of governmental experts at the UN level. The COP/MOP1 of Montreal in December 2005 decided to officially start the negotiations for the post-2012 phase. The EC will have to negotiate a new common agreement, along with future lines of the European 'bubble', for GHG emissions-reduction targets.

In order to negotiate and agree upon a common future GHG target at the EC level, the EU's old and new member states will have to be in the same international legal position in the context of the GCC agreements. Malta and Cyprus are currently listed as 'non-Annex I Parties' in the FCCC and therefore have different obligations compared to the other EU member states. Another issue of difference between new and old member states in terms of international climate regime obligations regards compliance with monitoring and reporting requirements under the FCCC and the Kyoto Protocol. These require the submission of national communications, annual GHG inventories and eligibility criteria needed to participate in the protocol's flexible mechanisms. The need for capacity building in CEECs to improve staff resources and infrastructure is still a priority. Nevertheless, the 25 EU member states have started to assume the same obligations within the international climate regime; this was one of the objectives upon which the EU accession requirements were negotiated.

The BSA and the influence of such obligations on the shape of the EU's foreign policy on GCC reflect the systemic theories described by Barkdull and Harris (2002: 63–91). The influence of the international system on European GCC policy and on European foreign climate policy towards the CEECs is manifested in this case. Article 4 of the Kyoto Protocol regulates the distribution of powers and obligations for the European Community in the context of international GCC regime; it guided the distribution of GHG emissions reduction commitments at the European level and European policies toward the new member states in terms of GCC mitigation policies and measures. However, the case of European GCC policy suggests a potential weakness of the systemic approach. The rigidity of the influence of the international system towards European climate policy is stressed in Article 4.4 of the Kyoto Protocol. This obliges the European Community to quantify targets for participating countries (Article 4.1) and to maintain this obligation until the end of the first commitment period (Article 4.3), even if there is a change in the composition of the regional organization (Article 4.4) – and this was exactly the case following the 2004 EU enlargement. Furthermore, Articles 4.5 and 4.6 regulate the responsibility of individual members in terms of compliance with the obligations agreed.

In terms of the GCC foreign policies of the CEECs, following the sixth conference of the parties (COP6) in the Hague (see Chapter 1), these countries set up their own negotiating group, namely 'Central Group 11' (CG11), composed of the 10 acceding states (with Malta and Cyprus as observers) together with Croatia. Supposedly based on the principles of consensus and the division of tasks, CG11 facilitated the formation of a common strategy within the international GCC negotiations and helped assess the advantages and disadvantages of Kyoto Protocol implementation in these countries. The CG11 collaborated with the EC negotiating group in order to produce joint submissions or statements at the COPs and to develop co-ordinated positions in the international GCC negotiations. The group was dismissed in 2003 following inclusion of the CEECs into the EU negotiating group.

CEECS AND EUROPEAN CLIMATE POLICY: THE CASE OF EMISSIONS TRADING

Annex I (developed country) parties to the Kyoto Protocol have the possibility to acquire emissions units from other Annex I parties, thereby allowing them to meet their emissions targets through the implementation of International Emissions Trading. The emissions units may be in the form of assigned-amount units, removal units, emission-reduction units (ERUs, in the context of JI) and certified emission reductions (CERs, in the context of the CDM). As a precondition, Article 17 of the protocol states that any trading must be 'supplemental to domestic actions for the purpose of meeting' reduction commitments. The EU has replied to this requirement by establishing the EATD, with which all EU member states shall comply.[9] The adoption of the EATD is an example of the extent to which the international regime has influenced Community policy on GCC (or vice-versa; see, for example, Chapter 14). This validates the systemic approach to EFP described in Barkdull and Harris (2002). The relevance of the rules established under the Kyoto Protocol and the solutions (e.g., flexible mechanisms) proposed at the international level represent one of the keys to understanding European GCC policy and related legislation.

The EATD is a domestic, compulsory and entity-based system compatible with states' international commitments under the Kyoto Protocol. Installations included in the system cover energy activities, production and processing of ferrous metals, mineral industries and other activities. Each installation shall hold a permit to operate under the EATD. Installations with a CO_2 emissions levels below their assigned amounts are allowed to sell emission rights to other installations in danger of exceeding their quotas. The system covers large plants in key sectors, such as power generation, heating industries, oil refineries, coke ovens, steel works and cement, glass, tile, ceramics and pulp and paper factories. Installations not in compliance with the requirements of the EATD are sanctioned through financial penalties. The first trading period started in 2005, 3 years before the start of IET. During the 'learning by doing' phase (2005–2008), the EU member states will be able to test the functioning of the market itself in order to be ready for the implementation of the international system in 2008 (Lefevere 2002: 41; Zapfel and Vainio 2002: 4).

The European Parliament amended the original Commission emissions trading (ET) proposal in several areas. Generally, emissions allowances in a trading system can be allocated in three different ways: auctioning, when national authorities sell permits to parties; grandfathering, when permits are allocated free of charge according to past emissions levels or to fuel consumption levels; and based on output, much like grandfathering but based on current or very recent activity (emissions are considered per unit of output). The Council common position of December 2002 changed several of the Parliament's preferences (see EU Council 2002b: 6–8). The possibility to opt out of the scheme, allowed only during the first phase of trading and depending on the Commission's approval, was granted to activities and sectors, not only to installations as requested by the Parliament.[10] Exempted installations, as well as installations covered by the scheme, have to reduce emissions and respond to verification requirements. An adverse outcome produced by this decision may be the opting out of entire sectors, which could have a negative influence on market prices and liquidity. A German proposal on a pooling system for grouping together installations of the industrial sector in order to distribute allowances to the different sectors instead of allocating them to single installations, was in part accepted. Concerning the method of allocation of permits, mainly distributed for free, the Council reduced the possibility for allocating allowances by auctioning in the second period to 10 per cent instead of the total amount as proposed by the Parliament,[11] while only 5 per cent of permits can be auctioned up to 2008 (for the first phase, the Parliament had proposed allocating 70 per cent of allowances for free and 30 per cent by auctioning).

The allocation of allowances is a key issue in the trading system because it affects several different interests and areas. In this regard, the European Commission released an official 'non-paper' clarifying the main open questions (European Commission 2003: 4–14). The inclusion in the scheme of sectors such as aluminum, chemicals and other gases was postponed to 2008 onwards, instead of starting in 2005 as preferred by the European Parliament (member states may apply for exceptions to the Commission, which will maintain a right to veto). The option to include CDM and JI in the scheme was confirmed (firms could buy credits from JI and CDM projects to make up allowance shortfalls), while the Parliament opposed this possibility in the first trading phase. In addition, as noted above, the Directive establishing a link between credits generated through JI/CDM to the EATD was adopted in 2004, allowing for direct trading of CERs and ERUs within Europe from 2005 and 2008 respectively.

In sum, the key issues, which characterized the final deal reached in the Parliament, were:

- voluntary allocation of permits through grandfathering, with the possibility to opt for auctioning (5 per cent in the first commitment period, 10 per cent in the second);
- additional activities and gases to be included in the system, either on the initiative of the Commission (by the end of 2004) or the member states (in the second period);
- only installations could be opted out of the system, according to the approval of the Commission;
- a pooling system based on groups of installations of the same sector;
- no links between different credits (JI/CDM to EATD).

EMISSIONS TRADING AMONG CEECS

The EU accession requirements obliged CEECs to implement the European Allowance Trading Directive by the date of accession. All CEECs (with the possible exception of Slovenia) are very probably far below their Kyoto targets and thus will not use emissions trading as a measure to comply with their international obligations. Rather, for such states emissions trading is an instrument to gain financial benefits for both public and private sectors. This rationale is quite important because the type of emissions cap is to be decided by the competent authority in each state on the basis of its perspective and overall goals on GCC, including choosing between counting emissions according to current or historic data, or considering probable future emissions. The amount of the emissions cap of the system, on the base of which the allocation of allowances is defined, represents a key decision with important political consequences for the participants. Considering the base year agreed within the FCCC (in most cases diverging from the usual base year of 1990) and the principle of 'common but differentiated responsibilities' (Article 3.1 of the FCCC), CEECs have a GHG emissions *surplus* that puts them in the position of being net sellers of allowances within the scheme. Despite the fact that the exact amount of such surplus is uncertain (the EU accession requirements in the environmental field could in fact be generating further reductions in GHG emissions thanks to energy efficiency rules and other EC measures), it is beyond doubt that this emissions benefit will be enough to cover emissions allowances demanded by the EU.

The question of the emissions surplus is very important when considering the allocation of permits. The definition of the CEECs' national allocation plans and the distribution of extra emissions units have to comply with EU competition law and state aid rules in order to avoid distortion in the Community internal market. Both national governments and private companies could see economic benefits through the selling of units, and these economic gains could be used as a way to attract stakeholders into the mechanisms (Jilkova and Chmelik 2001: 13). A competition issue could arise when permits are allocated for free. The allocation must be issued according to a business-as-usual emissions level and financial aid to industries must be limited as indicated in the EU scheme or when offering incentives to parties to participate in a voluntary ET scheme (OECD, IETA and IEA 2002a).[12] In any case, the European Commission leads the process on the allocation of permits: it has the power to accept or refuse the national allocation plans and it is obliged to provide official guidelines for the preparation of them (see European Commission 2003: 4–14).

The regulatory uncertainty that surrounded the EATD until the very last moment may have caused a few problems for the CEECs, which had to establish, in a relatively short timeframe, an adequate trading infrastructure and improve their institutional capacity as well as data collection and monitoring, reporting and verification facilities. On data verification and monitoring, for instance, Decision 280/2004/EC of the Council and the European Parliament establishing a new monitoring mechanism of Community GHG emissions and implementation of the Kyoto Protocol proposed new monitoring and reporting requirements covering areas such as the flexible mechanisms, including registries set out under the protocol. That decision forced the CEECs not only to set up national programs to implement GCC policies, but also demanded that they set up adequate monitoring mechanisms that would enable them to provide detailed GHG inventories. A similar monitoring system is already indicated by the IPPC Directive requiring member states to report annually on the emissions.[13]

Before adoption of the EATD, Slovakia, the Czech Republic and Poland were very active in the process of moving their national systems towards implementation of an emissions trading scheme (ETS), while the rest of CEECs were quite inactive in this sense. The domestic trading schemes studied in these three countries were designed to mimic as closely as possible the EU system (OECD, IETA and IEA 2002b: 5–18). Legal issues, political concerns, capacity building and technical problems represented the greatest challenges for these countries, which have to design their own rules for an ETS.

National authorities had to decide between a voluntary or a mandatory scheme, the method of allocating permits, and the range of entities and gases to be covered by the scheme. They had to set up national inventory and registry systems, establish a cap for the system, and define the type of unit to be traded. Finally, they had to establish adequate rules and guidelines for monitoring, reporting and verification, as well as penalties in the case of non-compliance.

The CEECs had the opportunity to ask for and negotiate possible exemptions or transition periods with regard to the implementation of the EATD, but this option was not considered. What could have forced some CEECs to ask for such special arrangements was the lack of capacity at the governmental as well as the industry level, even if only Lithuania and Latvia seemed to be candidates to opt for a transitional period before the accession (PointCarbon 2003: 2). At the second meeting of the Interim Committee, the CEECs asked for 'further clarification' and time to examine the directive, stressing the fact that the economic and environmental policy reasons for the implementation of such a procedure, especially in the first phase, are still lacking within the region. Hungary asked for further discussions on the matter and Latvia and Malta asked for further information on the way the proposed directive would be implemented. Given that economic changes in most accession countries after 1990 reduced their GHG emissions well below the targets agreed at Kyoto in 1997, a further reduction through ET would leave these countries in a difficult situation ahead of the next round of climate negotiations for the post-Kyoto period. Hungarian representatives claimed that emissions growth should have been allowed for the acceding countries, as it was for some other EU member states. It is important to stress that stakeholders in these countries were keen to see CEECs join the EU scheme from the very beginning, considering that this would be a chance to extend the size and liquidity of the single market.

CONCLUSION

The integration of the former candidate states into the EU is now complete, at least from the procedural and legislative point of view. This is true for all Community policies, including environmental policy. The considerable bulk of legislation that CEECs had to transpose into national legislation has contributed to improving infrastructure and to enhancing the quality of the environmental protection frameworks in these countries. With regard to the definition of EU policy on climate change, accession of the ten new member states did not greatly influence the European strategy on the reduction of GHG emissions in accordance with the EU and its member states' international obligations. Reduction targets under the Kyoto Protocol's first commitment period had already been negotiated and agreed within the framework of the Burden Sharing Agreement, and CEECs were not included in the EU GHG emissions bubble. This rigidity, both at the international level (ie Article 4 of the Kyoto Protocol) and at the European level (the BSA), was a 'systemic' constraint. The recognition of GCC at the global level has also functioned as a driving force for several states, as well as the European Community, towards the establishment of an advanced system of environmental protection leading to the reduction of GHGs.

The level of participation of the CEECs in the shaping of European GCC policy undoubtedly increased in the last years before enlargement, with their representatives raising concerns at different institutional meetings within the Council, the Parliament and the Commission. As to the acceding states' own GCC policies, the case of the European EATD shows the range of substantial difficulties these countries faced in order to implement European environmental legislation. Due to the complexity of the issue, either capacity building and stakeholder involvement is to be enhanced in the CEECs or institutional and financial support from the Commission should be increased.

This case study of the most recent phase of EU enlargement suggests that systemic approaches to environment and foreign policy may be the most suitable perspective from which to analyze European foreign policy on global climate change. The relationship between the international and regional systems (the latter being the EU), and the relationship between the EU and the national systems of member states (and member states-to-be), play a central role in defining roles and identities, as well as policies and measures at the local level, in terms of addressing the problem of global climate change.

NOTES

1. Romania and Bulgaria joined the EU in 2007. In the context of this chapter, the terms CEECs, acceding countries or new member states refers exclusively to Bulgaria, Czech Republic, Estonia, Hungary, Latvia, Lithuania, Poland, Romania, Slovakia, and Slovenia unless otherwise specified. Malta and Cyprus, though also now in the EU, have not been considered.
2. The analysis here may have useful lessons for future EU enlargement.
3. 'The conditions of admission and the adjustments to the Treaties on which the Union is founded, entailed by such admission, are set out in the Act annexed to this Treaty. The provisions of that Act shall form an integral part of this Treaty' (Treaty of Accession: Article 1.2).
4. In order to comply with the EU *Acquis*, CEECs were expected to spend €80–100 billion. Thus they would have to spend an average of 2–3 per cent of Gross Domestic Product (GDP) annually on environmental protection during the accession process.
5. Council Directive 97/11/EC of 3 March 1997 amending Directive 85/337/EEC of 27 June 1985 on the assessment of the effects of certain public and private projects on the environment and Council Directive 90/313/EEC of 7 June 1990 on the freedom of access to information on the environment.
6. Communication from the Commissions to the Council and the European Parliament on EU policies and measures to reduce GHG emissions: 'Towards a European Climate Change Programme', Brussels, COM 2000: 88; Communication from the Commission on the implementation of the first phase of the ECCP, Brussels, COM 2001: 580; 'Second ECCP Progress Report: Can we meet our Kyoto targets?', European Commission, April 2003.
7. The Council's special interim committee met for the first time on 27 February 2003 for consultations on the Common Position on the draft EU directive on packaging and packaging waste.
8. See EU Document 9702/98 of 19 June 1998 of the Council of the European Union reflecting the outcome of proceedings of the Environment Council of 16–17 June 1998. See also Annex I to the Council Decision of 25 April 2002.
9. See the Directive 2003/87/EC of the European Parliament and of the Council of 13 October 2003 establishing a scheme for GHG emission allowance trading within the Community and amending Council Directive 96/61/EC.
10. The opt-out clause was a key demand of the UK, which already had a domestic voluntary ETS in place
11. The original proposal from the Commission indicated that all allowances should have been allocated for free in the first phase; no indications were made concerning the second phase.

12. Annex III, point 5 of the EATD says the following: 'The plan shall not discriminate between companies or sectors in such a way as to unduly favor certain undertakings or activities, nor shall any installation be allocated more allowances than it is likely to need'.
13. See Council Directive 96/61/EC concerning integrated pollution prevention and control, OJ L 257, 10 October 1996. A decision on the implementation of the European Pollutant Emission Register (2000/479/EC) indicates how to proceed with reporting.

REFERENCES

Barkdull, John, and Paul G Harris (2002), 'Environmental change and foreign policy: a survey of theory', *Global Environmental Politics* **2**(2): 63–91.

Ecotec in association with EFTEC, IEEP, Metroeconomica, TME & Candidate Country Experts (2001), *The Benefits of Compliance with the Environmental Acquis for the Candidate Countries*, Brussels: Ecotec.

European Commission (COM) (1997), *Guide to Approximation of EU Environmental Legislation*, SEC(97) 1608, Brussels: European Commission.

European Commission (COM) (2001), *Handbook for Implementation of EU Environmental Legislation*. Brussels: European Commission.

European Commission (COM) (2003), *The EU Emissions Trading Scheme: How to Develop a National Allocation Plan*, Brussels: European Commission.

European Union Council (EU Council) (2002a) *Presidency Conclusions (12–13 December)*, Council Minutes (15917/02), Copenhagen: European Union Council, available at http://ue.eu.int/ ueDocs/cms_Data/docs/ pressData/en/ec/73842.pdf.

European Union Council (EU Council) (2002b), '2473rd EU Council Meeting Environment (9 December)', Council Minutes, 15101/02 (Presse 379): 6–8, Brussels: European Union Council, available at http://ue.eu.int/ueDocs/ cms_Data/docs/pressData/en/envir/73567.pdf.

Jilkova, Jirina and Thomas Chmelik (2001), *Domestic Emissions Trading System in the Czech Republic: Options for An Implementation Framework*, Prague: OECD and IEA.

Lefevere, Jürgen (2002), *Greenhouse Gas Emission Allowance Trading in the EU: A Background*, London: FIELD.

McGuinn, Jennifer (2003), 'EU funds may favour expensive solutions', *The Bulletin* **11**(4): 18–19.

Michaelowa, Axel and Regina Betz (2001), 'Implications of EU enlargement on the EU greenhouse gas, bubble and internal burden sharing', *International Environmental Agreements* **2**(1): 267–79.

Organization for Economic Cooperation and Development, International Emissions Trading Association and International Energy Agency (OECD/IETA/IEA) (2002a), *National Systems for Flexible Mechanisms: Implementation Issues in Countries with Economies in Transition*, Sentzendre, Hungary: OECD/IETA/ IEA.

Organization for Economic Cooperation and Development, International Emissions Trading Association and International Energy Agency (OECD/IETA/IEA) (2002b), *National Systems for Flexible Mechanisms: Implementation Issues in Countries with Economies in Transition. Working Group on Domestic Emissions Trading Systems*, Sentzendre, Hungary: OECD/IETA/IEA.

PointCarbon (2003), *Carbon Market Europe* (14 March), available at www.pointcarbon.com.

Steiner, Josephine and Lorna Woods (2001), *Textbook on EC Law*, London: Blackbone Press.

Zapfel, Peter and Matti Vainio (2002), *Pathway to European Greenhouse Gas Emissions Trading History and Misconceptions*, Working paper No. 85, Venice: Fondazione Eni Enrico Mattei.

14. The reflexivity of ideas in climate change policy: German, European and international politics

Lyn Jaggard

INTRODUCTION

Environmental issues have become integral to the national and global political arenas. Many of these environmental issues are transnational. The national and the international are often interlinked in environmental politics, and this is especially the case with global climate change (GCC). Climate change, with its anthropogenic causes and varying transnational effects, also involves injustices (see Chapter 15). The benefits of industrialization have accrued to those countries and people that are least vulnerable to GCC but most able to afford to implement adaptation policies. Conversely those areas of the world that are worst affected by GCC are those that tend not to have received the benefits of industrialization. The well-being of future generations will be fundamentally affected by the actions that are taken now. Therefore, questions of intergenerational justice are also relevant. As GCC is a phenomenon that is inherently global, it requires international negotiations and institutions to provide governance. To enable agreement on what form governance should take to carry out the necessary remedies requires complex discussions among states as well as scientists, politicians and citizens from around the world.

This chapter examines the evolution of GCC politics in Germany and its relationship to international GCC politics and policies (see also Chapter 2). Events at both national and global levels impact upon one another; there is a reflexive relationship. The chapter begins with a discussion of theory in this context, paying special attention to the role that certain ideas play in GCC-related politics and policy making. I then examine German and international environmental politics before showing how ideas embedded in them are reflected in foreign policy on global GCC.

CONSENSUS, DISCOURSE AND IDEAS IN THE POLICY PROCESS

The German political system is noted for its consensual nature. To achieve consensus, dialogue between the various parties involved in policy making is required. This feature of the political system, together with the imperatives of the international management of GCC, suggests that discourse theory – discourse ethics as propounded by Jürgen Habermas – may be helpful in explaining how and why Germany, and arguably other European states, responds to GCC the way they do.

According to Cavalier, Mellon and Ess, 'Habermas's central point [is that a] communicative action defines a rationality capable, through discourse, of arriving at universal norms' (Cavalier et al. 2002). Rationality is a key factor behind the belief that complex issues can be discussed and consensus found. The belief that human rationality can determine future paths of development can be traced back to the Enlightenment. Immanuel Kant philosophized on the subject. O'Neill (1989) explains Kant's view on reason, which was that individuals should think for themselves, think from the point of view of others, and think consistently and rationally. Kant proposed that by reasoning, people could act in ways that would be deemed proper for all people in that same situation, making such actions justifiable. By thinking in this way, principles can be agreed upon that are justifiable and universalizable. Kant calls this the Categorical Imperative.

McCarthy notes that:

> 'Habermas's discourse model represents a procedural reinterpretation of Kant's categorical imperative: rather than ascribing as valid to all others any maxim that I can will to be a universal law, I must submit my maxim to all others for purposes of discursively testing its claim to universality. The emphasis shifts from what each can will without contradiction to be a general law, to what all can will to be a universal norm' (McCarthy 1984: 326).

Thus, when consensus is reached, a universal norm is created. If all have agreed on an action it will be perceived as being a 'just' decision, therefore, all will be willing to abide by such a decision and any following actions. In other words, participatory justice will have been fulfilled. In some instances decisions arrived at consensually are institutionalized in law, that is, in rights and responsibilities. It follows that once rational discussion has brought about consensus on universal norms, such consensus can then be enshrined in rights. To quote Shue, 'A moral right provides the rational basis for a justified demand that the actual enjoyment of a substance be socially guaranteed against threats' (Shue 1996: 13).

Habermas discusses rights when he states that individual rights are important in understanding modern law: 'rights ('subjective rights' in German) fix the limits within which a subject is entitled to freely exercise her will. More specifically, they define the same liberties for all individuals or legal persons understood as bearers of rights' (Habermas 1996: 82). Habermas links the idea of individual rights, society and discourse:

> 'the procedure of democratic legislation must confront participants with the normative expectation of an orientation to the common good, because this procedure can draw its legitimating force only from a process in which citizens *reach an understanding* about the rules for them living together. In modern societies as well, the law can fulfill the function of stabilizing behavioral expectations only if it preserves an internal connection with the socially integrating force of communicative action' (Habermas 1996: 83–84).

Whilst it is not practical for every person within Germany or the international community to put forward their point of view, open and representative dialogue can be aspired to and worked towards.

As suggested above, justice – and, conversely, matters of injustice – pervade GCC politics (see Chapter 15). The definition of justice that is the starting point for this chapter is *the qualities and principles of fairness, even-handedness (i.e., equality) and moral decency*. Problems arise because terms such as moral decency and morally proper are open to contestation. Events and outcomes that are seen as fair and just by some people can be seen completely differently by another group of people.

Discourse-based theories with the aim of arriving at agreed policies, rights and responsibilities are apt. Habermas holds that during ethical discourse, 'every subject with the competence to speak and act is allowed to take part in a discourse; everyone is allowed to introduce any assertion whatever into the discourse; everyone is allowed to express his attitudes, desires and needs; [and] no speaker may be prevented, by internal or external coercion, from exercising his rights' (Cavalier et al. 2002). This idea of everyone being able to participate in discussions in order to find a consensus on the best course of action is a principle that can be seen, at least to some degree, to be in action within Germany. While not everyone has, nor could possibly have, equal access to discursive processes that result in decision-making, the fact that Germany has a consensually orientated political system does mean that a wide variety of opinions are considered (see Beuermann and Jäger 1996; Sturm 1996).

Although Habermas details the process through which universal norms can be determined, it is not the case that such norms are always achievable, as not all interests are generalizable. The main point is that ethical discourse be engaged in and that where a universal norm is not achievable, a consensually agreed upon compromise should be reached.[1] In this way the formation of democratic will can be achieved. It should be noted that compromize requires changing actions, while consensus can imply changing cognitions, interest patterns and options for actions. It should also be noted that ethical discourse is an ideal, as Stokke points out:

> 'Habermas does not deny that humans are able to act strategically, that is, in a goal-seeking manner towards objectified others, in order to realize a desired condition; the point is rather that their actions *cannot be reduced* to this purpose. There is something more going on which the actor himself cannot avoid or fail to notice: action expresses inner feelings and sentiments; it relates to social norms, either approvingly or rejectingly; and it communicates an understanding of the world' (Stokke 1998: 135).

One of the aspects of theory that is noted by Barkdull and Harris (2002: 84) as being 'largely unexplored' is the role of ideas in foreign policy-making. As is the case in all matters, the way GCC politics are conducted is influenced by the way in which the problem and its possible solutions are perceived. Discursive processes enable the values and ideas that are held by a large number of (in this case) German people to be considered and honed until a consensus is reached on the values and ideas that are reflected in policies. Thus values and ideas that are widely held across German society can be reflected in German foreign policy and therefore, have the potential to be reflected in international policy-making.[2]

Barkdull and Harris (2002) organize theory into three categories: systemic, societal and state-centric, within which they highlight the role of power, interests and ideas. Analysis in this chapter concentrates on the role of ideas. It shows the importance of ideas being expressed through public opinion, the media and social movements, which would fall under the societal category. The reflexiveness of domestic and international policy is also shown, with this dialectic relationship identifiable in the systemic category. Ideas have the ability to permeate all levels of discussions, thus paving the way for the reflexivity of policies between these various levels. It is not the intention in this chapter to argue that power and interests do not exist, but that the role of ideas is paramount and that discursive processes allow for the ideas of numerous people to be considered.

The next section looks at the evolution of environmental politics within Germany so that in the subsequent sections on international politics and German foreign policy we can see that the European and international policies on GCC are significantly reflective of domestic values. It is difficult to prove a causal link between the ideas of a particular person or group of people and the policy outcomes of another community. However, we can highlight correlations of opinions and probable connections and flows of influence (for further analysis, see Jaggard 2005).

GERMAN ENVIRONMENTAL POLITICS AND GLOBAL CLIMATE CHANGE

In 1969, Willy Brandt of the Social Democratic Party (SPD) was elected Federal Chancellor; he held this office until 1974. He was the first Chancellor to give environmental protection high priority, and in 1971 Germany's first Federal environmental program was adopted. Three principles formed the basis of this program and of all subsequent environmental policies: precautionary measures, polluter pays and industry-government cooperation. Each of these principles can be related to principles of justice. Precautionary measures are indicative of considerations of intergenerational justice. The polluter pays principle demands that the originators of the problem pay to resolve it. Paterson (2001) argues that is a form of retributive justice. As discussed earlier, participatory justice enables ideas to be heard and acted upon, thus enabling policy to be perceived as being just, resulting in acceptance of such policies, which enhances their potential for success. Industry-government cooperation can be seen as a part of this process.

Since they were introduced in Germany, both the precautionary and polluter-pays principles have been adopted by the international climate change regime (at least they are included in documentation if not always implemented). It can be argued that the international regime was significantly influenced by ideas emanating from Germany, with this transmission of ideas being achieved through what Habermas terms 'communicative rationality'. That is, the better argument prevailed, resulting in these principles entering into international environmental discourse.

Although this section is primarily about domestic politics some external influences should be noted, one of which is the United Nations (UN) Conference on the Human Environment held in Stockholm in 1972. It is certainly possible that the research, information available and the reflexive thought (ideas) that occurred around that time had an effect on politics within Germany. In addition to the first environmental program mentioned above, other German actions might have reflected sentiments expressed in the Stockholm process, such as Germany's convening of the Council of Environmental Experts in 1972 and the creation in 1974 of the Federal Environment Office (to research requirements for possible environmental regulations).

Von Weizsäcker (1994) argues that the policies introduced under Willy Brandt's leadership succeeded in producing cleaner rivers and air in industrial areas such as the Ruhr. However, by 1975 the global oil crisis was biting, both economically and in worsening unemployment and continued fuel cost rises. This had the effect of industrial leaders putting pressure on the government not to apportion further costs to industry in the form of environmental regulations. Thus environmental progress slowed in the following few years. However, Rowlands (1995) points out that during the oil crisis, although the use of coal increased, energy use in total decreased in West Germany. Economic growth did not suffer as perhaps would be thought; instead, energy use became more efficient. Rowlands points to this fact as the reason why German businesses were not violently opposed to proposed policies to reduce carbon dioxide (CO_2) emissions in the late 1980s. He argues that, instead, these policies were seen as being potentially beneficial in terms of efficiency and productivity. This is another instance when global issues have had an effect on domestic events and policies. The promotion of energy efficiency measures is a key component of contemporary foreign policy, with regard to GCC and sustainable development measures.

Inherent in German culture is the idea of forests being permanent features of the environment. As far back as the turn of the nineteenth century, lasting (sustainable) forestry was a declared aim. It is therefore not surprising that one issue that engendered popular concern was the widespread discovery in the early 1980s of *Waldsterben* (dying forests), the cause was attributed to air pollution. Beuermann and Burdick (1997) argue that due to this entrenched association with forests, during the 1980s and 1990s tropical deforestation and GCC were predominant in German thinking in relation to the concept of sustainability.[3] The need for emissions standards to reduce and control air pollution only came to the fore when the issue of *Waldsterben* was discovered and made public. This realization along with the strength of public concern was the catalyst that reinvigorated progress in German environmental policies. This is yet another issue that suggests linkages between values, ideas and policy.

In 1982 the government changed, having been led since 1974 by Chancellor Helmut Schmidt of the SPD. A coalition was formed from the Christian Democratic Union (CDU), the Christian-Social Union (CSU) and the Free Democratic Party (FDP). The new Chancellor was Helmut Kohl of the CDU; he remained Chancellor until 1998. Due to public concern over *Waldsterben*, the new government was vociferously called upon to take an active stance regarding environmental policies. It is also worth noting that members of the Green Party (which had formed from a social movement that had arisen in the 1970s due to popular concern over the environment and particularly over concern about nuclear safety [see Mayer and Ely 1998]) were voted into the *Bundestag* (the lower house) in 1983. This in effect sent a message to the government that the populace wanted environmental issues to be taken seriously; environmental ideas had to be taken on board and translated into policy. This sequence of events can be seen as an example of the ideas of the general populace being transmitted through the forming of a social movement and subsequently a political party. Barkdull and Harris's categorization of societal ideas fits this sequence of events, although it can be argued that this example has been so successful that it crosses into the category of state-centric ideas.

Under the Federal Emission Act, in 1983 the German government implemented strict emissions limits with regard to sulfur dioxide and large furnaces. Other environmental policies, such as the restriction of vehicle emissions, waste recycling programs and waste disposal regulations followed. Weale (1992) describes Germany as being active in pursuing international agreements with regard to pollution policies. He also argues that during the 1980s Germany became the environmental leader within Europe and that Germany's reunification in 1990 reinforced the idea of the need for environmentally protective action, as East Germany's dire environmental situation was realized.

The catalytic effect that *Waldsterben* had on German domestic environmental politics in the early 1980s transferred to Germany's efforts to push for air pollution policies to be introduced in Europe. Anderson and Liefferink (1997) argue that Germany's efforts were aided by the appointment of a German Environment Commissioner, Karl-Heinz Narjes, and by Germany holding the Presidency of the European Council in the first half of 1983, which had the effect of expediting a final proposal of an acidification policy that the European Commission was in the process of preparing. Similarly, Economy and Schreurs (1997) cite the success of the Green Party in Germany, and the subsequent uptake of green issues by mainstream political parties, arguing that this train of events continued on to the European level where 'Germany became a "primary force" in pushing the European Community on the introduction of a Large Combustion Plant Directive and later in the cases of stratospheric ozone depletion and global climate change' (Economy and Schreurs 1997: 9).

The Chernobyl nuclear accident, which occurred on 26 April 1986, was another exogenous event that impacted German Politics. Nuclear safety became the subject of much attention. The Interior Ministry had dealt with environmental issues since 1969, but Hatch (1995) argues that it inadequately handled events following Chernobyl. This led to the creation of the Federal Ministry for the Environment, Nature Conservation and Nuclear Safety (BMU). The Ministry of Transport handled Meteorological and GCC issues until late 1988. However, these agencies failed to provide adequate participation in the initial negotiations of the Intergovernmental

Panel on Climate Change (IPCC), and as a result responsibility for climate change was transferred to the BMU.

I have highlighted some events and ideas that have put pressure on the German government to follow certain domestic and foreign environmental policies. Values and ideas that were evidently held by a large proportion of the German populace were also those held by Klaus Töpfer, a member of the CDU and Environment Minister between 1987 and 1994.[4] According to Von Weizsäcker, Töpfer:

> 'soon brought both national and international recognition to the ministry. Helped by the [European] Commission and by France, Denmark and the Netherlands, he pushed the Large Combustion Plants Directive through the EC Council … He also took a leading role internationally on such topics as protection of the North Sea, the rain forests, global warming and the ecological reconstruction of Eastern Europe. At the Earth Summit in Rio de Janeiro he was certainly one of the leading figures representing the North and contributed much to the ultimate success of UNCED [United Nations Conference on Environment and Development]' (Von Weizsäcker 1994: 20).

This points to an important avenue by which ideas affected German environmental foreign policy and in turn influenced international environmental policy.

So far little has been said about the political system within Germany other than the fact that it is consensual in nature. Clearly this is pertinent to policy-making and so will be briefly explained (see also Chapter 2). The *länder* (regions, roughly equivalent to United States states) introduce some laws but they implement those made by both themselves and the federal government. Implementation is clearly an important factor and can obviously be effected with more or less vigor. Representatives of the *länder* constitute the *Bundesrat*, which is the upper house of parliament. The voting system is proportional and a second party list allows for people to express a preference for, in effect, the party that they wish to be in coalition with their main preference party. For all but 4 years of German post-WWII history there have been coalition governments. These characteristics allow for minority groups to have attention paid to their ideas and concerns. Thus, when environmental issues including GCC became fairly widespread concerns, the Green Party naturally gained support. This had the effect not only of the Green Party gaining *länder* government places and later federal government positions, but also of forcing the main parties to take on board environmental issues. The judiciary also plays an important part in German governance. Germany has a written constitution that allows for:

> 'A programmatic statement of principles [being] seen as an essential prologue to legislation and policy development, a practice that is probably reinforced

by the practice of coalition government in which political parties of different ideological persuasions have to come to some agreement [with] the emphasis on constitutionalism [having] the effect of making the policy process more formal' (Weale 1992: 82).

Weale points to the 1971 statement regarding the principles of precautionary measures, polluter pays and industry-government cooperation. These have effectively been institutionalized within the German policy-making system and hence are considered and worked towards in all environmental policy-making, both domestic and foreign. The institutionalization of these environmental principles in policy-making is reflective of the linkage between behavioral expectations, communicative action and the law that is identified by Habermas (see above).

Following the elections of September 1998, a Red-Green coalition formed a government led by Chancellor Gerhard Schröder. Jürgen Trittin of the Green Party was appointed as the Environment Minister. Another ministry important for GCC policy, the Economic Cooperation and Development Ministry, was headed by Heidemarie Wieczorek-Zeul, a representative of the SPD (the 'red' part of the coalition). Both the SPD and the Green Party are in favor of phasing out nuclear power. It is worth noting that during the 1980s the SPD adopted environmental policies that were in large part those already espoused by the Green Party. The adoption of people's ideas and concerns was a result of the political system that facilitates such inclusiveness.

It was a major intention of the Green Party, if not the whole government, to make their mark on Germany's environmental politics. In their first 2 years of government a number of measures were put in place to help reduce CO_2 emissions, including the following:

- Ecological tax reform, which envisages a gradual increase in energy prices in all segments in order to create incentives for the development of new technologies, coupled with the rational and economical use of energy.
- The Renewable Energy Act, which promotes the conversion of renewable energy sources into electricity.
- The market-launch program for renewable energy sources, which particularly benefits the use of solar panels, as well as rational energy use.
- The '100,000 roofs' program, which supports investments in photovoltaic systems.
- The promotion of low-sulfur and non-sulfur fuels to achieve a breakthrough in fuel-efficient, low-emission engine technology (BMU 2000: 7).

The BMU proposed a new GCC protection program, which was adopted on 18 October 2000. The summary of the program published by the BMU declares several ambitious targets for the federal government: to reduce emissions of carbon dioxide by 25 per cent by 2005 compared with 1990 levels, and to reduce emissions of the six greenhouse gases cited in the Kyoto Protocol by 21 per cent between 2008 and 2012, within the context of EU burden-sharing. The base year for CO_2, CH_4, and N_2O is 1990, and 1995 for H-CFC, CFC and SF_6 (CO_2 Carbon dioxide, CH_4 Methane, N_2O Nitrogen Oxide, H-CFC Hydro-chlorofluorocarbon, CFC Chlorofluorocarbon, SF_6 Sulfur hexafluoride). Furthermore, for the first time the federal government also set up specific targets relating to technology and energy resources: a doubling in the proportion of renewable energy sources by 2010 compared with the then current levels, and a further substantial increase in the proportion of renewable energy sources after 2010; the expansion of combined heat and power generation by means of set quotas, aimed at cutting CO_2 emissions by an additional 10 million tons by 2005, and by 23 million tons by 2010; a significant increase in energy productivity over the next few years; and, for the first time, setting of reduction targets for individual sectors (BMU 2000: 4).

More overarching intentions and beliefs are described in the BMU's GCC protection program, such as the importance of the EU and international negotiations and policies. Dedication to the continuance of GCC protection policy at both national and international levels is identified, as is the belief that the commitments of industrialized countries under the Kyoto Protocol's initial commitment period should be increased in future phases. It is believed, however, that developing countries with increasing emissions should be included in future commitment periods. A commitment to the continuance of support for GCC protection polices in developing countries is made, and it is stated that such commitments will be reviewed and updated (ibid).

Germany's leading domestic policies on GCC issues give them credibility in the push for international agreements; they are, after all, 'practicing what they preach' more than most other countries. Germany's leading domestic policies with regard to renewable energy are evidenced by the fact that Germany has over 11,000 wind turbines and is the world's leader in wind power production. In 2001 alone production capacity increased by 44 per cent (Bartlett 2002). The German government has agreed to phase out nuclear energy and is investing heavily in wind power. Germany's relative zeal for environmental action to be taken on a global scale is arguably due to the belief that such action is a necessity to ensure a fair and viable future for all – reflection of the values and ideas that are important to individuals being sought for all. It can also be argued that in many ways the adoption of domestic policies that impose strict environmental conditions necessitate a foreign policy that enthusiastically promotes similar policies to be adopted at an international level so as to prevent domestic isolation and competitive disadvantage.[5] If progressive policies are adopted on an international level, Germany is then at a competitive advantage, being in a position of technological advancement and thus able to sell its expertize abroad. The likelihood is that both factors have a bearing on German foreign policy. Whatever the exact reasons, domestic ideas and policies are being promoted at an international level.

The reflection of domestic policy in foreign policy can be seen in the aim of the German government to expand the use of solar power. In a report on a talk given by Jürgen Trittin in March 2002 at a conference on solar thermal electricity generation, the BMU asserts that Germany's research institutes and companies are world leaders in the development of solar power stations, and that, in Trittin's words, 'With the Government's help, solar power stations will become Germany's export hit of the future. This will benefit both climate protection and the economy' (Trittin quoted in BMU 2002a). According to the BMU report:

> 'Under its investment in the future program, the German Government will provide a total of 10 million Euro up to 2003 for research and development projects on solar thermal power stations … The Federal Environment Ministry has invested a total of 30 million Euros in renewable energies research. Jürgen Trittin believes that the investment in the future program should be continued until at least 2006' (BMU 2002a).

Since Germany is taking an active lead in renewable energy production, it will most likely be able to benefit economically from technology transfers; this would be in addition to environmental benefits that would not be limited to Germany.

I have tried to show that ideas of environmental protection, participative democracy/justice (discursive inclusion) are important within Germany. The following sections will argue that these ideas are reflected in Germany's foreign policy as well.

CLIMATE CHANGE AND INTERNATIONAL ENVIRONMENTAL POLITICS

The previous section on Germany's domestic policies included some references to international politics. This section will discuss in much more detail the linkages between German domestic politics and international environmental politics. The increasing interconnectedness of domestic and international politics has the effect of it being very difficult to completely separate the two levels.

In December 1990, the UN General Assembly resolved to establish the Intergovernmental Negotiating Committee (INC) for the Framework Convention on Climate Change (FCCC). The INC/FCCC first met in June 1991. The INC met every few months thereafter, but agreement was hard to achieve. Broadly speaking the US differed in their stance from the rest of what is commonly known as the North. The US was, and still is, far more strongly opposed to industrially restrictive agreements. Germany was much more progressive in its views, reflecting concerns that were important domestically. In the South views ranged from the hard stance of the oil-producing states, to a great deal of the developing world wanting to protect their rights to development by industrialization, through to the small island states that wanted far-reaching curbs on GHGs and a generally strong line taken against the anthropogenic causes of global warming.

Thirty-five of the world's small-island states joined together to push their concerns about GCC and sea level rise, forming the Alliance of Small Island States (AOSIS).[6] The AOSIS was active in INC/FCCC negotiations and pushed for the following principles, which are included in the FCCC: the principle of preventive action; the precautionary principle; the polluter-pays principle and State responsibility; the duty to cooperate; equity; the principle of common but differentiated responsibility; and a commitment to binding energy conservation and the development of renewal energy sources (AOSIS 2001b). These principles largely coincide with those supported by Germany and the inclusion of them in the agreement would have been due to consensus being reached by all involved in the negotiations and due to pressure from the delegations of many countries.

As noted above, the precautionary principle, which is seemingly ubiquitous in environmental discourse and modern environmental treaties, was first introduced in Germany (Von Seht and Ott 2000; Institute for European Environmental Policy 2004). During the INC's second session, a number of states submitted draft conventions (Ulbert 2003). Germany's draft included the precautionary principle, binding reduction targets and the call for immediate action. The US was against targets and 'doubted the usefulness of the precautionary principle on the basis of the lack of scientific certainty' (Ulbert 2003: 21). Nevertheless, the precautionary principle was included in the FCCC. This can be viewed as the result of discourse in which Habermasian communicative rationality prevailed. It is an example of the spread of ideas from the domestic level to the international level, or, to use Barkdull and Harris's terminology, from the state level (depicted in state-centric theories) to the systemic level (described by systemic-level approaches to environmental foreign policy).

Consensual politics as was mentioned in the previous section is a feature of German domestic politics; it is also necessarily a feature of international regimes. German delegations could be argued to be in a relatively strong position negotiating in such international situations because their political background gives them experience in similar conditions. During INC discussions, representatives of each country tried to negotiate the best possible deal in relation to their own interests, and it could be said that this reflects the realist tenets of self-help in an anarchic world order. It could also be argued that neo-liberal institutionalism better fits the facts because, although each country tried to maximize its own bargaining position, most, if not all, states had to compromise, albeit to varying extents.

Analysis that utilizes Habermas's discourse-ethics theory enables a greater understanding of events. It can be seen that interests of otherwise divergent states can converge over some matters. In climate-change related issues, states such as Germany and members of AOSIS have many similarities in aims and attitudes; as do some of the developing countries that are wishing to, or are already, rapidly industrializing, and the US. The INC process can be related to Habermas's discourse ethics because all interested parties were able to contribute their opinion in a rational discussion that led to agreement over the form that the FCCC would take, albeit with a continuing need for discussions aimed at achieving further agreements. Participatory justice led to an agreement that was largely perceived as just.

The FCCC was adopted on 9 May 1992, and it opened for signature during the June 1992 UN Conference on Environment and Development, subsequently entering into force on 21 March 1994 after being ratified by 50 states. The INC continued to meet until the First Conference of the Parties (to the convention), which was held in April 1995. At each Conference of the Parties (COP), various interest groups tended to present a largely unified viewpoint. The EU is one such grouping. At COP3 held in Kyoto in December 1997, AOSIS countries were first to propose a draft calling for a 20 per cent reduction in CO_2 emissions from 1990 levels, to be implemented by 2005. The Kyoto Protocol agrees different GHG reduction figures for different Annex I parties (developed countries, essentially those belonging to the Organization for Economic Co-operation and Development). These range from an 8 per cent cut for the EU to a 10 per cent increase for Iceland. Combining all Annex I countries targets gives a total cut of 'at least 5 per cent from 1990 levels in the period 2008–12' (UNFCCC 2005).

The EU can distribute its 8 per cent cut across the various countries as it sees fit. To enable the Kyoto Protocol to enter into force 55 parties to the convention, making up at least 55 per cent of 1990 emissions of Annex 1 parties (industrialized countries), had to ratify the Protocol. Without the US, which is the largest GHG emitter, ratification by at least the EU, Japan and

Russia had to occur. In December 2001 the German government adopted the bill on the enactment of the Kyoto Protocol, with the aim of finishing the parliamentary process by May 2002 and ratifying by June 2002. It was the fervent aim of Germany and the EU for the Kyoto Protocol to be ratified before the World Summit on Sustainable Development (WSSD) in August-September 2002. Germany and the rest of the EU ratified the Protocol on 31 May 2002, and Japan followed shortly after on 4 June 2002. However, it was not until 18 November 2004 that Russia ratified the Protocol, thus paving the way for it to come into force on 16 February 2005. The Kyoto Protocol targets are legally binding. This can be seen as an example that reflects the ideas of Henry Shue (1996: 13) and Jürgen Habermas (1996: 82–4) discussed above (see the section titled 'Consensus, Discourse and Ideas in the Policy Process'). The Kyoto Protocol was agreed through a discursively inclusive process that eventually resulted in the consensual outcomes of discussions being enshrined in rights and legally binding responsibilities.[7]

Although the Kyoto Protocol had been initiated at COP3 in 1997, many operational details were left to be worked out at subsequent conferences of the parties. At COP6 (Part II), held in Bonn during July 2001, the EU – Germany included – were instrumental in pushing for agreement to enable ratification of the Kyoto Protocol. Jürgen Trittin, Germany's Environment Minister, greatly influenced agreement at COP6 II. He persuaded the rest of the EU representatives to compromise and gain agreement.[8] This was seen as especially important as the US under George W. Bush's presidency declared the Kyoto Protocol 'fatally flawed' and had withdrawn from it in March 2001. Concessions were required in order to reach an agreement, but they were seen as better than effectively destroying the Protocol. The Bonn Agreement included recognition of the need for new and additional funding (although no specific amounts nor legally binding obligation to provide such funds were agreed) and the need for greater flexibility regarding sinks and clean development mechanisms.[9]

Negotiations and agreements at the EU level are viewed as somewhere between the domestic and the international levels. Reflexivity of ideas and policies can be seen between each of these levels. An example of this can be found in the adoption of the precautionary principle in the Treaty on European Union (otherwise known as the Maastricht Treaty), which came into effect on 1 November 1993. The principle had been adopted in Germany in 1971 and by the FCCC in 1992. It is pertinent to mention that the precautionary principle is not one that sits easily with all countries of the EU. As Wurzel (2002) explains, a standard-setting philosophy in Britain is the requirement for scientific proof. As all EU Member States agreed to the Treaty on European Union, this suggests that the adoption of the precautionary principle by the EU would have resulted from discourse in which all parties had the opportunity to contribute, all contributions were considered, and decisions were made through rational choice – that is, that the force of the better argument prevailed.

As mentioned at the start of this chapter, reflexivity can work in both directions. So far I have tried to show that ideas from within Germany percolated up to the EU and international levels. However, Germany (and other states) can be influenced by ideas that come from the international level; exogenous influences affect German policy (see Chapter 12). An example can be found in EU policy-making impacting on Germany in the context of the European Union (EU) Greenhouse Gas Emission Trading Scheme (ETS), which commenced on 1 January 2005. Companies are given a CO_2 emissions allocation. If this is exceeded a financial penalty is incurred, but if emissions are lower than the amount allocated, the excess can be sold. Much of German industry had fought against the implementation of such a scheme as it was felt that they had already taken large steps to reduce emissions and that they would be disadvantaged vis-à-vis industries in other countries. The ETS is seen by many as a forerunner for an international scheme, if this proves to be the case its influence will also be felt by industries around the world. The very nature of open and inclusive discourse that aims to bring about agreement through the rational choice of the most convincing argument allows ideas to flow in varying directions.

GERMAN FOREIGN POLICY ON GCC

On 8 December 2001, the German Environment Minister, Jürgen Trittin, spoke at the Berlin Conference on the Human Dimensions of Global Environmental Change. He said that Germany was working actively within the EU, the Council Presidency and the Commission, which negotiates at the global level, to influence GCC policy (Trittin 2001). He argued that such a united front enabled the continuation of the Kyoto process at the Bonn COP earlier that year, despite the US trying to block the process. He gave examples of environmental policy that had been implemented or decided upon within Germany, such as the decision to phase out nuclear power and the promotion of renewably sourced energy. He argued that these policies and others that make climate protection policy successful within Germany were instrumental in Germany's success at the Bonn negotiations in persuading other countries to agree that building nuclear power plants in developing countries would not be acceptable as offsetting measures under the FCCC's Clean Development Mechanism. This is an example of German ideas and values being reflected in foreign policy and ultimately in international environmental agreements, which in turn impact domestic policies of other countries. To use the terminology of Barkdull and Harris (2002), this is another example of the ideas at the state level becoming institutionalized at the systemic level.

On 19 December 2001, Hans Martin Bury, German Minister of State to the Chancellor, presented a draft national sustainability strategy entitled 'Prospects for Germany' (Federal Government of Germany 2001). Bury is reported as having said that environmental sustainability is 'the main theme of reform policy for this government ... Intergenerational justice, social solidarity, quality of life, and international responsibility are the coordinates of our strategy' (German Press and Information Office 2001). This statement can be seen as reflective of national values. It is clear that issues of justice and responsibility are important for Germany and that ethical considerations are not limited by national boundaries or to any large degree, by time.

During the formation of Germany's national sustainability strategy, systems were in place to enable multiple viewpoints to be considered. An Enquette Commission on Sustainable Energy Supplies in View of Globalization and Liberalization was established,[10] as was the Council for Sustainable Development, which was made up of members from different sectors of society. In addition to these consultation bodies and talks with major interest groups, an online forum to enable dialogue was set up by the Council for Sustainable Development. This avenue for public input was mentioned at Bury's presentation and in the related press release (German Press and Information Office 2001). Such publicity about this opportunity enables greater public participation in the policy-making process and is demonstrative of consensual politics in action. It also augurs well for the effectiveness of the policy because it has received widespread input and agreement. Sustainability is seen as requiring the participation of all sectors and members of society; each person and institution needs to consider the results of their actions. This is a reminder that, along with citizenship, which in itself confers the right or even duty to participate, there is correspondingly a certain amount of responsibility. This is an example of the importance of society's ideas being considered through participative dialogue in the formation of policy that is promoted on the international stage.

On 17 April 2002, the Federal Cabinet approved the final version of the national sustainability strategy, 'Perspectives for Germany: Our Strategy for Sustainable Development' (Federal Government of Germany 2002), which formed the German contribution to the WSSD. An article that was placed on the government website states that:

> 'The impetus for the federal government's sustainability strategy came from the 1992 UN Conference on Environment and Development in Rio de Janeiro,

> at which the international community of nations accepted the principle of sustainable development and established a global action program for the 21st century, Agenda 21. This calls upon the signatory states to develop a national sustainability strategy with the goal of achieving an economically powerful, socially equitable, and ecologically sound form of development' (Federal Government of Germany 2002).

The reflexivity between various levels, that is, local, national and international, is again evident, this time from the international level back down to the national level.

Some aspects of Germany's national sustainability strategy need to be considered here. In the foreword to 'Perspectives for Germany: Our Strategy for Sustainable Development' (Federal Government of Germany 2002) sustainable development is described as a way of directing globalization to enable social, ecological and economic well-being, and that by presenting its national strategy at the WSSD Germany is making an important contribution to this aim. The strategy is detailed, setting out objectives, priorities and measures to be taken to achieve sustainable development. It is stated that

> 'This strategy will be the basis for further political reforms and also for changes in the way businesses and consumers behave. Extending far beyond the ecological challenge, this strategy will also serve to provide guidelines for a comprehensive policy, able to meet the challenges of the future, to fulfill responsibility, which bridges across the generations, for devising a form of development which is economically, ecologically and socially sustainable' (Federal Government of Germany 2002: 3).

Evidently, ethical matters are seen as being important. The very notion of sustainable development implies intergenerational justice. This is made explicit in the report: 'The ethical basis of sustainability is that every generation must deal with its own problems and should not burden the coming generations with them' (Federal Government of Germany 2002: 5).

The German government's strategy includes rules for managing sustainability. Several areas of action are identified and are pertinent here (Federal Government of Germany 2002: 5). In the long term, consideration is given to the actual renewability of all 'renewable' natural resources that are used, and when using all non-renewable natural resources, such as fossil fuels, consideration is to be given to how these functions can be fulfilled by the use of an alternative. It is also foreseen that, in the long term, the products of such resource usage will not reach a level that is detrimental to the world's ecosystems. The decoupling of energy and resource usage from economic growth is seen as a necessity, as is the offsetting of the growing demand for energy, resources and transportation by increasing efficiency.

The sustainability strategy advocates that:

> 'international framework conditions are to be devized in such a way that people in all countries can lead a life of human dignity according to their own cultural principles and can participate in economic developments. Environment and development form a unity. In an integrated approach the fight against poverty is to be linked to respect human rights, to economic development, environmental protection, and also to responsible action on the part of Government' (Federal Government of Germany 2002: 10).

The strategy points out that since 1998 wind power capacity has trebled; that the future of wind energy is offshore; and that demand for solar panels is high (Federal Government of Germany 2002: 17). It is the government's policy to encourage the development of renewable energy. Government targets include the doubling of the proportion of renewably sourced energy in primary energy consumption by 2010 from 2000 levels, and for renewably sourced energy to reach approximately 50 per cent by the middle of the century (Federal Government of Germany 2002: 18). Nuclear energy is confirmed as being highly risky because waste products cause problems for generations. Therefore, nuclear energy is not considered a viable alternative to conventional forms of energy production. The phasing out of nuclear energy production within Germany is assured (Federal Government of Germany 2002: 19)

The report makes it clear that sustainability strategies must take into account the effects of domestic actions on other areas of the world. Support is expressed for the establishment of a UN World Commission for Sustainable Development that would ensure that continued globalization occurs within the confines of sustainability (Federal Government of Germany 2002: 35). With specific reference to the WSSD, the report says that Germany is committed to developing strategies to increase the worldwide use of renewable energy and to increase energy efficiency, and to the following:

- Strengthening the financial and economic basis for Sustainable Development in developing countries by means of trade facilitation measures and developmental collaboration;
- Joint initiatives with the German economy, trades unions and non-governmental organizations to take greater account of social and environmental factors and also of Sustainable Development principles in making direct foreign investments;
- Development of the structures of the UN in the area of Sustainable Development with the additional aim of involving developing countries more effectively in the future direction of the processes of globalization;

- Improved co-ordination of environmental activities in international institutions and within a framework of international agreements;
- Strengthening of the United Nations Environmental Program (UNEP) with a view to developing it into a world environmental organization (Federal Government of Germany 2002: 37).

This government document is arguing for intergenerational justice to be considered and implemented. It is not only a rallying cry to the German populace; it is also a policy statement of what the government intends to implement.

One of the aims of Germany – and indeed the EU – at the WSSD was to agree on targets for the increase of renewable energy. This did not happen. It has been argued that the US, Japan and the OPEC states blocked this possibility.[11] Chancellor Schröder responded by announcing that Germany would host a conference on renewable energy. A declaration of like-minded countries was submitted during the final stages of the Johannesburg summit which committed the signatory states to setting ambitious goals for increased use of renewable energies at global, regional and national levels, together with adopting schedules for the meeting of such goals. The declaration was born from a German initiative, it was submitted by the EU and it was signed by, amongst others, all EU Member States and candidate countries and AOSIS. The decision for Germany to host an international conference on renewable energy was made before the WSSD, and hence before the declaration of like-minded countries was initiated. Nevertheless, the conference on renewable energies (held in June 2004) was viewed by the Germans as an important step in the initiative's implementation (BMU 2002b). This is an example of the transmission of ideas between state and system levels.

Another example of the transferability of ideas through discourse between various political levels can be found in German and EU influence in agenda setting for the WSSD. According to Andreas Krämer of Ecologic,[12] the German government wanted energy to be on the agenda at the WSSD but that in the WSSD PrepComs it was not clear that this would be the case. The EU, led by France and Austria pushed for energy to be included in the summit. This initiative was strongly supported by Germany. The German Ministry for Development and Cooperation (BMZ) hosted a conference that was attended by, amongst others, Klaus Töpfer of UNEP. Krämer (2003) argued that the EU's initiative was strengthened by the BMZ conference (and the presence of Töpfer) and shortly thereafter energy was included on the agenda for the WSSD. Although a link between these events and inclusion of energy on the WSSD agenda cannot be proven, such a link is probable. At the very least, these events would have added extra weight to any argument for energy's inclusion on the WSSD agenda.

Much of German environmental foreign policy is channeled through the EU, which presents a united position at global conferences such as the WSSD and at GCC specific international negotiations such as those held annually by the Conference of the Parties to the FCCC. In order to arrive at a unified position prior to international negotiations, delegations from member states meet regularly together with representatives of the Environment Directorate General of the European Commission. Germany has representatives at all meetings; it has the resources – both financial and human – to be able to do so (this is not the case for all EU countries). Not only is Germany represented at all meetings, but it also takes an active part in them. It has been said that Karsten Sach of the BMU is a very active, well-respected and authoritative member of the climate change working group, and that people listen to him.[13] However, GCC policies have evolved over a number of years and it is the case that continuing discourse between the member states has resulted in there being a large degree of confluence of positions. This points to the consensual adoption of ideas through multi-participative discourse that can, as indicated in the introduction, result in changing ideas about the best policy to pursue.

To return more specifically to German foreign policy, during 2002 the BMZ placed on their website a report entitled 'From Rio to Johannesburg: a review of progress in selected areas of German development policy' (BMZ 2002). This report says that 'German development cooperation is committed to the guiding vision of sustainable development and its translation into tangible action through the agreements reached in Rio' (BMZ 2002). This is another instance that is demonstrative of the reflexivity of ideas and policies between the international level, the national level and indeed foreign policy. In the report's section on protecting the environment, it is acknowledged that the energy sector is a major source of greenhouse gases and that these cause damage to the atmosphere. It is also noted that globally in excess of two billion people are without secure energy supplies. It is argued that more people must have access to energy and that this increase should be met in a sustainable manner. The report states that Germany can make a substantial contribution to this challenge as it is amongst the world leaders in renewable energy technology innovation. Domestic policy allows international leadership and, it could be argued advantage, thus reinforcing Germany's foreign policy. It is also stated that Germany provides €102 million annually for promoting renewable energy sources in the developing world, and Germany is the third largest contributor to the Global Environment Facility,[14] which in turn is said to provide US$200 million per annum for climate protection in developing countries. Germany also expresses support for the Kyoto Protocol's Clean Development Mechanism (BMZ 2002).

Focal areas are identified for Germany's development cooperation policy with regard to energy, these are: the provision of policy advice, the promotion of renewable energy and the improvement of energy efficiency. The financing of sustainable development and the undertaking of industrialized countries made at Rio to increase their overseas development assistance (ODA) to 0.7 per cent of Gross National Product (GNP) is discussed. There has been a general trend of declining ODA, but it is noted that Germany began to reverse this trend in 1998 and that it will aim to contribute 0.33 per cent by 2006. The report also points out that Germany is the third largest provider of bilateral ODA and that it has a great deal of influence in multilateral development organizations (BMZ 2002). Germany is increasing support to partner countries that are planning to introduce national sustainability strategies (BMZ 2002). This foreign policy aim reflects domestic ideas, values and policies.

It is acknowledged that for sustainable development to be successfully implemented, widespread participation at all levels is needed. Developing countries need to be able to participate in international negotiations, the benefits of globalization and in the use of global resources. Citizen participation is also seen as a necessary requirement for successful policy implementation (BMZ 2002). The sentiments expressed in this report reflect those of multiple participation, consensual politics and considerations for justice. These values, both within German domestic politics and in German foreign policy, have been described throughout this chapter.

Following elections on 18 September 2005 and subsequent negotiations between the main political parties, a coalition agreement was reached on 11 November 2005 between the CDU, CSU and SPD. This agreement enabled the formation of a new government. The coalition agreement includes the environment, climate change and related issue areas.[15] It appears that the aims of the German government will remain largely the same as those under the SPD/Green coalition. This impression is reinforced by the facts that the BMU is headed by Sigmar Gabriel of the SPD, and the Economic Cooperation and Development Ministry identified earlier as important in GCC policy continues to be headed by Heidemarie Wieczorek-Zeul of the SPD. The long-term confluence of ideas and policy aims is consistent with consensual decision-making.

CONCLUSION

As noted in the introduction to this chapter, Barkdull and Harris (2002) suggested that insufficient analysis has been undertaken of the role of ideas in environmental foreign policymaking. This chapter is an attempt to add to this under-represented area of research. Barkdull and Harris (2002) classify different focal areas of foreign policy analysis: systemic (the international arena), societal (the domestic arena) and state-centric (government). A complete picture of the politics of GCC will not be given through the analysis of any one these areas; to look at the interconnections between approaches gives a more realistic picture.

This study has drawn on Habermas's work on discourse ethics to examine the role of ideas at these various levels, and at the interaction among these levels of analysis, and thus adds to the understanding of these complex processes. Discursive interactions facilitate the consideration of a multiplicity of ideas and the dissemination of those ideas that are considered to form the 'better argument'. In giving a brief overview of German domestic environmental politics and German interaction in international environmental politics, this chapter has tried to illustrate how ideas can be transmitted between society, the state and indeed the international level, although this process is not necessarily linear. It has been shown that German domestic politics and values place importance upon the environment and upon justice. These values are reflected in German foreign policy. Whilst it is extremely difficult to prove causal links between the ideas of a particular grouping of people and the policy outcomes of a wider community, it is possible to identify correlations between them and probable connections. Reflexivity occurs in both directions, but it seems that German foreign policy and German efforts in the international arena to influence GCC negotiations are more greatly influenced by domestic policy than the other way around.

NOTES

1. All affected parties having the opportunity to contribute to the discussion, all contributions being listened to, and decisions being made through rational choice, that is, that the force of the better argument prevails.
2. Examination here of this reflexivity is based on textual analysis of documentation and information gleaned from participants in this process.
3. The public became aware of massive tropical deforestation and the fact that forests act as 'carbon sinks' (they take in carbon dioxide and give off oxygen).
4. Töpfer went on to become Chairman of the UN Commission on Sustainable Development from 1994 to 1995. Since 1998 he has been UN Under-Secretary General and Executive Director of the United Nations Environment Programme and Director-General of the UN office at Nairobi.
5. It is in fact the case that while the Ministries for Finance and for Economics and Labor agree that climate change needs to be addressed, they prefer international agreements to Germany taking too great a lead with domestic policies, precisely because they do not want Germany to be put at a competitive disadvantage (Interviews with Thomas Frisch of The Federal Ministry of Economics and Labor [Berlin by telephone, 24 April 2003] and Jürgen Wenderoth of the Federal Ministry for Finance [Berlin, 8 April 2003]).

6. Membership of AOSIS has increased to 43 countries (four of which are observers) and is described as 'a coalition of small-island and low-lying coastal countries that share similar development challenges and concerns about the environment, especially their vulnerability to the adverse effects of global climate change' (AOSIS 2001a). (The AOSIS has a website at www.sidsnet.org/aosis/main-aosis.htm, to which Germany has contributed funding.)
7. Not all international agreements are reached through such discursively inclusive processes and not all result in legally binding commitments.
8. Interview with Dietrich Brockhagen of the German Advisory Council on Global Change, Berlin, 3 March 2003.
9. For further information, see IISD (2001).
10. Enquette Commissions are explained by Beurmann and Burdick as advisory bodies 'to the federal government. One half of the members are politicians, the other are scientists appointed by parties represented in parliament. Enquette Commissions are established to give policy advice on complex political issues' (Beuermann and Burdick 1997: 102).
11. A number of newspaper reports argued this position. See *Independent on Sunday* (2002).
12. Ecologic is an institute for international and European environmental policy.
13. Interview with a member of the Directorate General of the Environment of the European Commission, Brussels, 21 February 2004.
14. The Global Environment Facility was established as a result of a Franco-German initiative (BMZ 2002).
15. For further information see CDU, CSU, and SPD (2005).

REFERENCES

Alliance of Small Island States (AOSIS) (2001a), *Alliance of Small Island States (AOSIS): The Alliance*, available at www.sidsnet.org/aosis/main-aosis.htm.

Alliance of Small Island States (AOSIS) (2001b), *Alliance of Small Island States*, available at www.sidsnet.org/aosis/background.html (November).

Anderson, Mikael S and Duncan Liefferink (1997), 'Introduction: the impact of the pioneers on EU environmental policy', in Mikael S. Anderson and Duncan Liefferink (eds), *European Environmental Policy: The Pioneers*, Manchester: Manchester University Press.

Barkdull, John and Paul G Harris (2002), 'Environmental change and foreign policy: a survey of theory', *Global Environmental Politics* **2**(2) (May): 63–91.

Bartlett, Patrick (2002), *Germany Boosts Wind Power*, available at http://news.bbc.co.uk /1/hi/world/europe/176054.stm (January).

Beuermann, Christiane and Bernhard Burdick (1997), 'The sustainable transition in Germany: some early stage experiences', in Timothy O'Riordan and Heather Voisey (eds), *Sustainable Development in Western Europe: Coming to Terms with Agenda 21*, London: Frank Cass.

Beuermann, Christiane and Jill Jäger (1996), 'Climate change politics in Germany: how long will any double dividend last?', in Timothy O'Riordan and Jill Jäger (eds), *Politics of Climate Change: A European Perspective*, London: Routledge.

Bundesministerium für Umwelt, Naturschutz und Reaktorsicherheit (BMU) (2000), *Germany's National Climate Protection Programme Summary*, Berlin: BMU.

Bundesministerium für Umwelt, Naturschutz und Reaktorsicherheit (BMU) (2002a), *Jürgen Trittin: Solar Power Stations: Germany's Export Hit of the Future*, available at www.bmu.de/english/news/pressrelease020314.php.

Bundesministerium für Umwelt, Naturschutz und Reaktorsicherheit (BMU) (2002b), *Declaration of Like-Minded Countries on Increasing the Share of Renewable Energies*, available at www.bmu.de/english/download/b_jointdeclaration.php.

Bundesministerium für wirtschaftliche Zusammenarbeit und Entwicklung (BMZ) (2002), *From Rio to Johannesburg: A Review of Progress in Selected Areas of German Development Policy*, Berlin: BMZ, available at <www.bmz.de/include/cgibin/druck.pl?default>.

Brockhagen, Dietrich (2003), *German Advisory Council on Global Change. Interview* (3 March), Berlin: German Advisory Council on Global Change.

Cavalier, Robert, Carnegie Mellon and Charles Ess (2002), *Introduction to Habermas's Discourse Ethics*, available at http://caae.phil.cmu.edu/cavalier/Forum/meta/background/HaberIntro.html.

CDU, CSU and SPD (2005), *Working Together for Germany with Courage and Compassion: Coalition Agreement between the CDU, CSU and SPD* (11 November), available at www.bundesregierung.de/Anlage951220/Coalition+Agreement.pdf.

Economy, Elizabeth and Miranda A Schreurs (1997), 'Domestic and International Linkages in Environmental Politics', in Miranda.A. Schreurs and Elizabeth Economy (eds), *The Internationalization of Environmental Protection*, Cambridge: Cambridge University Press.

Federal Government of Germany (2001), *Prospects for Germany*, available at http://eng. bundesregierung.de/documente/Artikel/ix_7632.htm.

Federal Government of Germany (2002), *Perspectives for Germany: Our Strategy for Sustainable Development*, available at http://www.bundesregireung.de/frameset/index.jsp.

Frisch, Thomas (2003), *Interview by telephone* (24 April), Berlin: Federal Ministry of Economics and Labor.

German Press and Information Office (2001), 'Minister of State Bury Presents Draft of National Sustainability Strategy Containing 21 Objectives for the 21st Century', *Press Release No. 588/01*, available at http://eng.bundesregierung.de/dokumente/ Pressemitteilung/ix_65804_2726.htm.

Habermas, Jürgen (1996), *Between Facts and Norms: Contributions to a Discourse Theory of Law and Democracy*, translated by William Rehg, Cambridge: Polity Press.

Hatch, Michael T (1995), 'The politics of global warming in Germany', *Environmental Politics* **4**(3) (Autumn): 415–40.

The Independent (2002), 'Earth Summit: After days of Intense Negotiations Leaders Settle on a Blueprint to Keep the Planet Alive', *Independent on Sunday* (3 September), available at http://news:independent.co.uk/world/environment/story.jsp?story=329998.

Institute for European Environmental Policy (2004), *Manual of Environmental Policy: The EU and Britain*, available at http://www.mep-online.com/chapter2/toc_2_1.html.

International Institute for Sustainable Development (2001), 'COP-6.bis Final', *Earth Negotiations Bulletin* **12** (30 July): 176.

Jaggard, Lyn D (2005), *Germany's International Relations of Climate Change*, Thesis for the degree of Doctor of Philosophy, UK: University of Birmingham.

Krämer, Andreas (2003), *Ecologic*, Interview (12 February), Berlin.

Mayer, Margit and John Ely (eds) (1998), *The German Greens: Paradox Between Movement and Party*, Philadelphia: Temple University Press.

McCarthy, Thomas (1984), *The Critical Theory of Jürgen Habermas*, Cambridge: Polity Press.

O'Neill, Onora (1989), *Constructions of Reason: Explorations of Kant's Practical Philosophy*, Cambridge: Cambridge University Press.

Paterson, Matthew (2001), 'Principles of justice in the context of global climate negotiations', in Urs Luterbacher and Detlef Sprinz, *International Relations and Global Climate Change*, Cambridge, MA: MIT Press.

Representative of the Environment Directorate General of the European Commission (2004), *Confidential interview* (24 February), Brussels.

Rowlands, Ian (1995), *The Politics of Global Atmospheric Change*, Manchester: Manchester University Press.

Shue, Henry (1996), *Basic Rights: Subsistence, Affluence and US Foreign Policy*, Princeton: Princeton University Press.

Stokke, Olav S (1998), 'Understanding the formation of international environmental regimes: the discursive challenge', in Arild Underdal (ed.), *The Politics of International Environmental Management*, Dordrecht: Kluwer Academic Publishers.

Sturm, Roland (1996), 'Continuity and Change in the Policy-Making Process', in Gordon Smith, William E. Paterson and Stephen Padgett (eds), *Developments in German Politics 2*, Basingstoke: Macmillan Press Ltd.

Trittin, Jürgen (2001), 'The role of the nation state in international environmental policy', Speech and paper presented in Berlin (8 December).

Ulbert, Cornelia (2003), 'Ideas and Discursive Structures: US and German Positions in Negotiating Framework Convention on Climate Change', Paper presented at the 2003 Hamburg Conference *Does Discourse Matter? Power, Discourse and Institutions in the Sustainability Transition* (11–13July).

United Nations Framework Convention on Climate Change (UNFCCC) (2005), *Kyoto Protocol*, available at www.unfccc.int/resource/docs/convkp/kpeng.html.

Von Seht, Hauke and Hermann E Ott (2000), *EU Environmental Principles: Implementation in Germany*, Wuppertal: Wuppertal Institute for Climate, Environment and Energy.

Von Weizsäcker, Ernst U (1994), *Earth Politics*, London: Zed Books.

Weale, Albert (1992), *The New Politics of Pollution*, Manchester: Manchester University Press.

Wenderoth, Jürgen (2003), *Interview* (8 April), Berlin: The Federal Ministry for Finance.

Wurzel, Rüdiger (2002), *Environmental Policy-Making in Britain, Germany and the European Union: The Europeanisation of Air and Water Pollution Control*, Manchester: Manchester University Press.

15. Sharing the burdens of global climate change: international equity and justice in European policy

Paul G Harris*

INTRODUCTION

Preceding chapters have analyzed how and why several European countries and the European Union (EU) have started to act on global climate change (GCC). In this chapter, Europe's policies and actions on GCC are examined in the context of global justice generally and international environmental equity (IEE) or fairness in particular. The concept of IEE refers to the fair and just sharing of the burdens associated with environmental changes (see Harris 2001b: 25–43). The questions I want to look at here are, first, what role do and should considerations of international (social and distributive) justice and IEE play in the climate change regime? Secondly, in what ways might ideas about global justice and IEE have shaped European policies on GCC? Finally, are European countries, and is the EU as an organization and a community, doing enough to share the burdens of GCC?[1] We can address these questions from both practical and normative-ethical perspectives. From a practical perspective, it is reasonable to assert that Europe should do its part to act on the provisions of the Framework Convention on Climate Change (FCCC) calling for stabilization of atmospheric greenhouse gas (GHG) concentrations at a level that avoids 'dangerous anthropogenic interference' with the Earth's climate (UNFCCC 1992: Article 2). From a normative perspective, we might argue that Europe should meet a number of commonly accepted ethical standards of international burden sharing. In the real world, of course, practical matters and ethical ones often overlap. Thus the international negotiations leading to the FCCC, and subsequent measures to implement it, have been suffused with discussions about how to prevent GCC while fulfilling ethical demands that those who cause the problem act first and those who suffer from it most (and are least able to act) are afforded special consideration and assistance.

In their survey of analytical approaches to environmental foreign policy (EFP), Barkdull and Harris (2002) describe some of the scholarly work on the role played by ideas. From these perspectives, which often fall under the rubric of constructivism, ideas and values can shape the identities of states and in turn shape or define the international system, especially if those ideas are pushed by major international actors (e.g., hegemonic states or influential intergovernmental organizations). Ideas and values can cross-pollinate among states or emerge from international organizations and regimes, in turn having impacts on state behavior in particular issue areas. Here I am interested in the degree to which IEE has started to permeate the GCC process and shape European behavior, the degree to which Europe has fostered these values, and even the degree to which IEE is becoming 'embedded' in the climate change regime – much as economic liberalism has (for better or worse) become entrenched in the global trade regime (see Barkdull and Harris 2002: 73; cf Ruggie 1998: 62–84). This 'construction' of IEE, justice and fairness has arguably occurred in the context of GCC. International environmental equity has contributed more to the GCC policies of European states and the EU than to the policies of other countries. The EU is in turn driving the construction of IEE more broadly as it becomes a European *interest* and a shaper of its internationally relevant domestic policies (e.g., on energy use and conservation) and foreign policies on GCC.[2]

The next section of this chapter introduces the notion of IEE in the context of the GCC agreements. The third and fourth sections look at how these notions of burden sharing and IEE arguably should and do apply in this issue area. The fifth section suggests that Europe has been a leader on IEE in the GCC negotiations over the last decade and more, and the sixth section points to what European states and the EU have done to take on some of the burdens of GCC (the preceding chapters offer rich details on European policies and actions in this regard). These actions are then assessed from practical and especially normative perspectives in the seventh section of the chapter. The upshot is that Europe is doing more than any other part of the world to address GCC and to share the burdens associated with it. However, while it can be relatively proud of its actions compared to some other countries, notably the United States (US), it is not doing nearly enough. Both practical and normative imperatives point to the need for much more urgent action by Europe to share the burdens of GCC. The eighth section of the chapter looks at the theoretical implications of Europe's *relatively* equitable response to GCC before I conclude. The aim here is not only to show (or at least suggest) how ideas at the various levels have shaped European GCC policies – they have to an extent, as the previous chapter argues – but also to show that ethical considerations in this context exist, that they can serve as one measure of Europe's responses to GCC, and that they ought to guide future action.

In this chapter I discuss 'European' policies and obligations with regard to GCC. I am referring to the EU organization, including the Commission, as well as the member states.[3] That is, I am generally referring to the EU as a conglomerate.[4] Especially with regard to many important issues of concern, such as energy policy, there is much that distinguishes EU states from one another. However, to an increasing degree, on environmental issues generally and particularly on collective policies with regard to GCC, Europe does have a kind of common foreign policy.[5] When it comes to GCC, the Commission has sought to place the EU in a leadership position, which helps explain European states' actions relative to other developed countries.[6]

INTERNATIONAL EQUITY IN THE CLIMATE CHANGE AGREEMENTS

A generation ago, the world's governments came to recognize the need for international cooperation to address global warming and associated changes to the Earth's climate system. Toward that end, in 1992 most of them, including those in Europe, signed the FCCC. The core objective of the FCCC is the:

> 'stabilization of greenhouse gas concentrations in the atmosphere at a level that would prevent dangerous anthropogenic interference with the climate system. Such a level should be achieved within a timeframe sufficient to allow ecosystems to adapt naturally to climate change, to ensure that food production is not threatened and to enable economic development to proceed in a sustainable manner' (UNFCCC 1992: Article 2).

Developed countries agreed to a non-binding target of reducing their GHG emissions to 1990 levels by 2000. They failed to achieve that goal.

The FCCC's third conference of the parties (COP), held in Kyoto in December 1997, was an effort to turn the voluntary commitments into binding ones. Diplomats negotiated the Kyoto Protocol to the FCCC, which requires developed country parties to reduce their emissions of GHGs by about 5 per cent below 1990 levels by 2012. The EU countries agreed to reduce their aggregate emissions by 8 per cent. At the late-1998 fourth conference of the parties in Buenos Aires, and in subsequent meetings, the developed countries agreed upon a set of actions for implementing the Kyoto Protocol. However, even with full implementation, as agreed so far the FCCC and its Kyoto Protocol will result in reductions of less than 5 per cent of the developed countries' emissions of GHGs. Global emissions will continue to rise precipitously. Climate change will continue, virtually unabated, short of new and much more aggressive cuts in GHG emissions. Indeed, scientists say that emissions of carbon dioxide (CO_2) and other GHGs must be cut by at least 60 per cent just to stabilize their concentrations in the atmosphere and to prevent chaos in the global climate system (WRI 1996; BBC World Service 2005).[7]

According to the FCCC, countries should protect the world's atmosphere from GCC 'on the basis of equity and in accordance with their common but differentiated responsibilities and respective capabilities', with the economically developed countries taking the lead in addressing the problem and its effects (UNFCCC 1992: Article 3[1]). Explicit in the treaty was the understanding that the economically less developed countries would not be required to undertake commitments until the developed countries first do so in earnest. This notion of *common but differentiated responsibility* was agreed because the developed countries are most responsible for historic GHG emissions and their potentially adverse consequences,[8] and because they are less vulnerable to the impacts of climate change and more able to pay for necessary action to cut GHGs. What is more, per capita emissions in the industrialized world are much higher than in most of the developing countries, with considerations of IEE suggesting that those countries with the higher per capita emissions do more than those with lower emissions per person (Ashton and Wang 2003: 14).

As the poor and less affluent countries of the world develop economically, they are becoming major sources of GHGs. China is already the second largest emitter of GHGs, and it will take first place in the next few decades (WRI 1996: 315–25). Consequently, in the long run the developed countries cannot by themselves solve this problem; comprehensive participation of the developed countries *and* the major developing countries is required. Some countries – the US, in particular – see the immediate answer to global warming in cutting GHG emissions from the developing world, especially China and India. However, the world's poor countries are unlikely to substantially reduce their emissions simply because the wealthy countries want them to do so. Developing countries have said that they will not join any future international climate protection agreement – another binding protocol to the FCCC or side agreement to the Kyoto Protocol – that is not fair and equitable based on commonly accepted notions of which country or countries are most responsible for the problem – namely, the US, Europe and other developed countries (Harris 2001a; Muller 2002). As Shukla puts it, 'The basic concern of the developing countries is not whether or when to initiate the mitigation actions, but how the mitigation burden will be distributed among nations. This is a justice issue, concerned with an equitable distribution' (Shukla 1999: 157). Consequently, for international climate change negotiations to be successful in coming years – for them to result in actual, robust and widespread overall cuts in GHGs – the developed countries must *demonstrate* that they acknowledge their 'guilt' for damaging GHG emissions and that they are acting to substantively reduce those emissions. In addition, in accordance with the FCCC, the developed countries should assist developing countries in their efforts to deal with this problem.

This difficult situation shows that normative-ethical considerations like IEE, fairness, and responsibility – notions not commonly perceived as essential to serious international discourse (or of most foreign policy practices and analyses) – are absolutely central to efforts to address GCC. Most economically developed countries – particularly those in Europe – now at least recognize this proposition (Harris 1997b). It is of course important to ask exactly what the largest polluters (especially on a per capita basis) ought to be doing to address global warming and the effects of climate change, especially for the world's most vulnerable people and countries.[9] With only about 5 per cent of the world's population, the EU produces about 15 per cent of global GHG emissions (DGE 2002: 8), and it arguably has the wealth to spare to help those who suffer from climate change. In the European Commission's words, 'Richer, industrialized nations have a duty to take the lead in cutting the emissions that could have the most serious impact on developing countries' (DGE 2002: 8–9).[10]

The EU's Climate Protection Program explicitly acknowledges that equity is 'fundamental to the climate challenge' for four reasons (CPP 2004: 15): First, it is a legal imperative based on the FCCC's obligations to act based on equity and common but differentiated responsibilities. Secondly, it is a moral imperative: 'Citizens of the global community face a moral compulsion to engage on the basis of justice and equity. As global interconnectedness grows through globalization and shared environmental and geopolitical challenges, the moral imperative becomes further strengthened' (CPP 2004: 15). Thirdly, it is a political imperative because the nature of the problem requires some countries to take the lead because 'countries will only participate if they perceive the climate regime to be equitable' (CPP 2004: 15). Fourthly, it is a practical imperative because 'the challenge of climate change may only be practically resolvable if equity – in its strongest sense – is addressed. Both pillars of addressing climate change – mitigation and adaptation – rely on a fundamental recognition of equity and sustainable development. The practical imperative inextricably merges the sustainable development goals of the South with the global climate challenge' (CPP 2004: 15). The upshot is that IEE has now permeated the GCC regime, and it has started to shape GCC policies of major countries, in particular those of a number of European states and the EU itself.

SHARING THE BURDENS OF CLIMATE CHANGE: SOME NORMATIVE CONSIDERATIONS

What obligation do states have to share the burdens of GCC?[11] That is, to what extent are they ethically obligated to address the causes and consequences of GCC by cutting their emissions of GHGs and compensating those who have and will suffer from climate change caused by those emissions? It is impossible to state definitively what is a 'fair' or 'equitable' distribution of climate change burdens. There are too many disagreements about who is to blame and what should be done. Having said this, it is possible to approach an answer (cf Linnerbooth-Bayer 1999). Regardless of the specifics, it seems unassailable, to start with, that there ought to be a fair and equitable sharing (or distribution) among countries of the burdens (and decision-making authority and possibly even some potential benefits) associated with climate change. One can say 'unassailable' because it is difficult to envision any reasonable person or government arguing that there ought to be an *unfair* or *inequitable* (as opposed to possibly *unequal*) sharing of burdens associated with GCC.

Ethical philosophers have endeavored to define what is fair and equitable, both within and between countries, in the context of GCC, and the United Nations' Intergovernmental Panel on Climate Change (IPCC) has grappled with the question (see Bruce et al. 1996: 79–125; Odingo et al. 1994). In simple terms, equity means the quality of being fair, impartial or even-handed in dealings with others (Pearsall and Hanks 1998). People will, of course, disagree about the precise definition or content of fairness and equity. Indeed, that is arguably much or even most of what the climate change negotiations over the last almost-two decades have been about (see Hampson and Reppy 1996; Harris 1996, 2000).[12] In the final analysis, what constitutes a fair and equitable sharing of GCC burdens will be the result of political bargaining among the states – and other influential actors, such as corporations and non-governmental organizations (NGOs) – involved. This will not suit ethical purists, but reality and ethics – what is or is not right or wrong – will collide. Power does not make right in this circumstance any more than others, but bargaining and, yes, power, plays a role in determining which normative principles actually shape outcomes. However, at the very least, as Shue argues, fairness and equity mean doing no additional harm to other countries and other people, and 'the fundamental focus of equity is the protection of an adequate minimum for those unable to protect their own' (Shue 1995: 392). At the most basic level, fairness demands that Europe (among other developed parts of the world[13]) act to limit and ultimately end the harm it does by way of GCC, giving due consideration to the least well-off countries and people.

Discussions of fairness and equity often refer to two general categories of issues: procedures (or how decisions are made) and outcomes (or consequences). Procedural equity requires that basic rights of individuals or states be respected in decision-making, and that those affected by decisions be allowed to participate in the formulation of those decisions. The fairness or equity of outcomes demands that there be a fair distribution of burdens and a fair allocation of benefits (Banuri et al. 1996). These conceptions of course apply both within and between generations and within and between states, with many pointing out that they also apply between humans and other species.[14] However, again, what might we mean by 'fair' or 'equitable'? Among the many conceptions of international fairness and equity, six are considered here. They are premised on rights, causality and responsibility, utilitarianism, Kantian ethics, Rawlsian justice and impartiality.[15]

From a rights perspective, individuals have inherent rights, such as minimum nutrition, freedom from torture, and so forth, simply because they are human beings. Shue (1996) argues that individuals have at least the right to subsistence, for without it no other rights can be exercized. Thus, insofar as the burdens of climate change inhibit subsistence rights – as they will do for many as climate change manifests itself in droughts, floods, the spread of pests and other detrimental effects on agriculture, notably in the poorest parts of the world (see IPCC 2001) – the distribution of benefits and burdens is not, at present and into the foreseeable future, fair and equitable. Conceptions of fairness and equity based on causality or responsibility assert that those responsible for causing harm are responsible for ending and ultimately righting that wrong. According to Shue, 'the obligation to restore those whom one has harmed is acknowledged even by those who reject any general obligation to help strangers … this is because one ought even more fundamentally to do no harm in the first place' (Shue 1995: 386). The European states, the US and other developed countries acknowledged when they joined the FCCC that they deserve the bulk of the blame for climate change and that they have some responsibility to aid other countries that will be affected by it. Utilitarians argue that benefits and burdens should be distributed in such a way as to achieve the greatest good for the greatest number of individuals (cf. Warnock 1962; Bentham 1970). Thus, from this perspective, the burdens of climate change ought to be distributed in such a way as to maximize overall human 'happiness'.[16]

From a Kantian perspective of equity and fairness, people should not be treated as means to one's own ends, but as ends in themselves (Kant

1948). It is unfair to exploit other people because they would not freely choose to be exploited. Thus if the burdens of climate change are imposed on countries against their free will, then that distribution is not fair and equitable. A Rawlsian conception of what is a fair and equitable distribution of burdens can be derived from assessments of self-interest decided in an 'original position' behind a 'veil of ignorance'. Inequalities in distribution are acceptable insofar as they benefit the least advantaged in society, because that is what those in the original position (of, for example, not knowing what country or social class one would be born into) would choose (Rawls 1971).[17] Applied to climate change, one might say that, insofar as the requirements of weak, small-island and poor coastal countries, which are subject to sea-level rise and increased storms caused by global warming, are seriously considered in the context of GCC, then fairness and equity are being furthered. However, so far, their demands have received too little serious attention by the developed countries, at least compared to the scale of suffering they are likely to endure with greater frequency. A conception of fairness and equity based on impartiality suggests that we ought to assess what is fair and equitable based on what is reasonable (Barry 1995). It is not reasonable to expect an equal relationship between (for example) Europe and Vanuatu. European countries should aid Vanuatu to cope with climate change (using the words of Brown in an analogous context), 'not because it is in [their] interest to do so but because justice as impartiality suggests that the case for such aid cannot be reasonably denied' (Brown 1992: 181).

Biermann has summed up the 'essence of a general concept of justice in international environmental policy' (from the perspective of practice and international law) as (1) more capable states shall accept more duties (stricter regulations); (2) more capable states shall assist others; and (3) equal participation shall be guaranteed and participation shall be transparent (Biermann 1999: 163–9). His characterization points quite strongly to the evolving practice of rich states aiding poor states (even if they are not responsible for the suffering of the latter). Ringius et al. (2002: 17) argue that for any burden-sharing rule in the context of climate change to be acceptable it must at least be compatible with three principles of equity: responsibility, capacity and need (including basic need). Beyond that, the rules are and have been open to substantial negotiation, including that influenced by national interests and power (Ringius et al. 2002: 17). The upshot is that, while debates about the precise content of IEE in the context of GCC continue, equity considerations are already being integrated into the GCC regime and the relevant policies of a number of major developed states, notably in Europe more than elsewhere.

SHARING THE BURDENS OF CLIMATE CHANGE: SOME ISSUES OF IMPLEMENTATION

These varied conceptions of fairness and equity are germane to the international politics – and associated domestic burdens – of GCC in a number of ways, including (1) the distribution of GHG emissions, (2) the transfer of financial resources and technology and (3) compensation for harm resulting from climate change.[18]

Distribution of Emissions

A fair and just solution to the climate change problem requires addressing the distribution of GHG emissions reductions and the costs that accompany those reductions. The disproportionate responsibility of European and other developed countries for causing climate change – that is, the present concentration of GHGs in the atmosphere – has been invoked by most countries during international negotiations. Parties to the FCCC essentially agreed that the primary and initial burdens of emissions reductions should be borne by the developed countries. The preamble to the FCCC notes that most current and historical emissions of GHGs have originated in the developed world, that 'per capita emissions in developing countries are still relatively low and that the share of global emissions originating in developing countries will grow to meet their social and developmental needs', and that actions to address global warming should first consider the 'legitimate priority needs of developing countries for the achievement of sustained economic growth and the eradication of poverty' (UNFCCC 1992: Preamble). The convention calls on all parties to share the burdens, but to do so in such a way that those most responsible for historical emissions and those most capable of taking action (generally the same countries) bear the bulk of the burden.

Article 3(1) of the FCCC declares: 'The Parties should protect the climate system for the benefit of present and future generations of humankind, on the basis of equity and in accordance with their common but differentiated responsibilities and respective capabilities. Accordingly, the developed country Parties should take the lead in combating climate change and the adverse effects thereof'. The developed countries affirmed the principle of common but differentiated responsibility in the Berlin Mandate, agreed at the 1995 first conference of the FCCC parties. They pledged to act first to reduce their GHG emissions before requiring developing countries to do so. Thus negotiations for the subsequent Kyoto Protocol were premised on the common-but-differentiated-responsibility principle. The developed countries accepted this standard because they knew developing countries would not – and in many cases could not – otherwise join the climate regime and eventually limit their emissions.

Not surprisingly, there has been much dispute regarding how to allocate GHG emissions. This was true in the (relatively simple) case of the Kyoto Protocol, but applies even more to follow-on agreements for the much greater cuts that will be required in the future to have a substantial impact on GCC. The most obvious allocation of GHG emissions, one proposed by developing countries and their advocates, was to an equal global per capita allotment of allowable emissions. The rich countries would then have to pay the poor countries for the use of their allotment. While the GCC negotiations so far have arrived at bargains short of codifying equal per-capita rights to the atmosphere, Biermann believes (along with developing country diplomats) that only equal per capita allotments, which 'have an inherent appeal due to their link to basic human rights of populations in both South and North', are 'likely to have the normative power to grant the climate governance system the institutional stability it needs in the decades and centuries to come', not least because, 'in the unequivocal words of the Indian prime minister at the 2002 conference of the parties to the climate convention, it does not seem that "the ethos of democracy can support any norm other than equal per capita rights to global environmental resources" (Vajpayee 2002)' (Biermann 2005: 20).[19] In contrast, Hourcade and Grubb argue that 'for political and economic reasons such a proposal has no chance of being accepted by developed countries because it leads to unacceptable costs for them' (Hourcade and Grubb 2000: 199).

At some point the most advanced developing countries will have to come on board, with limits and, in the cases of the largest polluters, eventually even cuts of their own. However, there is little chance that they will agree to GHG cuts that let the developed countries continue their disproportionate emissions or do not require them to help the poor countries (especially the least developed among them) to continue growing economically. As suggested above, the need for Europe and other developed regions to share more of the burden is manifest; it is a necessary prerequisite for much more action by China, India and other large developing countries that are growing sources of the problem itself.

There is also the important question of how any emissions limitations should be distributed among developing countries themselves. Some of the newly industrialized countries and those with substantial wealth, such as Saudi Arabia, Singapore and South Korea, as well as rapidly industrializing countries, such as Brazil and China, will presumably be expected to act long before the very poorest countries are seriously expected to do anything more than receive aid for mitigation of their suffering from climate change (Ott et al. 2005: 91; see CPP 2004).

Financial Resources and Technology Transfer

Financial and technological transfers to the developing countries will be required to help them adapt to climate change and to assist them in eventually limiting their GHG emissions. These transfers are often justified based, again, on the North's disproportionate responsibility for the problem. The FCCC demands that industrialized countries help the developing countries by providing finance and technology to meet treaty objectives. Developing countries joined the convention only after it was agreed that their development prospects would not suffer in the process. There was an implicit understanding that some sort of international fund would be established to compensate developing countries for the costs of participation in the convention (Mott 1993). Article 11 of the FCCC describes the financial mechanism envisioned to provide financing 'on a grant or concessional basis, including for the transfer of technology', to help poorer signatories fulfill treaty commitments (Article 11[1]). The Global Environment Facility was designated as the funding mechanism for the FCCC (see Mott 1993).

In several articles the climate convention calls on participants to provide developing countries with aid to assist them in fulfilling their information and reporting requirements (Article 4[3]). Article 4 calls for new and additional resources delivered from the developed countries in an 'adequate' and 'predictable' fashion to assist developing countries in complying with their obligations under the convention (Article 4[3]). Developed countries are to take steps to 'promote, facilitate and finance, as appropriate, the transfer of, or access to, environmentally sound technologies and know-how', especially to developing countries (Article 4[5]). The same article declares that developing countries' effective implementation of the treaty 'will depend on the effective implementation by developed country Parties of their commitments … related to financial resources and transfer of technology and will take fully into account that economic and social development and poverty eradication are the first and overriding priorities' of the developing countries (Article 4[7]).

By signing the treaty, developed country governments accepted these principles. The Bonn and Marrakech Accords, which lay out measures for implementing the Kyoto Protocol, established several new funding mechanisms to help poor countries adapt to climate change: the Special Climate Change Fund, the Least Developed Country Fund, and the Kyoto Protocol Adaptation Fund (see Ott 2001b; Huq 2002; Vrolijk 2002; Dessai and Schipper 2003). Importantly, the parties to the FCCC have agreed that *new and additional* funding, beyond that in existing aid programs, would be needed, and the EU and its member states, along with Canada,

Iceland, New Zealand, Norway and Switzerland, agreed to donate €$450 million annually (some of it new money) to these new funding mechanisms by 2005 (IEA 2001: 8; UNFCCC 2001, cited in Ashton and Wang 2003: 10, 21; IIED n.d.).[20] The developed countries (and their industries) are of course looking for ways to achieve the objectives of the FCCC at minimum cost. Transfers were already part of joint implementation projects that permitted developed country firms to save money by reducing emissions at facilities in developing countries. These projects were endorsed by the Kyoto Protocol and clarified in subsequent international negotiations, and they have increased in number and scale in recent years (see below). Joint implementation projects and tradable permit schemes will go some way toward actualizing financial and technology transfers, thereby assisting the developing countries, in the process making emissions reductions less costly in the developed countries. However, the developed countries have been slow to act on these principles, and the sums of money and transfers of technology have been only modest so far.

Compensation

If one accepts several of the notions of fairness and equity outlined above (particularly those based on causality and responsibility), the developed countries ought to compensate those countries that will suffer from the effects of climate change. The low-lying island and coastal states, for example, will feel the effects of rising seas (and may already be feeling them).[21] They want to be – and ethically ought to be – compensated for these adverse effects. However, this is one area where both the treaty and the stated intentions and actions of the developed countries fall far short of many principles of IEE. The most the treaty says about this is Article 4(4): 'The developed country parties and other developed parties … shall also assist the developing country Parties that are particularly vulnerable to the adverse effects of climate change in meeting costs of adaptation to those adverse effects'. The developed countries implicitly acknowledged that they ought to compensate those countries that will suffer from climate change, but in the foreseeable future there will be only modest actions toward this end. 'In other words', as Paterson points out, 'compensation is only formulated [in the FCCC] as a vague principle without any concrete implementation scheme' (Paterson 1997: paragraph 5.4.3.3). What is more likely is additional financial assistance from the wealthy countries to the poorer ones to help them *adapt* to climate change. This is the direction that Europe has moved in. This very different from compensation for harm, however.

EUROPE'S RESPONSE TO GCC: LEADERSHIP IN INTERNATIONAL NEGOTIATIONS

What has Europe done to equitably share the burdens of GCC? Several European states and the EU have arguably been global environmental leaders, with the EU in particular 'championing the Kyoto Protocol on climate change despite US opposition to the agreement' (Axelrod et al. 2005: 208) along with other international efforts to deal with climate change (Schreurs 2004: 222). In the run-up to the Kyoto COP, the EU 'tried to position itself as the global leader on climate change issues', after being pushed by some 'progressive' member states (Collier 2002:180). Porter, Brown and Chasek have defined the EU as a 'lead state' on climate change issues because it was consistently ahead of other countries in calling for action to combat the problem through GHG cuts (Porter et al. 2000: 114, 119). To be sure, not everyone would call Europe the GCC leader, but when contrasted with other state actors it has been relatively so.[22]

The Run-up to Kyoto

By 1990, the European Commission, especially the environment commissioner, decided that Europe should be a leader in the GCC negotiations before the 1992 Rio Earth Summit (McCormick 2001: 281). Doing so was made easier by GHG emissions cuts adopted at the national level by some member states (McCormick 2001: 281). Formal international negotiations that would lead to the FCCC started in 1991, and by 1992 members of EU committed themselves to stabilizing their GHG emissions at 1990 levels by 2000 (IEA 1994, cited in Betsill 2005: 106). At the Earth Summit the EU sought agreement among other countries to establish binding targets, by dates certain, for reducing GHG emissions (Schreurs 2004: 211).[23] The FCCC did not include these firm commitments. According to Yamin, in the early climate change negotiations, the EU tried to balance the US demand that developing countries commit themselves to GHG emissions limits and those same countries' refusal to do so (Yamin 2000: 62). On the one hand, the EU 'supported the G-77 and China's opposition to new commitments for developing countries [and] dismissed demands for DC [developing country] commitments as premature because it argued the principle of common but differentiated responsibility required effective leadership by Annex I [developed country] Parties' (Yamin 2000: 62). On the other hand, it argued that 'regulation of DC emissions in the long term was an environmental necessity' (Yamin 2000: 63). In the end, of course, the developing countries were able to avoid any commitments on the latter, and developing country emissions limitations will for the time being depend largely on projects in the context of the Clean Development Mechanism (CDM) and the like, often financed by Europe.[24]

In the period between the signing of the FCCC in 1992 and the Kyoto Protocol in 1997, the Europeans continued their efforts. They called 'for an international agreement based on binding emissions targets and timetables and premised on the understanding that the industrialized states were to act first by cutting domestic emissions' (Schreurs 2004: 213). According to Schreurs (2004: 213), 'Germany was a particularly strong advocate of action', hosting the first COP in 1995, from which the Berlin Mandate emerged. Even before the conference, the EU's Council of Environmental Ministers was pushing for diplomats there to adopt a protocol to the FCCC that would set targets and timetables for reducing GHGs after 2000 (Haigh 1996: 182). Alongside environmental NGOs and the Alliance of Small Island States (AOSIS) – whose members feared the devastating effects of sea-level rise – the EU was noteworthy for strongly arguing for binding targets and timetables for lowering GHG emissions (Betsill 2005, 112). Given that the only other group of states supporting emissions cuts were from the developing world, the EU states' position put them at the forefront of pledges to act on climate change by reducing GHGs (see Betsill 2005, 108).[25] Describing climate change negotiations of the mid-1990s, Leggett argued that 'The European Union has for a long time now been the most progressive force at the climate negotiations, outside the Alliance of Small Island States. If there were to be any prospect of a protocol with cuts in Kyoto, the EU would have to make the first move. They did so', calling for a 15-per cent cut in emissions of three main GHGs (CO_2, methane and nitrous oxide) by 2010 (Leggett 2001: 259).[26]

By early 1996, Germany was calling for CO_2 emissions cuts by developed countries of 10 per cent from 1990 levels by 2005 and 15–20 per cent by 2010, falling between the AOSIS proposal of cutting CO_2 emissions 20 per cent by 2005 and a British proposal of 5–10 per cent by 2010 (Leggett 2001: 236). According to Leggett, the EU

> 'made it clear, in partial support of Germany, that Europe supported action going beyond "no regrets". What this should mean, the EU suggested, was stabilization of atmospheric CO_2 concentrations at less than 550 parts per million – a doubling of atmospheric CO_2. This target should guide "limitation and reduction efforts", which translated implicitly meant that the long-term goal would have to be deep cuts' (Leggett 2001: 236–7).[27]

At March 1997 international GCC negotiations, the EU opposed US efforts to subject developing countries to emissions targets: 'Developed countries should lead the way, they said, and indeed had already agreed to do so in the Berlin Mandate itself' (Leggett 2001: 260). In Attfield's view 'In the negotiations leading up to the Kyoto Protocol agreement the European Union played a valuable role in persuading other developed nations to join a new greenhouse gas regimen on principles acceptable to the Third World' (Attfield 1999: 199). This may be a slight exaggeration of the degree to which the agreement fits the preferences of the developing world, but it at least suggests further that the EU was considerate of developing country perspectives on equity in the context of GCC.

Differences between European and other developed states, notably the US, were especially evident during the Kyoto negotiations. Shortly before the Kyoto conference, with the US opposing GHG emissions cuts, top European diplomats said that they were 'sick with disappointment that the USA was being so backward' (Leggett 2001: 280). Schreurs notes that, while the US was demanding that the developing countries undertake 'meaningful measures' to limit their GHGs, the EU was a strong supporter of 'the idea that the main responsibility for global warming lay with the developed states and that therefore they should set an example for the rest of the world by making sharp emissions cuts at home through a mix of regulations, incentives, and voluntary measures' (Schreurs 2004: 215–216). At Kyoto the EU followed through, pushing for a 15 per cent cut in GHGs by 2010 and putting pressure on other countries, in Elliott's reckoning giving the Europeans the 'moral upper hand' (Elliott 2004: 88). During the Kyoto COP and in subsequent GCC negotiations, the Europeans also defied the US in pushing for the majority of emissions cuts to be made at home, whereas the US wanted a larger role for 'flexible mechanisms,' notably emissions trading between states.[28] As Betsill points out, whilst the US and industry representatives were pushing for flexible instruments for GHG emissions reductions during negotiations for the Kyoto Protocol, 'the EU, most developing countries, and environmental groups objected, arguing that extensive reliance on such mechanisms would allow rich countries to buy their way out of making any meaningful commitments domestically, thereby violating the "polluter pays" principle' (Betsill 2005, 114–15). Throughout the GCC negotiations during the 1990s, leading up to and through the 1997 Kyoto conference, the EU led on GCC and 'officially recognized its global responsibility in this area' (Collier and Loftstedt 1997: 2). However, progress toward *achieving* this goal was only modest (at best) during the 1990s.

After Kyoto

The Europeans seem to have taken GHG cuts even more seriously following the signing of the Kyoto Protocol. According to a May 1999 European Commission report on EU preparations for implementing it:

'Developing countries understandably oppose constraints on their economic development. A dialogue on the role of developing countries in the fight against climate change will therefore have to take into account the principles of equity and of common but differentiated responsibilities. ... Furthermore, the EU is fully supporting the view that industrialized countries must show the way in addressing the problem. ... In order to convince the "non-Annex I" developing countries to participate in an open and constructive debate, the EU and other industrialized countries will need to show demonstrable progress [by the] 2005 [target date]' (European Commission 1999: 19).

The European Commission proposed, and European negotiators followed, a firm EU position on a number of issues at the sixth COP at The Hague during November 2000. They demonstrated substantial – even remarkable, relative to some other developed countries (notably the US) – dedication to equity and fairness in the developed world's response to climate change. The Commission proposed, and the EU demanded at the COP, that the use of 'flexible mechanisms' by developed parties be severely limited, thus requiring them to largely undertake their pledged emissions cuts (European Commission 2000; Ott 2001a: 287). The Commission recommended that at the COP European diplomats:

'ensure that industrialized countries take real action at home to reduce emissions by securing a primary role for domestic policies and measures ... limit the extent to which "sinks" that absorb carbon, such as forests, can be used by industrialized countries to offset emissions ... establish a comprehensive and tough regime to oversee Parties' compliance with the Protocol, including effective penalties with a clear economic impact for non-compliance by industrialized countries with their emission targets [and] help developing countries meet their Convention commitments [reporting requirements] and adapt to climate change through capacity building, transfer of clean technologies and financial assistance' (European Commission 2000: 2).

This sounds amazingly like a recipe for climate equity. The Commission acknowledged that large developing countries needed to be brought on board the climate change regime if it is to be effective in the long term, but it also argued that the FCCC accepts that 'the extent to which developing countries will implement these commitments in practice depends on industrialized countries making good on commitments ... to provide financial resources and transfer of technology' (European Commission 2001: 11). According to the Commission, the EU wanted 'to see COP6 develop a broad process to cater to the adaptation needs of developing countries, particularly the least developed and most threatened ones. The EU is sensitive to developing countries' concerns and wants to work constructively to get a better understanding of their needs and how to address them' (European Commission 2000: 12). This falls short of meeting IEE obligations, but the concepts underlying EU policy are clearly (to this observer) guided by, or at least fit, those obligations to a significant degree.

Some observers of the international negotiations on climate change had assumed that Europe would join the US in rejecting the Kyoto Protocol following President Bush's rejection of it in March 2001. However, the EU was able to persuade other parties to the FCCC that the Kyoto Protocol should be kept alive (DGE 2002). The EU immediately sent a letter to President Bush declaring that, for the EU, an agreement 'leading to real reductions in greenhouse gas emissions is of the utmost importance … The global and long-term importance of climate change and the need for a joint effort by all *industrialized* countries in this field makes it an integral part of relations between the USA and the EU' (cited in Gelbspan 2004: 94–95; my emphasis). As Schreurs describes it, the EU:

> 'surprised the Bush administration. Rather than joining the US in abandoning the agreement, the EU's resolve to move forward with ratification and win the support of other nations to do the same intensified … Prior to the Bush announcement, cracks were evident in the EU regarding climate change. The Bush decision, however, so angered the Europeans that they were able to overcome internal disagreements and present a strong united block supporting the Kyoto Protocol. They called upon the Bush administration to do the same. Seldom in history has the EU criticized the US as forcefully as it has over this issue' (Schreurs 2004: 209).

Indeed, at the late-2005 joint COP11 and first meeting of the parties to the newly ratified Kyoto Protocol, the EU joined with most world governments in aggressively opposing US efforts to derail the negotiations generally and the protocol in particular (see ENB 2005). Admittedly, given the utter failure of the US to show any resolve in addressing GCC, it is relatively easy to declare the EU the leader on this issue. Having said that, even without the US comparison, the EU was at least at the forefront in the developed world – and even very far ahead of many developing countries' wishes (e.g., the oil producers).

EUROPE'S RESPONSE TO GCC: CUTTING GREENHOUSE GASES

While the EU and its member states' implementation of GHG cuts has been less spectacular than the EU's plans, they have nevertheless been leaders in 'long-term target setting and driving the international climate policy process' (Michaelowa 2004: 23). The EU has faced difficulties in acting on GCC, to be sure. A case in point is the failed effort to actualize a union-wide energy tax and other measures during the 1990s (see Collier

1997b: 50–58; McCormick 2001: 282–4; Zito 2000: 167–95; see Chapter 11). Nevertheless, the EU has been much more active than other countries in moving to cut GHG emissions, and it is doing more to use the Kyoto mechanisms (e.g., joint implementation, the CDM, emissions credits) and is providing more capacity-building funds (Michaelowa 2004: 23). Indeed, EU efforts to limit European GHG emissions began in the 1980s, and by 1990 the EU had set a goal of stabilizing CO_2 emission at 1990 levels by 2000 (Schreurs 2004: 210). In mid-1990 the European Council called for the implementation of 'strategies and targets' for cutting GHGs, and a joint meeting of environmental and energy ministers later in the year agreed that CO_2 should be 'stabilized at 1990 levels by the year 2000' (McCormick 2001: 281). Some EU member states went further. For example, in 1990 Germany stated its plan to cut its CO_2 emissions by '25 per cent of 1987 levels' by 2005 (Schreurs 2004: 210; see Social Learning Group et al. 2000, cited in Schreurs 2004: 228, and Chapter 2). This was a reflection of the degree to which the precautionary principle – that preventative action to protect the environment should be taken even if scientific certainty does not obtain – and the belief that polluters should pay for pollution prevention (the 'polluter pays' principle) had already permeated European thinking and its EFP (Schreurs 2004: 210–11; see Harris 2004).

In 1996 the EU Council of Ministers agreed that 'global average temperatures should not exceed 2 degrees C above [the] pre-industrial level' (European Commission 2005c: 3). This is the measure against which the Commission sets its emissions objectives to meet the mandate of the FCCC. This mandate to keep to the 2-degree maximum temperature rise was confirmed in March 2005 (European Council 2005: 15–16). Toward this objective, the European Commission has proposed and implemented over the last decade and more a number of programs designed to actually reduce GHG emissions across the EU. These efforts have met with some success (see Collier 1997: 50–58). They have included the European Climate Change Program, the Renewable Energy Directive, the Environmental Technologies Action Plan and energy-efficiency guidelines, among a plethora of others.[29] In 2004, the EU adopted an Action Plan on Climate Change and Development (see European Council 2004). Arguably, the most serious development has been the EU-wide internal emissions-trading system that took effect at the start of 2005 (see DGE 2002: 19; Klepper and Peterson 2005; Rosenzweig et al. 2005; European Commission 2005b; Chapter 12). The scheme covers thousands of large factories, power production facilities and other major polluters. Member states set emissions limits for these facilities, and those that lower their emissions are permitted to sell the surplus. It is too early to assess the effectiveness of this scheme, but already the Commission has demonstrated willingness, and some capacity, to fend off efforts by some member states to go easy on some of the affected industries (JIQ 2005b: 2). For example, in early 2006 it rejected a request from the UK to increase the amount of CO_2 it is permitted to emit from 2005 to 2007 under the EU emissions-trading scheme (BBC News 2006).

The EU as a whole was on a path of GHG reductions in the 1990s, but without additional policies its emissions were projected in 2002 to drop only 4.7 per cent by 2010, 3.3 per cent short of the EU's collective Kyoto target (EEA 2002: 5, 74; see Duwe 2004: 10). However, a 2000 report on the European Climate Change Program projected that the EU 'could exceed the target with *additional policies and measures*' (ECCP 2003: iv; see also European Commission 2003b, 2004). European Union-wide actions are only now starting to have a real impact (Duwe 2004: 6), but planned measures are likely to be enough for Europe to exceed its Kyoto obligations (Duwe 2004: 29), although most of this would be the result of actions by Britain, Germany and Sweden (EEA 2002: 74; see also EEA 2004).[30]

Some members states have put in place domestic policies that are bringing their GHG emissions down, although some of their conditions are unique, as with the UK's switch to natural gas, whereas others are acting later in accordance with EU-wide measures (Duwe 2004: 6). By 2000, half of the EU was moving toward exceeding their share of EU-wide (bubble) GHG emissions, notably Austria, Belgium, Denmark, Greece, Ireland, Italy, the Netherlands, Portugal and Spain (ECCP 2003: 10). Spain has been one of the greatest laggards in the EU, with its emissions in 2000 almost 34 per cent higher than a decade earlier. Only a few EU member states are on track to cut their GHG emissions in line with the internal burden agreement to reach Europe's –8 per cent Kyoto target, although several (Denmark, Germany, France, Luxemburg, Sweden, UK and the EU-15 as a group) have cut emissions relative to 1990 (Klepper and Peterson 2005: 5).[31] Some countries have been especially ambitious, notably Germany, which was the first developed country to take 'far-reaching steps towards a stringent climate policy', and the Nordic states have followed its lead (Ott 2001a: 278). By 2000, Germany was on track to achieve its –21 per cent commitment within the EU burden-sharing agreement, already lowering its emissions by more than 19 per cent (ECCP 2003: 10). While some of Germany's emissions cuts at first resulted from economic declines in eastern Germany (relative to West Germany), about 15 per cent of the cuts (in the first half of the 1990s) came from overhauls of infrastructure in what was East Germany, and the government has implemented a relatively aggressive set of climate change-related policies, including energy taxes, cogeneration of heat and power, and measures to promote renewable energy (Ott 2001a: 293; see Chapter 2).

The latest assessment on EU GHG emissions from the European Environment Agency (released in late June 2005) shows that EU-wide emissions increased 1.5 per cent in 2003 after falling in 2002 (EEA 2005). The EU-15 saw an overall increase of 1.3 per cent relative to 2002 (EEA 2005). This increase was attributed to increased coal use for power production and a colder than normal winter. Nevertheless, overall emissions of the EU-15 countries were down 1.7 per cent relative to 1990 (down 2.9 per cent if averaged over the preceding 5 years). While many of the new member states have much lower emissions than in 1990, as expected, several major EU-15 states were also down (Germany –18.5 per cent, UK –13.3 per cent, and France –1.9 per cent) (EEA 2005). Consequently, by 2005 the European Commission was reporting that the EU had reduced its emissions by 3 per cent below 1990 levels (European Commission 2005c: 10). Further cuts were expected from EU and national policies, including emissions trading (from January 2005) and carbon sequestration projects (EEA 2005), and recent measures of the European Climate Change Program that were not in effect until 2003 (European Commission 2005a). As of June 2005 the EU expected the EU-15 states to meet their overall Kyoto target of –8 per cent (European Commission 2005a), with the new member states likely to cut much more.

Most EU member states have joined Germany in adopting 'climate protection programs with more or less far-reaching policies' (Ott 2001a: 293; see German Federal Environment Ministry 1997). The plans by several member states for future (post-2012) emissions cuts are ambitious: the UK announced a goal of –60 per cent by 2050, calling on other developed countries to do likewise; France referred to a target of –75 per cent by 2050, also calling on other developed countries to undertake such a commitment; Germany announced a target of –40 per cent by 2020 (contingent on the whole EU cutting emissions by 30 per cent in the same period); the Netherlands proposed –30 per cent by 2020 (Special Committee on a Future Framework for Addressing Climate Change 2004: 13). The European Commission acknowledged 'the need for further emission reductions beyond 2012, by suggesting an EU target to reduce emissions by an average of 1 per cent per year up to 2020 and a global target of 20 to 40 per cent reduction by 2020, both from 1990 levels' (EEA 2002: 7). In a paper laying out post-2012 strategies on GCC, the European Commission reminded members that achieving the goal of limiting average global temperature increases to 2 degrees Celsius above pre-industrial levels would require significant cuts in emissions of GHGs (JIQ 2005a: 2).

Importantly, the Commission anticipated that the EU's share of global GHG emissions would drop to below 10 per cent (from 15 per cent now) – whereas emissions from the developing world would rise above 50 per cent

(JIQ 2005a: 2). In March 2005, the EU Ministers of the Environment and the Heads of State and Government affirmed their desire to push for deeper cuts in GHGs by developed countries (JIQ 2005a: 2). The European Council of Environment Ministers proposed that developed countries reduce GHG emissions 15–30 per cent below 1990 levels by 2020 and 60–80 per cent by 2050 (JIQ 2005a: 2). The EU Heads of Government seconded the former sentiments,[32] with an effort to start codifying early proposals for 20–40 per cent cuts in GHGs by 2020 (compared to 1990) and total cuts of 70 per cent in the long term.[33]

EUROPE'S RESPONSE TO GCC: A PRELIMINARY ASSESSMENT

Given that we can reasonably expect international negotiations and policy evolution on climate change to continue for a century or more, any assessment of Europe's response to the problem must be preliminary and based on the early stages of the process.[34] That Europe has started to recognize its responsibilities is demonstrated by repeated official declarations, its diplomacy over more than a decade, and actions on the part of several member states to reduce and, in the case of many laggards, at least limit their business-as-usual GHG emissions. However, recognizing one's responsibility for harm to others is only the first step. The stated *policies* of Europe – particularly a few member states and the Commission in particular – have often fit, to varying degrees, with several conceptions of fairness and equity. The actual *behavior* of European governments and the EU, while clearly moving in the right direction, is more difficult to assess, in part because it has only just started.

In this section, I want to consider this question: Have Europe's policies and actions with regard to GCC been fair and equitable? Put another way, has Europe started to take on its fair share of the GCC burden? We can start to find answers by looking to the conceptions of international fairness and equity introduced earlier, namely those premised on rights, causality and responsibility, utilitarianism, Kantian ethics, Rawlsian justice and impartiality. The upshot is that, although European actions were fairer and more equitable than those of other major countries, most notably the US under the George W Bush administration, more must be done before one could comfortably say that Europe's sharing of climate change burdens has been adequately and appropriately fair and equitable.

Before doing this, we can begin by asking whether Europe's actions are enough to achieve the *practical* objectives of the FCCC, objectives that of course have ethical importance given the scale of suffering that GCC will bring. On the matter of emissions reductions needed to prevent 'dangerous anthropogenic interference' to the Earth's climate – the central feature of the FCCC – Europe has started to act, but it has hardly adopted a sufficiently robust set of policies. The cuts so far are simply nowhere near what is needed. While the developed states as a group agreed to an overall 5.2 per cent cut in GHG emissions under the Kyoto Protocol, the EU accepted a bigger cut of 8 per cent, to be met by burden sharing among member states. Even if the EU cuts are fully realized, which now seems likely but is not certain, they are on the order of only one-tenth of the effort that is required of Europe. Looked at another way, Europe's actions are more than any other state or group of states has agreed to undertake (see UNFCCC 1997, Annex B). Europe has focused much more on actually cutting its emissions – subscribing to the polluter pays principle – than other regions and states. That is, it is not only seeking cost-effective ways of taking credit for cuts it helps bring about outside Europe, but it is also implementing emissions cuts within the EU (although the bubble allows several of the member states to increase their GHG emissions). Without the involvement of the EU and its member states, the FCCC and Kyoto Protocol, limited as they are, would be much weaker – and the FCCC might never have been agreed (Haigh 1996: 182). However, despite Europe's actions and pronouncements on cutting its GHG emissions, the degree to which it lives up to its own rhetoric, let alone the requirements of global environmental equity, is hindered by continuing subsidies to the coal industry, funding for road transport and export credits that add to climate change (Duwe 2004: 26). Europe's GHG limitations are important first steps, but given the scale of the problem they would have to be *much* greater to actually protect the Earth's climate system.

What can we say of Europe's actions from a normative-ethical perspective? Europe arguably acts much more ethically than do its developed-country counterparts. From the perspective of rights, Ott and Santarius characterize the equity considerations this way: 'equity in its meaning of fundamental human rights will take center stage – the right to life, health and the protection of property' (Ott and Santarius 2003: 8). It seems that threats to these basic rights are recognized by, and are concerns of, the EU, which has increasingly invoked the welfare of those vulnerable to GCC as objects of European concern and justification for its GHG cuts and development assistance to aid adaptation in the developing world. Europe has also said and done much more to demonstrate that it accepts substantial *responsibility* for the causes of climate change. The EU has acknowledged repeatedly that, along with other major developed economies, its historic emissions are the cause of ongoing and much of the anticipated changes to climate. For example, the EU's environment commissioner stated emphatically that:

> 'action has to be based on the principle of *common but differentiated responsibility*, as enshrined in the Climate Convention. We can not expect the developing countries to do something that many industrialized countries in the world, with all their research and technological capabilities, have not been able to do. The EU does not think this is realistic or fair, taking into account historical responsibility, current capability and actual per capita emissions' (European Commission 2001c, original emphasis).

Instead, the stated EU policy, and that of member governments, is to reduce EU-wide GHG emissions and to assist developing countries through aid and technical know-how to promote sustainable development (European Commission 2001a: 1). Toward that end, at the World Summit on Sustainable Development and the 2002 New Delhi COP, EU diplomats tried to push for greater commitments on the part of developed countries to undertake GHG reductions beyond those in the Kyoto Protocol. The EU's actions, limited as they are compared to the problem and associated ethical obligations, carry more weight (morally, arguably) because the European Commission has declared that it believes that carbon sinks that would enable developed countries to offset emissions cuts 'should not become a loophole enabling rich countries to avoid cutting domestic pollution' (DGE 2002: 18).

From the perspective of utilitarianism, Europe has recognized that suffering in the developing world from climate change will outweigh the 'happiness' derived from pollution-intensive lifestyles in Europe. Europe has started to cut its own emissions, recognizing that the world's poor must enjoy additional development, and it is starting to aid developing countries to reduce the suffering from GCC that will be experienced there. According to the Commission, climate change should be addressed 'in a way that is coherent with the overarching objective of poverty reduction' in the developing world (EuropaWorld 2003). From a more Kantian-like perspective, Europe shares the US view that the developing world is a means to an end: protection of the Earth's climate for all. However, unlike the US, it sees the developing world as more than an instrument for promoting its own ends. In 2003, the European Commission adopted a strategy for helping developing countries to cope with climate change, including a myriad of programs to assist with adaptation to, and mitigation of, the effects of climate change within poor countries (see European Commission 2003a). The strategy is quite explicit in stating more than once that 'climate change concerns needs to be fully mainstreamed into EU development cooperation and development staff made aware of the disproportionate impact climate change is likely to have on the poorest countries and on the poorest people in all developing countries' (European Commission 2003a: 13).[35] What the developing world is experiencing seems to matter to Europe.

Considering again Rawlsian conceptions of justice, the EU increasingly seems to accept that the inequitable distribution of rights to use the global atmosphere are unfairly hindering the world's least well off, and this seems to be one motivation for it having started to act on GCC. The European Commission has argued, in effect, that those who are suffering from Europe's global pollution need to be aided to cope with the consequences: 'developing countries are the most vulnerable given their high dependence on these climate-sensitive economic sectors [agriculture, fisheries, forestry and tourism] and their low capacity to adapt. Strengthening their adaptive capacity would contribute to their development' (European Commission 2005c: 7). Margot Wallström, when she was the EU's environment commissioner, said, 'developing countries are the most vulnerable to climate change and therefore deserve our full support in addressing this threat' (EuropaWorld 2003). In 2003 the EU's policy was 'giving high priority to the issue of adaptation to the adverse effects of climate change, in addition to mitigation, in recognition of the fact that developing countries, in particular least developed countries and Small Island Developing States, are more vulnerable to climate change and adaptation is critical for them' (Europa 2003). Upon adoption of its strategy to aid developing countries affected by GCC, Poul Nielson, the EU Commissioner for Development and Humanitarian Aid, said that the adverse effects of climate change:

> 'will disproportionately affect poorer countries with economies predominantly based on natural resources. What's more, the ability of developing countries to adapt to climate change is undermined by a lack of financial resources, adequate technology and stable and effective institutions. The Commission is committed to assisting developing partners in reconciling their legitimate needs for economic development with the protection of the environment and sustainable use of resources. We believe the best way to do this is by addressing climate change concerns within EU development cooperation activities in complete coherence with the overarching objective of poverty reduction' (EuropaWorld 2003).[36]

Europe also seems to accept that fundamental notions of what is fair preclude it from justifiably continuing business as usual. Impartiality suggests that Europe must change its ways. An EU minister is reported to have said that 'The EU believes that it is not realistic to ask the developing countries to reduce or limit their emissions if we cannot show that we, as the biggest emitters, have done something ourselves' (Bhattacharya 2001). This seems obvious, but it is something that is rarely stated by the rich countries.

The EU has been more forthcoming than other countries with regard to recognizing responsibilities and rhetorically supporting the equity demands of developing countries. Preisitz describes (and promotes) three elements of European leadership on climate change: early ratification of the

Kyoto Protocol, domestic implementation of the Kyoto commitments, and addressing the 'needs and interests' of developing countries (Preisitz 2002: 62–6). While she does not put it in the terms of international environmental equity, that is what she describes: recognizing the harm, acting to bring that harm to an end (or at least starting to reduce it), and aiding those who have suffered from the harm. However, it would be an overstatement to say that Europe has been *very* forthcoming in acting upon these sentiments. Actions by the EU countries over the last decade to reduce environmental impacts were greater than those of other major economies, and the EU certainly has been more supportive of developing country demands in international environmental negotiations. However, many EU member countries, much like the US, continue to increase their emissions of global pollutants. Nevertheless, the Europeans are now committed to taking firmer (first) steps toward sharing the burdens of climate change, much more so than the US. What is more, some EU states, such as Germany, have started taking very serious and concrete measures to fulfill and exceed their GHG reduction obligations under the Kyoto Protocol.

Of course the assertion here that Europe has started to recognize these normative considerations and to act upon them is based on an interpretation of statements and policies, and therefore somewhat subjective. However, it seems safe to say that Europe is more willing to do the right thing than other developed countries. Without overstating European actions, Europe has gradually developed a more equitable attitude toward the developing world than have most developed countries, notably the US, including in the environmental issue area (see Harris 2004).[37] The EU 'is seen as more just for showing sympathy for developing countries' (Papadakis 2002: 14). Like the US, Europe wants the large developing countries to join in limiting GHG emissions, simply because doing so is critical to addressing the problem in the long term. However, the EU and some of its member states are setting examples and following the principle of common but differentiated responsibility – whereas the US still expects much poorer countries, at least the largest ones, to act first. Clearly the European perspective and policies are much fairer by almost any ethical measure. That other wealthy countries, notably the US and Australia, sought to first constrain emissions of developing countries has 'reinforced the perception that the EU occupies the higher moral ground and shows more understanding for developing nations' (Papadakis 2002: 4).

If one somehow concludes from what has been said here that Europe has adequately responded to the burdens of GCC, one should be disabused of this notion. Europe has started to take on burdens in this issue area, and it has recognized its responsibility to do so, consistent with practical and normative-ethical arguments. It has certainly been a leader in this respect compared to the US and most other developed countries. Yet it has *not* done enough, and it has very far to go before its own actions make a big enough contribution to robustly reducing GHG emissions, addressing the suffering from climate change in the developing world, and helping the developing countries advance in ways that will be less harmful to the planet in the future. The upshot is that some of the countries of Europe and the EU itself have done more to take on the burdens of climate change than the US, Australia and other developed country parties to the FCCC, but *far* less than what is required of them from both practical and ethical perspectives. Thus we are left with a mixed bag: while Europe has failed to act sufficiently to meet the aims of the FCCC, and it has not fully lived up to a range of ethical standards that appropriately apply in this circumstance, compared to other global actors it has done the most to implement IEE principles in the context of global climate change. If it is not acting ethically, it is certainly acting less *un*ethically.

INTERNATIONAL EQUITY IN EUROPEAN GCC POLICY: THEORETICAL CONSIDERATIONS

Where does Europe's apparent concern for international environmental equity, at least relative to other major countries (e.g., the US) come from? How can an idea like IEE permeate and become part of European GCC policy, and why does this seem to be happening? Looking at EFP theories summarized by Barkdull and Harris (2002), the role of IEE in Europe's GCC policies fits approaches focusing on ideas, particularly those theories arguing that objectives of foreign policy and even national interests are constructed from powerful ideas. One way of conceiving this is in terms of worldviews. Johnson (1998) has argued that the climate change policies of the EU, Japan and the US can be explained by the fact that 'dominating national positions are related to worldviews, and that these differences go beyond national interests' (Johnson 1998, cited in Linnerbooth-Bayer 1999: 47). According to Johnson 'in dealing with matters of international concern, every country approaches the negotiating table with different ideas of what are realistic and desirable outcomes. Such differences in outlook can be ascribed in large part to differences in institutional cultures, which influence how a country will frame an issue, who will play the most significant role in shaping the debate, what the policy making process will be like, and which instruments will be chosen to achieve the desired result' (Johnson 1998: 242).

That equity would start to influence European foreign policy on GCC fits a pattern. Lumsdaine (1993) and Noel and Therien (1995) have shown that countries that are generous at home – where people have a sense of responsibility toward one another and social welfare programs are broadly supported – are more likely to want to aid those abroad who are in need of help. Social democratic 'welfare states', like those of Western Europe, have the most generous foreign aid programs focused on the poorest among the developing countries, while conservative welfare states, best characterized by the US, tend to be least generous with foreign aid. For example, Noel and Therien show that the values and principles that are central features of social democratic institutions within countries had an effect on the foreign aid regime that developed during the second half of the twentieth century. Redistribution of financial resources based on need, not just based on markets, is a characteristic of many domestic societies *and* of international relations (Noel and Therien 1995: 551) In short, those countries that are generous at home tend to be generous abroad. Looked at this way, Europe's relative willingness to take on its share of the burdens associated with GCC makes sense: it fits a pattern, which has developed over half a century, of increasing concern for those less well off at home *and* abroad. Europe's worldview includes concern for those who suffer in places far away.

This challenges major 'realist' conceptions of foreign policy, which tend to discount the power of principled ideas and values in shaping state behavior, at least externally. Conversely, it supports constructivist approaches to international relations, which argue that states have different national identities that can profoundly affect foreign policy and state behavior. According to Boekle et al. (2001: 3)

> 'EU members states' different national identities generate different European and foreign policies as long as the social construction of collective identity takes place on the national rather than the European level. Conversely ... a European foreign policy takes shape inasmuch as identity is constructed on the European level and as the EU gains "actorness" in international politics'.

Arguably, there is a European identity on climate change, increasingly bolstered by the European Commission, which is permeated with considerations of international environmental equity.

This 'social construction' of IEE in GCC policies has been bolstered by the 'public space' of international GCC negotiations in which 'policy-relevant knowledge or ideas are made legitimate, motivate collective action and symbolize a common identity' (Smith 2004: 124). Biermann describes what he calls a 'climate governance architecture', which requires the establishment of 'universally accepted basic norms and problem frames among states' (Biermann 2005: 8). He points out that international relations research over the last decade has shown that 'political behavior of states cannot be explained merely through plain calculations of material interests and power, as earlier theories in the framework of political realism had posited. Instead, states are guided in their behavior by international norms that prescribe and prohibit certain types of behavior and that create an international society that "socializes" states' (Biermann 2005: 8–9).[38] While there are several norms that must guide responses to climate change, considerations of equity in general – and the notion that the developed countries, including Europe, have special responsibility to act and aid the developing world in this issue area – are now established and infused throughout European policy on GCC, even if other, increasingly anachronistic, norms based on old (and narrow and self-defeating) measures of national self-interest still compete.

As Wiegandt has argued:

> 'Perceptions of [social] justice are an integral part of the design of climate change policy, which seeks to alter human influence on atmospheric processes without further aggravating differences in welfare among different countries and social groups. Part of the rationale is to foster compliance (the notion being that countries will only agree to do what they believe is fair); part of it comes from broader underlying notions of equity that describe overall societal relations' (Wiegandt 2001: 147).

The former is nothing new and fits with realist views of the world; that the latter would take on so much importance in Europe is noteworthy. Even when equity considerations do not guide policy – as I believe sometimes is the case for some European governments and for the EU – they nevertheless limit policy choices to certain kinds of compromises, as Wiegandt suggests: 'compromises on policy may be achieved because they lead to compatible ethical outcomes. Knowing, however, that policy alternatives are incompatible with deeply held principles of justice in different societies will, conversely, alert negotiating parties to the impossibility of reaching agreement on those actions' (Wiegandt 2001: 147). The domestic attachment to equity and social justice – and its effects on foreign policy – as described by Lumsdaine (1993) and Noel and Therien (1995) explains why 'unjust' GCC policy alternatives are less palatable in Europe than in, say, the US. As Smith puts it:

> 'if a foreign policy identity can be manifested as shared language, understandings, procedures and collective foreign policy actions, then Europe does exhibit such an identity. Here there is the possibility of a true constructivist interpretation, by which EU states reconstitute themselves in line with common values and create a new collective identity in the process' (Smith 2004: 124).

International environmental equity is part of that new identity, at least when it comes to climate change.

Burchell and Lightfoot (2004: 334) describe Manners' (2002) argument that understanding EU actions requires recognizing the normative foundation of an organization that evolved historically through the development of common policies and treaties. Manners said that, because the 'EU has been constructed on a normative basis', it is predisposed 'to act in a normative way in world politics' (Manners 2002: 252, cited in Burchell and Lightfoot 2004: 334). Increasingly, European interests in the GCC policy field have been at least partly constructed from, or to include, international equity and justice. This has had a material impact on real-world policies, suggesting that ethical ideas can shape policies of member states and the EU. Europe's relative embrace of equitable burden sharing has also (arguably) contributed to its integration into the environmental foreign policies of other major actors in this issue area, which through feedback loops can and should further bolster IEE in Europe's GCC policies. This is difficult to prove, but the indicators, some introduced above, are strong. Regardless, equity and fairness are now well-established objectives of Europe's GCC policies. More could and should be done to implement IEE in the context of GCC, but it now seems nigh impossible that we would return to the days when the EU and most of its member states would deny that the *differentiated* aspect of 'common but differentiated responsibility' applies to them and requires them to move forward, not to dig in their heels as the US and some other developed countries continue to do.

CONCLUSION

In June 2005, the US Senate passed a non-binding resolution calling for 'mandatory, market-based' policies to 'slow, stop and reverse' GHG emissions by the US (although during the Senate debate the basic GCC science was questioned) (US Senate 2005). Despite this development, there is little likelihood that the House of Representatives – or even the Senate itself, let alone the president – would accept more than the ongoing US policy on climate change: pronouncements on voluntary measures and more research, but no action to speak of. Toward this end, the following month President Bush reiterated his opposition to the Kyoto Protocol and any requirement that the US cut its emissions of GHGs. Even two powerful hurricanes that battered the US's southern states in mid-2005, the destructive power of which atmospheric scientists increasingly believe likely resulted from global warming, seem unlikely to spur significant action by the US (see Rosenthal 2005). This sharply contrasts with European policies, which by themselves seem modest at best but when set alongside US policies demonstrate a remarkable willingness to act upon, and more equitably share the burdens of, global climate change. The upshot is that European *policy* and *rhetoric* have moved into line with common conceptions of fair and equitable burden sharing. Europe's *actions* began moving in that direction, albeit in a modest way when compared to the continent's responsibility for the problem and the potential magnitude of GCC's effects. Europe has taken action at home and will very likely take substantially more in future. It has started to take on its fair share of the burdens of climate change.

Given the complex institutional character of the EU, with member states competing among themselves and with the organization, and still less than explicit responsibility and competence on foreign policy issues, it is especially remarkable that considerations of equity and fairness have been able to have a significant impact. On one hand, GCC equity is an issue around which member states can generally agree; it is very difficult for member states to stand up, as the US has done, and proclaim IEE a non-starter. Also, as Edwards argues, 'external rhetoric has to be backed up with implementation … The link between the internal and external aspects of the EU is therefore central to the question of [GCC] leadership' (Edwards 2002: 50). Once the EU and most of the members agreed to take on some of the burden of climate change, and particularly once it was established how this would be shared within Europe, a European foreign policy of GCC burden sharing became somewhat entrenched. While it has been imperfect from both practical and ethical perspectives, in the area of climate change Europe has shown substantial 'directional' leadership (Metz et al. 2001: 168): it has pushed the idea of IEE in this context and implemented it domestically to a significant degree (see Gupta and Ringius 2001). As European states internalize IEE as an important national and EU objective, it becomes more pronounced and it in turn affects international affairs and responses to global environmental changes in particular. Europe becomes the shaper of the GCC regime (much as the US shaped the post-war free-trade regime).

What should Europe do in the future? It ought to do and it can do more. It can of course go much further with existing GHG cuts and aid to developing countries. However, entirely new GCC policies are justifiable. For example, one important issue that Europe has an obligation to consider and to transform into new policy is that of GCC-related immigration. Since Europe is responsible for much of the suffering from climate change that will be experienced in the developing world, European states are ethically obligated to take in migrants who would otherwise experience that hardship. Europe would be obliged to take in at least tens of thousands – and more appropriately hundreds of thousands – of *additional* migrants from those countries (and especially areas within them) most adversely affected by climate change,[39] such as low-lying states suffering from sea-level rise and

African states experiencing increased drought. This should be a policy specifically geared toward, and identified as, mitigation of suffering from Europe's atmospheric pollution.

Given the total failure of the US to respond as it should to GCC, and the slow response in other countries where action is being taken, both the practical and normative burdens will have to be borne by Europe. Europe will have to set the example and send the strongest possible signal – through its own GHG cuts and its aid to the developing world – that the developed countries are taking on their share of the burden. Only then are we likely to see major action by the large developing countries to limit their own emissions and hopefully to avert disaster. As Collier puts it:

> 'With most greenhouse gases originating in the industrialized "Western" capitalist societies, it is appropriate that it is from these regions of the world that the initiatives for change should come. If reasonably stable unions of peoples such as the EU cannot find ways of addressing the climate change issue it is difficult to see how it can be addressed at all' (Collier 1997a: 199).

For better or worse, as things stand now, Europe is the only realistic hope – a realistic hope that derives from its nascent embrace of international environmental equity.

NOTES

* The author wishes to acknowledge support from Lingnan University through a Social Sciences Research Grant and the assistance of Steve Chung in gathering documentary materials. Many of ideas in this chapter appear in Harris 2006.

1. Burden sharing is examined on the global level, in particular Europe's practical and ethical burdens relative to the developing world. Burden sharing within the EU is not examined in this chapter (see Chapter 9, Aidt and Greiner 2002 and Bode 2004).
2. I have argued elsewhere at some length (Harris 2001b) about the degree to which equity considerations have permeated international environmental politics and started to shape EFP. For a discussion of the EU's role in the spread of environmental policy instruments through policy learning and transfer, see Jordan, Wurzel and Zito (2003).
3. The European Parliament arguably is more radical on these (and many other) issues than are the Commission and the member states. I have not looked at the Parliament as a separate force or 'pressure group', but someone ought to. (I am grateful to Brian Bridges for pointing this out to me.)
4. 'European Union', as used here, encompasses the European Community, one of the EU's core pillars in the context of the Treaty of Rome and the legal entity that signed the FCCC. See Haigh (1996: 156).
5. For a study of common policies on environmental issues, see Jordan and Liefferink (2004).
6. At least following agreement of the FCCC; it was the member states, not the European Community/EU, that were heavily involved in negotiations resulting in the FCCC (Haigh 1996: 181). For discussions of EU leadership in the context of GCC, see Edwards (2002), Gupta and Ringius (2001), Oberthür (1999) and Preisitz (2002). See Chapter 4 (Table 4.1) for a typology of leadership modes in this context.

7. Some scientists, notably the Lamont-Doherty Earth Observatory's Wallace Broeker, argue that GHGs must be brought to zero later in this century (BBC World Service 2005).
8. The industrialized countries are collectively responsible for 85 per cent of historic GHG emissions and 65 per cent of current emissions (Ashton and Wang 2003: 14; see IPCC 2001). On common but differentiated responsibility, see Biermann (1996) and Harris (1999a).
9. I have tried to do some of this in Harris (2001b). To put per capita emissions into perspective, consumption in the EU is very roughly twice that in China, but roughly half that in the US (Shukla 1999: 151).
10. It might be worth asking rhetorically whether one could possibly imagine the current US government saying anything close to this. Words go only so far, of course, but the degree to which the EU and many of its member states have at least recognized their ethical obligations with regard to GCC, and that they have actually started to act upon them, is *relatively* very profound.
11. I first exercised many of the ideas in this section, and applied them to the US case, in Harris (1999b).
12. Consistent with common usage, the term equity, as used here, refers to *distributive* justice, that is, the 'fairness' or 'rightness' of distributing benefits and burdens.
13. This begs the questions of cosmopolitan justice, notably the obligations of *individuals* to one another. Arguably well-off people, wherever they may reside, have obligations to act in ways analogous to the oblations of well-off states. To get individuals to act is very difficult and often requires government action, but this does eliminate the obligation (c.f. Singer 1996 and Paterson 1996: 190).
14. While considerations of fairness and equity between generations (intergenerational equity) and between species (interspecies equity) are important, here I focus on considerations of equity as they relate to contemporaneous sharing or distribution among states (*intra*-generational equity). For discussions of all three types, see Cooper and Palmer (1995). For a rare examination of intra-national justice in the context of GCC, see Pan (2003).
15. Here I do not pretend to take on the burdens of philosophical exegesis. This typology mirrors Paterson's framework (see Paterson 1996, Banuri et al. 1996, Brown 1992 and Harris 1997a). In contrast, Jamieson (2002: 298) believes that global environmental justice, which includes obligations to the environment and other species as well as (potentially) to other people, 'does not lend itself naturally to the application of "big picture" theories of justice', such as Rawls (1971) and Nozick (1974).
16. This begs the question (to use one example that comes to mind) of who is happier, an affluent person with a big new car spewing CO_2, or a very poor person who finally has enough to eat – and what value we place on different kinds of happiness.
17. Rawls argues that his theory does not apply much to international relations. Nevertheless, several scholars have argued otherwise (see, for example, Beitz 1979).
18. See Paterson (1996, 2001) and Shue (1992, 1995, 1996). There are a number of other important measures of equity, such as participation in decision-making and procedural fairness, which are important but not included here due to space limitations. For several additional papers on applying equity and fairness considerations in the context of GCC, see Toth (1999) (see also Rayner et al. 1999 and Wiegandt 2001). These other perspectives are germane. For example, the EU has as one of its objectives helping poor countries to participate in climate change negotiations (see European Council 2004: 15).
19. Baer (2002) also argues that equal per capita emissions rights are the only ethical option, noting that this has the practical benefit of offering options for developing-country emissions limits in the future.
20. For a description of EU funding in the field of GCC, see European Commission (2001d: 147–61).
21. The anticipated adverse impacts of GCC for very poor countries are described in Huq et al. (2003).
22. For a proposal on how the EU could be a stronger leader on GCC, see Ott and Oberthür (2001).

23. I recall that at the time some Americans were saying that the Europeans were doing this knowing full well that their efforts would fail. In other words, it was perceived to be an empty gesture. Subsequent events have shown that it was not.
24. Wiegandt characterizes the CDM as 'the convergence of two related principles: equity and efficiency', achieving the former by shifting 'resources from developed to developing countries, which will reduce the economic disparities inherent in the process of achieving global emissions reductions' (Wiegandt 2001: 138).
25. Indeed, environmental NGOs pushing hardest for strict targets and timetables for GHG emissions cuts aligned themselves with negotiating blocks from the EU, AOSIS and G-77 and China (Betsill 2005, 110), showing how the EU position was viewed as the most sympathetic within the developed world toward the developing world.
26. This was a compromise between a Dutch proposal (supported by AOSIS and environmental NGOs) for making the cuts by 2005 (see Chapter 4) and opposition to that deadline from Britain (see Chapter 3) and France. Cuts were envisioned in an EU-wide 'bubble', meaning some would make larger cuts to allow other members to increase emissions (see Chapter 1). This idea for an EU bubble was opposed by other developed countries, as was the EU plan to limit cuts to only three GHGs (Leggett 2001: 259).
27. This aim to hold CO_2 concentrations at no more than double the pre-industrial historic norm was intended to prevent 'dangerous' GCC, in accordance with the core objective of the FCCC. See Leggett (2001: 237).
28. For a discussion of EU positions regarding flexible mechanisms, see Christiansen (2003), and emissions trading in particular, see Christiansen and Wettestad (2003) and Cass (2004). The position of the EU remains that the majority of emissions reductions should come from domestic cuts, rather than through trading and carbon sinks (Gelbspan 2004: 97; see below) – although trading *within* the EU is now a major component of its GHG-reduction policies.
29. See European Commission (2001d) and Swedish NGO Secretariat on Acid Rain (2004) for a list and summary of the various strategies and legislation. Some of these are significant. For example, one EU directive requires 22.1 per cent of EU electricity to come from renewable sources by 2010 (Strauss 2005). For an analysis of the European Climate Change Program, see European Commission (2001b), Preisitz (2002) and ECCP (2003).
30. Contrary to an oft-heard view, CO_2 emissions reductions in Britain and Germany are *not* necessarily mostly the result of special circumstances (i.e., the former's switch from coal to natural gas when energy markets were liberalized, and the latter's restructuring of East Germany following reunification) (Eichhammer et al. n.d.). The European Environment Agency determined that EU GHG emissions dropped 4 per cent between 1990 and 1999 (DGE 2002: 11), ranging from remarkable reductions during this period of 43.3 per cent in Luxembourg and 18.7 per cent in Germany to increases of 23.2 per cent in Spain and 22.4 per cent in Austria (European Environment Agency statistics cited in DGE 2002: 12). By way of comparison, during the same period US emissions increased by 12 per cent (Schreurs 2004: 226).
31. These comments apply to the EU-15; a number of new member states have seen emissions reductions simply because their economies declined relative to the 1990 base year.
32. The 2050 goal was left out at the insistence of Austria and Germany (JIQ 2005a: 2).
33. According to the EU's sixth environment action plan (DGE 2002: 14).
34. If suggesting that the GCC negotiations may continue for more than a century seems far-fetched, bear in mind that much easier international negotiations on trade (GATT, WTO) are far from finished after more than half a century. The question is whether the plight of some small-island and low-lying states, among others most adversely affected by climate change, will have a speedier influence on negotiations than has the economic plight of poor countries on the trade regime.
35. See European Commission (2003a) for a long list of relevant development projects.
36. A private report from the US subsequently reached the same conclusion. See Kuzma and Dobrovolny (2004).

37. The Lomé Convention, which aimed 'to improve conditions of some of the world's poorest and most troubled countries' (Lister 1998: 33), comes to mind.
38. The literature to which Biermann (2005) refers includes March and Olson (1998), Finnemore (1996), Hall and Taylor (1996), Barnett and Finnemore (1999) and Finnemore and Sikkink (1998). For a skeptical view of the efficacy of moral discourse and norms in international environmental affairs, see Boehmer-Christiansen and Kellow (2002).
39. Drawing roughly on an assessment by Byravan and Rajan (2005).

REFERENCES

Aidt, Toke and Sandra Greiner (2002), 'Sharing the Climate Policy Burden', HWWA Discussion Paper 176, Hamburg: Hamburg Institute of International Economics.

Ashton, John and Xueman Wang (2003), 'Equity and Climate Change in Principle and Practice', Working draft (July) in the *Beyond Kyoto* series of papers, Washington: Pew Center on Global Climate Change.

Attfield, Robin (1999), *The Ethics of the Global Environment*, Edinburgh: Edinburgh University Press.

Axelrod, Regina S, Norman J Vig and Miranda A Schreurs (2005), 'The European Union as an environmental governance system', in Regina S Axelrod, David Leonard Downie and Norman J Vig (eds), *The Global Environment: Institutions, Law and Policy*, Washington: CQ Press.

Baer, Peter (2002), 'Equity, greenhouse gas emissions, and global common resources', in Stephen H Schneider, Armin Rosencranz and John O Niles (eds), *Climate Change Policy: A Survey*, Washington, DC: Island Press.

Barkdull, John and Paul G Harris (2002), 'Environmental change and foreign policy: a survey of theory', *Global Environmental Politics* **2**(2) (May): 63–91.

Banuri, Tariq, Karl Goran Maler, Michael Grubb, Harold K Jacobson and Farhana Yamin, (1996), 'Equity and social considerations', in JP Bruce, H Lee, and EF Haites (eds), *Climate Change 1995: Economic and Social Dimensions of Climate Change*, Cambridge, England: Cambridge University Press.

Barnett, Michael N and Martha Finnemore (1999), 'The politics, power and pathologies of international organizations', *International Organization* **53**(4): 699–732.

Barry, Brian (1995), *Justice as Impartiality*, Oxford: Oxford University Press.

BBC News (2006), 'Britain's Emissions Plan Rejected' (22 February), available at <http://news.bbc.co.uk/1/hi/uk/4740878.stm>.

BBC World Service (2005), *Global Business* (2 January).

Beitz, Charles R (1979), *Political Theory and International Relations*, Princeton: Princeton University Press.

Bentham, Jeremy (1970), *An Introduction to the Principles of Morals and Legislation* Burns and H.L.A. Hart, (eds) London: Athlone Press.

Betsill, Michele M (2005), 'Global climate change policy: making progress or spinning wheels?', in Regina S Axelrod, David Leonard Downie and Norman J Vig, *The Global Environment: Institutions, Law, and Policy*, Washington: CQ Press.

Bhattacharya, Jayanta (2001), 'Kyoto protocol: rich nations must make sacrifices', *The Statesman* June 10: 2.

Biermann, Frank (1996), '"Common concern of humankind": the emergence of a new concept of international environmental law', *Archiv des Volkerrechts* **34**(4) (December): 426–81.

Biermann, Frank (1999), 'Justice in the greenhouse: perspectives from international law', in Ferenc L Toth (ed.), *Fair Weather?: Equity Concerns in Climate Change*, London: Earthscan.

Biermann, Frank (2005), *Between the United States and the South: Strategic Choices for European Climate Policy*, Global Governance Working Paper No. 17, Amsterdam: Global Governance Project, available at www.glogov.org.

Bode, Sven (2004), 'European Climate Policy: Burden Sharing after 2012', HWWA Discussion Paper 265, Hamburg: Hamburg Institute of International Economics.

Boehmer-Christiansen, Sonja and Aynsley Kellow (2002), *International Environmental Policy: Interests and the Failure of the Kyoto Process*, Cheltenham, UK: Edward Elgar.

Boekle, Henning, Jorg Nadoll and Bernhard Stahl (2001), 'European foreign policy and national identity: detecting the link', Paper presented at the Fourth Pan-European International Relations Conference, Canterbury, UK, 6–10 September (draft).

Brown, Chris (1992), *International Relations Theory*, Oxford: Columbia University Press.

Bruce, James P, Hoesung Lee and Erik F Haites (eds) (1996), *Climate Change 1995: Economic and Social Dimensions*, Cambridge: Cambridge University Press.

Burchell, Jon and Simon Lightfoot (2004), 'Leading the way? The European Union at the WSSD', *European Environment* **14**: 331–41.

Byravan, Sujatha and Sudhir Chella Rajan (2005), 'Before the flood', *International Herald Tribune*, 11 May 2005: 8.

Cass, Loren (2004), 'Issue Framing and Preference Change: The Evolution of the European Union Position on International Emissions Trading', Paper prepared for the International Studies Association Annual Conference, Montreal, 17–20 March 2004.

Christiansen, Atle C (2003), 'The role of flexibility mechanisms in EU climate strategy: lessons learned and future challenges?', *International Environmental Agreements* **4**: 27–46.

Christiansen, Atle C and Jorgen Wettestad (2003), 'The EU as a frontrunner on greenhouse gas emissions trading: how did it happen and will the EU succeed?', *Climate Policy* **3**: 3–18.

Climate Protection Programme (CPP) (2004), 'South-North Dialogue on Equity in the Greenhouse: A Proposal towards an Adequate and Equitable Global Climate Agreement', Wuppertal, Germany: Wuppertal Institute for Climate, Environment and Energy, available at www.wupperinst.org/download/1085_proposal.pdf.

Collier, Ute (1997a), 'Comparative analysis and conclusions', in Ute Collier and Ragnar E. Loftsedt, *Cases of Climate Change Policy: Political Reality in the European Union*, London: Earthscan.

Collier, Ute (1997b), 'The EU and climate change policy: the struggle over policy competences', in Ute Collier and Ragnar E. Loftstedt (eds), *Cases in Climate Change Policy: Political Reality in the European Union*, London: Earthscan.

Collier, Ute (2002), 'EU Energy policy in a changing climate', in Andrea Lenschow (ed.), *Environmental Policy Integration: Greening Sectoral Policies in Europe*, London: Earthscan.

Collier, Ute and Ragnar E Loftstedt (eds) (1997), 'The climate change challenge', in Ute Collier and Ragnar E Loftstedt (eds), *Cases in Climate Change Policy: Political Reality in the European Union*, London: Earthscan.

Cooper, David E and Joy A Palmer (eds) (1995), *Just Environments: Intergenerational, International and Interspecies Issues*, London: Routledge.

Dessai, Suraje and Emma Lisa Schipper (2003), 'The Marrakech Accords to the Kyoto Protocol: analysis and future prospects', *Global Environmental Change* **13**: 149–53.

Directorate-General for the Environment of the European Commission (DGE) (2002), *EU Focus on Climate Change*, Luxembourg: Office for Official Publications of the European Communities.

Duwe, Matthias (2004), 'Climate Policy in the European Union: A Brief Overview of EU-wide Policies and Measures', Mimeo (30 November). Brussels: Climate Action Network Europe, available at www.aef.org.uk/downloads/EU%20and%20Climate%20-%20Duwe.pdf.

Earth Negotiations Bulletin (ENB) (2005), 'Summary of the Eleventh Conference of the Parties to the UN Framework Convention on Climate Change and First Conference of the Parties Serving as the Meeting of the Parties to the Kyoto Protocol: 28 November–10 December 2005', *Earth Negotiations Bulletin* **12**, 291 (12 December), available at www.iisd.ca/vol12/enb12291e.html.

Edwards, John Huw (2002), *Has the European Union Exercised Leadership in the International Climate Change Regime Since the Hague Conference?*, Thesis for the degree of Master of European Studies, Brussels: College of Europe, Bruges Campus.

Eichhammer, Wolfgang, Frank Gagelmann, Joachim Schleich and John Chesshire, n.d. 'Reasons and Perspectives for Emissions Reductions in Germany and the UK', available at www.isi.fhg.de/publ/downloads/isi02a08/emission-reduction.pdf.

Elliott, Lorraine (2004), *The Global Politics of the Environment*, Houndmills, UK: Palgrave Macmillan.

Europa (2003), 'EU Presidency Statement: Environment and Sustainable Development', available at http://europa-eu-un.org/articles/cs/article_2922_cs.htm.

EuropaWorld (2003), 'Commission to Help Developing Countries Meet the Challenge of Climate Change', 21 March, available at www.nieuwsbank.nl/en/2003/03/18/r014. htm.

European Climate Change Program (ECCP) (2003), *Second ECCP Progress Report: Can We Meet Our Kyoto Targets?* (April), available at europa.eu.int/comm/environment/ climat/pdf/second_eccp_report.pdf.

European Commission (1999), 'Preparing for Implementation of the Kyoto Protocol', Commission Communication to the Council and the Parliament, Brussels: European Commission.

European Commission (2000), 'The EU's Position for COP6', European Commission Briefing Paper (3 November), Brussels: European Commission, available at http://europa.eu.int/ comm/environment/climat/eupositions.htm.

European Commission (2001a), 'Commission Reacts to US Statements on Kyoto Protocol', Brussels, March 29: 1, available at http://europa.eu.int.

European Commission (2001b), 'Communication from the Commission on the Implementation of the First Phase of the European Climate Change Programme' (23 October), Brussels: European Commission.

European Commission (2001c), 'EU Reaction to the Speech of US President Bush on Climate Change', (June 12), Brussels: European Commission (mimeograph).

European Commission (2001d), 'Third Communication from the European Community under the UN Framework Convention on Climate Change', Commission Staff Working Paper (30 November), Brussels: European Commission.

European Commission (2003a), 'Climate Change in the Context of Development Cooperation', Communication from the Commission to the Council and the European Parliament (11 March), Brussels: European Commission.

European Commission (2003b), 'Report from the Commission under Council Decision 93/389/EEC as Amended by Decision 99/296/EC for a Monitoring Mechanism of Community Greenhouse Gas Emissions' (28 November), Brussels: European Commission.

European Commission (2004), 'Catching Up with the Community's Kyoto Target' (20 December), Report from the Commission, Brussels: European Commission.

European Commission (2005a), 'Climate Change: More Coal Use Pushes Up EU Greenhouse Gas Emissions in 2003', Press Release IP/05/767 (22 June), Brussels: European Commission.

European Commission (2005b), 'EU Emissions Trading: An Open Scheme Promoting Global Innovation to Combat Climate Change', Brussels: European Commission Publications Office.

European Commission (2005c), 'Winning the Battle Against Global Climate Change', Communication from the Commission to the Council, the European Parliament, the European Economic and Social Committee and the Committee of the Regions (9 February), Brussels: European Commission.

European Council (2004), 'Climate Change in the Context of Development Cooperation: Council Conclusions' (24 November), Brussels: European Council.

European Council (2005), 'European Council Brussels 22 and 23 March 2005: Presidency Conclusions' (23 March), Brussels: European Council.

European Environment Agency (EnEA) (2002), *Greenhouse Gas Emission Trends and Projections in Europe: Are the EU and the Candidate Countries on Track to Achieve the Kyoto Protocol Targets?*, Copenhagen: European Environment Agency.

European Environment Agency (EnEA) (2004), *Greenhouse Gas Emission Trends and Projections in Europe 2003*, Copenhagen: European Environment Agency.

European Environment Agency (EnEA) (2005), *Annual European Community Greenhouse Gas Inventory 1990–2003 and Inventory Report 2005: Submission to the UNFCCC Secretariat*, Technical Report No. 4/2005, Copenhagen: European Environment Agency.

Finnemore, Martha (1996), 'Norms, culture, and world politics: insights from sociology's institutionalism', *International Organization* **50**(2): 325–47.

Finnemore, Martha and Kathryn Sikkink (1998), 'International norm dynamics and political change', *International Organization* **52**(4): 887–917.

Gelbspan, Ross (2004), *Boiling Point*, New York: Basic Books.

German Federal Environment Ministry, ed. (1997), *Causes of the Trend in CO_2 Emissions in Germany between 1990 and 1995*, Berlin: German Federal Environment Ministry.

Gugele, Bernd, Kati Huttunen, and Manfred Ritter (2003), *Annual European Community Greenhouse Gas inventory 1990–2001 and Inventory Report 2003*,

Brussels: European Environment Agency, available at http://reports.eea.eu.int/technical_report_2003_95/en.
Gupta, Joyeeta and Lasse Ringius (2001), 'The EU's climate leadership: reconciling ambition and reality', *International Environmental Agreements* **1**: 281–99.
Haigh, Nigel (1996), 'Climate change policies and politics in the European Community', in Tim O'Riordan and Jill Jager (eds), *Politics of Climate Change: A European Perspective*, London: Routledge.
Hall, Peter A and Rosemary CR Taylor (1996), 'Political science and the three institutionalisms', *Political Studies* **XLIV**: 936–57.
Hampson, Fen Osler and Judith Reppy (eds) (1996), *Earthly Goods: Environmental Change and Social Justice*, Ithaca: Cornell University Press.
Harris, Paul G (1996), 'Considerations of equity and international environmental institutions', *Environmental Politics* **5**(2) (Summer): 274–301.
Harris, Paul G (1997a), 'Affluence, poverty and ecology: obligation, international relations and sustainable development', *Ethics and the Environment* **2**(2) (Fall): 121–38.
Harris, Paul G (1997b), 'Environment, history and international justice', *Journal of International Studies* **40** (July): 1–33.
Harris, Paul G (1999a), 'Common but differentiated responsibility: the Kyoto Protocol and United States policy', *Environmental Law Journal* **7**(1) (1999): 27–48.
Harris, Paul G (1999b), 'Sharing the costs of climate change: an assessment of American foreign policy', *Cambridge Review of International Affairs* **12**(2): 289–310.
Harris, Paul G (2000), 'Defining international distributive justice: environmental considerations', *International Relations* **15**(2) (August): 51–66.
Harris, Paul G (2001a), 'Assessing climate change: international cooperation and predictions of environmental change', *Politics* **21**(1) (February 2001): 11–21.
Harris, Paul G (2001b), *International Equity and Global Environmental Politics*, Aldershot: Ashgate.
Harris, Paul G (2004), 'International development assistance and burden sharing', in Norman J Vig and Michael G Faure (eds), *Green Giants?: Environmental Policies of the United States and the European Union*, Cambridge, MA: MIT Press.
Harris, Paul G (2006), 'Europe and environmental change: sharing the burdens of global warming', *Colorado Journal of International Environmental Law and Policy* **17**(2) (Spring), 309–55.
Hourcade, Jean-Charles and Michael Grubb (2000), 'Economic dimensions of the Kyoto Protocol', in Joyeeta Gupta and Michael Grubb (eds), *Climate Change and European Leadership: A Sustainable Role for Europe?*, London: Kluwer Academic Publishers.
Huq, Saleemul (2002), 'The Bonn-Marrakech Agreements on funding', *Climate Policy* **2**: 243–46.
Huq, Saleemul, Atiq Rahman, Mama Konate, Youba Sokona, and Hannah Reid (2003), *Mainstreaming Adaptation to Climate Change in Least Developed Countries (LDCs)*, London: International Institute for Environment and Development.
Intergovernmental Panel on Climate Change (IPCC) (2001), *Climate Change 2001: Impacts, Adaptation, and Vulnerability*, Cambridge: Cambridge University Press.
International Energy Agency (IEA) (1994), *Climate Change Policy Initiatives*, **1**: OECD Countries, Paris: International Energy Agency.

International Energy Agency (IEA) (2001), 'Climate change and the Bonn Agreement: questions and answers from the IEA', Paris: International Energy Agency, available at www.iea.org/textbase/papers/2001/quesbonn.pdf.
International Institute for Environment and Development (IIED), n.d. 'Adaptation to climate change and least developed countries', available at www.iied.org/docs/climate/LDC_ flyer_final.pdf.
Jamieson, Dale (2002), *Morality's Progress*, Oxford: Clarendon Press.
Johnson, Anne K (1998), 'The influence of political culture on the formation of pre-regime climate change policies in Sweden, the United States and Japan', *Environmental Values* **7**: 223–44.
Joint Implementation Quarterly (JIQ) (2005a), 'EU council proposes ambitious climate policy', *Joint Implementation Quarterly* **11**(1) (April): 2.
Joint Implementation Quarterly (JIQ) (2005b), 'EU emissions trading: an overview of the first three months', *Joint Implementation Quarterly* **11**(1) (April): 12–14.
Jordan, Andrew and Duncan Liefferink (2004), *Environmental Policy in Europe: The Europeanization of National Environmental Policy*, London: Routledge.
Jordan, Andrew, Rudiger Wurzel and Anthony Zito (2003), 'How "new" environmental policy instruments (NEPIs) spread in the European Union: an analysis of the role of the EU in shaping environmental governance', European Union Studies Association Biennial Conference, 27–29 March 2003.
Kant, Immanuel (1948), *The Moral Law*, London: Hutchinson.
Klepper, Gernot and Sonja Peterson (2005), *Emissions Trading, CDM, JI, and More: The Climate Strategy of Europe*, Paper for the Climate Change Modelling and Policy Series, Fondazione Eni Enrico Mattei (April), available at www.feem.it/Feem/Pub/ Publications/WPapers/default.htm.
Kuzma, Jennifer and Lydia Dobrovolny (2004), *The Global Climate and Economic Development*, Minneapolis: Humphrey Institute of Public Affairs.
Leggett, Jeremy K (2001), *The Carbon War: Global Warming and the End of the Oil Era*, New York: Routledge.
Linnerbooth-Bayer, Joanne (1999), 'Climate Change and Multiple Views of Fairness', in Ferenc L. Toth (ed.), *Equity Concerns in Climate Change*, London: Earthscan.
Lister, Marjorie (ed.) (1998), 'Europe's new development policy', in Marjorie Lister (ed.), *European Union Development Policy*, Houndmills, UK: Macmillan.
Lumsdaine, David H (1993), *Moral Vision in International Politics: The Foreign Aid Regime, 1949–89*, Princeton: Princeton University Press.
Manners, Ian (2002), 'Normative power Europe: a contradiction in terms?', *Journal of Common Market Studies* **40**(2): 235–58.
March, James G and Johan P Olson (1998), 'The institutional dynamics of international political order', *International Organization* **52**(4): 943–69.
McCormick, John (2001), *Environmental Policy in the European Union*, New York: Palgrave.
Metz, Bert, Marcel Berk, Marcel TJ Kok, Jelle G van Minnen, Andre de Moor and Albert Faber (2001), 'How can the European Union contribute to a COP-6 Agreement?: An Overview for Policy Makers', *International Environmental Agreements* **1**: 167–85.
Michaelowa, Axel (2004), 'Can the EU provide credible leadership for climate policy beyond 2012?', *Kyoto Protocol: Beyond 2012*, Jakarta: Pelangi Indonesia.
Mott, Richard (1993), 'The GEF and the conventions on climate change and biological diversity', *International Environmental Affairs* **5**(4) (Fall): 299–312.

Muller, Benito (2002), *Equity in Climate Change: The Great Divide*, Oxford: Oxford Institute for Energy Studies.

Noel, Alain and Jean-Philippe Therien (1995), 'From domestic to international justice: the welfare state and foreign aid', *International Organization* **49**(3) (Summer), 523–53.

Nozick, Robert (1974), *Anarchy, State and Utopia*, New York: Basic Books.

Oberthur, Sebastian (1999), 'The EU as an international actor: the protection of the ozone layer', *Journal of Common Market Studies* **37**(4): 641–59.

Odingo, Richard Samson, AL Alusa, F Mugo, JK Njihia and A Heidenreich (eds) (1994), *Equity and Social Considerations Related to Climate Change*, Nairobi: ICIPE Science Press.

Ott, Hermann E (2001a), 'Climate change: an important foreign policy issue', *International Affairs* **77**(2): 277–96.

Ott, Hermann E (2001b), 'The Bonn Agreement to the Kyoto Protocol: paving the way for ratification', *International Environmental Agreements* **1**(4) (December): 469–76.

Ott, Hermann E and Sebastian Oberthur (2001), *Breaking the Impasse: Forging an EU Leadership Initiative on Climate Change*, Berlin: Heinrich Boll Foundation.

Ott, Hermann E and Tilman Santarius (2003), 'Developments in international and European climate policy in 2003', Paper for the Wuppertal Institute for Climate, Environment and Energy.

Ott, Hermann E, Bernd Brouns, Wolfgang Sterk and Bettina Wittneben (2005), 'It takes two to tango: climate policy at COP 10 in Buenos Aires and beyond', *Journal for European Environmental and Planning Law* **2**: 84–91.

Pan, Jiahua (2003), 'Emissions rights and their transferability: equity concerns over climate change mitigation', *International Environmental Agreements* **3**(1): 1–16.

Papadakis, Elim (2002), 'Challenges for global environmental diplomacy in Australia and the European Union', National Europe Centre Paper No. 21, Paper presented to conference on The European Union in International Affairs, Australian National University, 3–4 July 2002.

Paterson, Matthew (1996), 'International justice and global warming', in Barry Holden (ed.), *The Ethical Dimensions of Global Change*, New York: St Martin's Press.

Paterson, Matthew (1997), 'Equity or justice', in Detlef Sprinz and Urs Luterbacher (eds), *International Relations and Global Climate Change*, PIK Report No. 21, Potsdam: Potsdam Institute for Climate Impact Research, available at www.pik-potsdam.de/pik_web/ publications/pik_reports/reports/reports/pr.21/pr21_8.htm.

Paterson, Matthew (2001), 'Principles of justice in the context of global climate change', in Urs Luterbacher and Detlef F. Sprinz (eds), *International Relations and Global Climate Change*, Cambridge, MA: MIT Press.

Pearsall, Judy and Patrick Hanks (eds) (1998), *The New Oxford English Dictionary*, Oxford: Oxford University Press.

Porter, Gareth, Janet Welsh Brown and Pamela Chasek (2000), *Global Environmental Politics*, 3rd edn, Boulder: Westview Press.

Preisitz, Ute (2002), 'European Union's climate change policy towards the implementation of the Kyoto Protocol and its role as an international leader', Master's Thesis, Graz: University of Graz.

Ringius, Lasse, Asbjørn Torvanger and Arild Underdal (2002), 'Burden sharing and fairness principles in international climate change policy', *International Environmental Agreements* **2**: 1–22.
Rosenthal, Elisabeth (2005), 'Bush takes heat on global warming', *International Herald Tribune* (2 September), available at www.iht.com/articles/2005/09/01/news/warm. php.
Rawls, John (1971), *A Theory of Justice*, Cambridge, MA: Harvard University Press.
Rayner, Steve, Elizabeth L Malone and Michael Thompson (1999), 'Equity issues and integrated assessment', in Ferenc L. Toth (ed.), *Fair Weather?: Equity Concerns in Climate Change*, London: Earthscan.
Rosenzweig, Richard, Dirk Forrister and Rob Youngman (2005), 'A Paper on European Greenhouse Gas Performance, the European Union Emissions Trading Scheme, and Technology Policy', Paper for the Trans-Atlantic Conference on Energy Policy (28 March) Washington, DC.
Ruggie, John Gerard (1998), *Constructing the World Polity*, London: Routledge.
Schreurs, Miranda A (2004), 'The climate change divide: the European Union, the United States, and the future of the Kyoto Protocol', in Norman J Vig and Michael G Faure (eds), *Green Giants?: Environmental Policies of the United States and the European Union*, Cambridge, MA: MIT Press.
Shue, Henry (1992), 'The unavoidability of justice', in Andrew Hurrell and Benedict Kingsbury (eds) *The International Politics of the Environment*, Oxford: Oxford University Press: 373–97.
Shue, Henry (1995), 'Equity in an international agreement on climate change', in Richard Samson Odingo, AL Alusa, F Mugo, JK Njihia, and A Heidenreich (eds), *Equity and Social Considerations Related to Climate Change*, Nairobi: ICIPE Science Press.
Shue, Henry (1996), *Basic Rights*, Princeton: Princeton University Press.
Shukla, PR (1999), 'Justice, equity and efficiency in climate change: a developing country perspective', in Ferenc L. Toth (ed.), *Fair Weather?: Equity Concerns in Climate Change*, London: Earthscan.
Singer, Peter (1996), 'Famine, affluence, and morality', in William Aiken and Hugh LaFollette (eds), *World Hunger and Morality*, Upper Saddle River, NJ: Prentice Hall.
Smith, Michael (2004), 'Institutionalization, policy adaptation and European foreign policy cooperation', *European Journal of International Relations* **10**(1): 95–136.
Social Learning Group, William C Clark, Jill Jaeger, Josee van Eijndhoven and Nancy M Dickson (eds) (2000), *Learning to Manage Global Environmental Risks, Vol 1: A Comparative History of Social Responses to Climate Change, Ozone Depletion, and Acid Rain*, Cambridge: MIT Press, 2000.
Special Committee on a Future Framework for Addressing Climate Change (2004), *Sustainable Future Framework on Climate Change*, Interim Report, Japan: Global Environmental Sub-Committee, Industrial Structure Council (December).
Strauss, Michael J (2005), 'Wind spins its way into the energy equation', *International Herald Tribune* (6 July): 19.
Swedish NGO Secretariat on Acid Rain (2004), 'EU on climate change: targets, strategies and legislation', *Acid News* **1** (March): 11–14.
Toth, Ferenc L (ed) (1999), *Fair Weather: Equity Concerns in Climate Change*, London: Earthscan.

United Nations Framework Convention on Climate Change (UNFCCC) (1992), *ILM* **31**, 849, Article 2, available at http://unfccc.int/essential_background/convention/items/2627.php.

United Nations Framework Convention on Climate Change (UNFCCC) (1997), *The Kyoto Protocol to the Convention on Climate Change*, Bonn: UNFCCC.

United Nations Framework Convention on Climate Change (UNFCCC) (2001), 'Report of the Conference of the Parties on its Seventh Session Held at Marrakech from 29 October to 10 November 2001', Addendum 1 (FCCC/CP/2001/13/Add.1), 21 January, available at http://unfccc.int/resource/docs/cop7/13a01.pdf.

United States Senate (2005), 'Energy Policy Act of 2005', *Congressional Record* (22 June): S6980–S7063.

Vajpayee, Atal Bihari (2002), 'Speech by Prime Minister Shri Atal Bihari Vajpayee at the Eighth Session of the Conference of the Parties to the UN Framework Convention on Climate Change' (30 October).

Vrolijk, Christiaan (2002), 'A new interpretation of the Kyoto Protocol: outcomes from the Hague, Bonn and Marrakesh', Sustainable Development Programme Briefing Paper No. 1, London: Royal Institute of International Affairs.

Warnock, M (ed.), (1962), *Mill: Utilitarianism and Other Writings*, Glasgow: Collins.

Wiegandt, Ellen (2001), 'Climate change, equity, and international negotiations', in Urs Luterbacher and Detlef F. Sprinz (eds), *International Relations and Global Climate Change*, Cambridge, MA: MIT Press.

World Resources Institute (WRI) (1992), 'The difficulty of stabilizing emissions', *Research Topics*, available at http://population.wri.org/pubs_content_text.cfm?ContentID=792, Washington, DC: World Resources Institute.

World Resources Institute (WRI) (1996), *World Resources 1996–97*, Oxford: Oxford University Press.

Yamin, Farhana (2000), 'The role of the EU in climate negotiations', in Joyeeta Gupta and Michael Grubb (eds), *Climate Change and European Leadership: A Sustainable Role for Europe?*, London: Kluwer Academic Publishers.

Zito, Anthony R (2000), *Creating Environmental Policy in the European Union*, Houndmills, UK: Palgrave.

CONCLUSION

16. Explaining European responses to global climate change: power, interests and ideas in domestic and international politics

Paul G Harris

INTRODUCTION

This volume has described and explained several aspects of European and European Union (EU) responses to global climate change (GCC). We have examined the policies of, and politics within and among, several European states, especially Germany, the Netherlands, Norway, Poland, Spain, Sweden and the United Kingdom (UK). We have also endeavored to explain how and why the EU formulated and implemented its GCC policies, and we have sought to illuminate the influences of member states and the international GCC regime on the EU's internal and external policies in this issue area. We have also tried to show how and why ideas, including principled (normative) beliefs, have influenced national and regional responses to global climate change. Our main aims have been (1) to analyze these 'whats', 'hows' and 'whys' of European countries' and the EU's GCC policies, and (2) to add to existing understandings of these policies by explicitly thinking more or less in terms of environmental foreign policy (EFP) and by applying EFP theories to them. Analyzing and thinking about European and EU GCC policies from this 'foreign policy perspective' by definition highlights the crossovers and interactions between domestic and international politics and policy, including the actors, institutions and forces at both levels, thereby increasing our knowledge of how and why Europe has responded to GCC the way it has, and helping us to anticipate how European actors might behave in the future.

A RECAP OF THEORY

How can we explain Europe's domestic, regional and international responses to global climate change? The contributors to this volume have built upon and added to existing answers to this question. A common reference point for their case studies has been Barkdull and Harris's (2002) survey of theoretical approaches to EFP. As summarized in Chapter 1, Barkdull and Harris synthesized much of the EFP research into a typology of approaches categorized by three levels of analysis – the international system, domestic society (notably politics and culture) and the state (government and the people working therein) – and three core variables – power, interests and ideas (or cognition). This chapter highlights some of our explanations for Europe's GCC policies in the context of Barkdull and Harris's typology (see Figure 16.1, p. 404). While these explanations sometimes fall neatly into one cell in the matrix (e.g., distributions of power in the international system might explain state behavior), it is important to note that consideration of more than one approach – a combination of levels of analysis and/or variables – is most often found to be necessary for arriving at satisfactory explanations for European and EU GCC policies and practices. Before summarizing some of our findings, we can first recap Barkdull and Harris's summary of EFP theory.

In the context of the international system, *power*-based theory is usually associated with 'realism', which asserts that international regimes are a consequence of the power of one or more 'great powers'. Theoretical approaches taking this perspective assert that 'foreign policy is about acquiring power, and the state's position in the international hierarchy strongly influences which foreign policies will be rational for the state (and hence by assumption will be adopted)' (Barkdull and Harris 2002: 70). Power-based approaches to foreign policy may also focus our attention on distributions of power in domestic society, for example that of non-governmental actors, such as private industry. Here we are often interested in the role of direct and indirect influence on policy wielded by elites (often economic ones). We can also find explanations for state behavior by looking at executive-legislative relations and the relative power within government of leaders, policymakers or bureaucrats, particularly if we can identify 'which actor(s) in the state dominates the policy process' (Barkdull and Harris 2002: 80).

Thinking systemically, *interest*-based theory is often associated with liberal internationalism, which points to states creating and supporting international regimes because it is in their collective interest to do so (or at least in the interest of enough states to get the regime up and running and to pay its operating costs). As Barkdull and Harris (2002: 68) note: 'Interest-based theory might focus on which material interests in society different groups promote. Environmental policy might be understood as the outcome of bargaining and compromise toward a common solution in regard to a given problem. No actor or group would be expected to be permanently dominant; rather, temporary coalitions would negotiate for policies that would be broadly acceptable to all affected parties'. Theories focusing on interests in government (the state) often ask, 'What are the interests of the state, and what is the role of interests within the state? The first approach directs attention to state interests that exist apart from society … The premise is that the state's interests cannot be reduced to interests in society, such as group interests … The second approach … says that the foreign policy agencies have certain interests that they pursue in the foreign policy process [and] outcomes are explained as the product of bureaucratic bargaining and compromise …' (Barkdull and Harris 2002: 81–82). *Ideational* theories point to the manner in which new ideas or knowledge shape state behaviors or come to constitute them; state identities and national interests can be 'constructed' over time from influential ideas. For example, the work of Finnemore (1996): 'shows that ideas and values generated at the domestic level can lead to the creation of international organizations and international policies that are reflected in new institutions in other states. The international structure Finnemore investigates is one of "meaning and social value" rather than power' (Barkdull and Harris 2002: 73). According to Finnemore (1996: 2), 'States are *socialized* to want certain things by the international society in which they and the people in them live'. Ideas in domestic society, such as ideology, belief systems, stereotypes, myths, and public opinion, can help explain policy and behavior (Barkdull and Harris 2002: 77–78).

Last but not least (because, admittedly, the typology depicted by Barkdull and Harris of course does not exhaust the possible approaches to EFP), we can ask how ideas operate within government. Ideas that can affect policy range from 'world views, principled beliefs, and causal beliefs' serving as 'road maps' for policymakers, to scientific understanding, disseminated by scientists but also often used by 'knowledge brokers' to 'mold environmental foreign policy so that it reflects their own political and economic interests' (Barkdull and Harris 2002: 82–83). From this EFP perspective, we might also be interested in the psychology of leaders and decision-making groups, as well as the role that societal groups (e.g., non-governmental organizations [NGOs]) play in 'shaping policymakers' values and even scientific understanding of environmental and resource issues' (Barkdull and Harris 2002: 83).

Bearing these variables and the analytical approaches in mind, the next section summarizes some of the theoretical findings of the case studies in earlier chapters.

EXPLAINING EUROPE'S RESPONSES TO GCC: THEORETICAL FINDINGS

Based on what we have learned from our case studies, how can we explain Europe's responses to global climate change? What do the chapters say about which theories have the most utility in this respect? In Chapter 2, Hatch showed that the process of domestic politics interacting with regional and international relations can shape GCC policy. While power and the way it is wielded by societal interests and bureaucratic actors is important, he shows that this power-play does not provide a satisfactory explanation. We must also bear in mind that 'societal and bureaucratic influences are mediated through such historically developed institutions as national styles of policymaking or the organization of the political system itself – electoral arrangements and the network of political parties that shape the electoral arena'. Hatch showed how domestic forces determined German policy, which in turn fed into the German government's efforts to shape EU, and in turn international, GCC policy. While some of these forces, such as those exerted by the coal industry, were unfavorable to action on GCC, others – such as those from industries hoping to capitalize on demand for energy-efficient technologies and political parties that had taken up the environmental mantel – allowed Germany to have a more proactive policy, with material impacts on EU EFP and the GCC regime. 'The result', Hatch finds, was the 'adoption of ambitious reduction targets and a set of measures that reflect a search for consensus among various bureaucratic actors, powerful economic interests, and influential environmental NGOs'.

Cass's chapter, while also demonstrating the importance of domestic politics, finds that ideational factors in both domestic society and the international system were especially influential in shaping GCC policy in the UK. Ideas helped form national preferences, which in turn linked 'domestic ideational dialogue and international normative debates'. Political norms in the UK are often at odds with those in the rest of Western Europe, thereby limiting congruence between British and continental environmental policies and pressuring the UK government to 'internationalize' domestic norms, in an effort to limit the cost of policy responses within the UK. Cass demonstrates how 'normative debates' about the UK's response to climate change took place, often simultaneously, at domestic, EU and international levels. Cass also points to ways in which interests within the UK government were influential in policy formulation: 'Within the normative parameters surrounding climate policy, successive British governments strategically maneuvered in what could be described as a "three-level game" to pursue what were perceived to be British national interests related to GCC policy. The government linked climate policy to other domestic and European political initiatives and manipulated the negotiations to achieve important political objectives' that served UK interests. Overall Cass's case study shows that the state has an expected central role in shaping GCC policy, but it also shows that there is 'a significant role for ideational variables in establishing the parameters within which domestic and environmental foreign policy must be pursued'.

Domestic politics and institutions, and ideas and interests competing in society, are normally important shapers of foreign policy on climate change, as Kanie's case of the Netherlands shows. Powerful ideas are particularly important because, once they are formulated through a process of interest-interaction within domestic society (acting upon state policy), they can be wielded by the state and other actors (e.g., environmental NGOs) at the international level. Kanie finds that once the Netherlands' foreign policy on GCC was formulated at the domestic level (partly in concert with European partners), power became particularly important: 'Although policies, institutions and ideas still matter in explaining EFP and behavior of a country, the power-based approach is important for explaining the influence of a particular state in the process of building international environmental regimes'. The Netherlands was able to wield power through 'unilateral action combined with social persuasion', a kind of 'instrumental leadership' that can be asserted by small- and middle-power states. 'In other words', according to Kanie, 'the size of the power does not matter in exerting domestic resource-based multilateral leadership. In terms of approaches to foreign policy study, this means that interests and cognition/ideas account more for middle power leadership than does a power-based approach to foreign policy'.

Ideas in society are also important in Bokwa's case study of Polish GCC policy. However, the influence of ideas was not all pro-environmental, instead being the consequence of communist-Stalinist ideas perpetuated through time. Thus the most important variable in the case of Poland has been policy imposed largely by fiat from external hegemons. First Poland was (involuntarily) driven by Soviet prescriptions and preferences, which were environmentally harmful, and after the fall of communism Poland has been (voluntarily) driven by EU prescriptions and preferences, which are intended to bring the country into compliance with environmental protection objectives, including in the context of GCC. In sharp contrast to the case studies that precede it, the one in Chapter 5 shows that ideas shaping policy can be 'sharply detached from the beliefs and opinions widely held in society'. Indeed, during the communist period Poland did not, and does not now, have a politically influential environmental lobby that can push government or industry. Thus, Bokwa finds, 'Poland's climate foreign policy is not rooted in a domestic *environmental* policy so much as outside political stimuli', and its 'GCC policy will remain a matter of external conditions instead of national debate based on public awareness of the issue' because the demands of EU membership will continue to drive policy to a substantial degree.

Marshall's case study of Sweden's GCC diplomacy suggests that the 'key' to understanding it is found in domestic politics. Having said that, she points out that Sweden's GCC policies have been influenced to an extent by international power distributions. Much as in the case of the Netherlands visited in Chapter 4, Sweden's traditional power resources (e.g., economy and military) are limited, meaning that what influence it does wield comes from 'soft power and norm-based diplomacy'. Sweden's 'environmentally friendly' norms would ordinarily be expected to guide its EFP, but those norms failed to do so because of the way the policy process at all three levels of analysis (the state, society and the international system) interacted. The resulting 'tensions played out at the systemic level where Sweden deviated from its usual norm-based foreign policy, akin to the focus on ideas in Barkdull and Harris' (2002) typology, to a foreign policy that protected sovereignty, akin to the focus on interests in the typology'. Interestingly, Marshall concludes that 'an idea-based foreign policy can be inconsistent, just as a power or interest-based foreign policy can be. Norm influence is the primary method for Sweden's exercise of international power, but even when all the factors that support norm influence were present, a state can fail to engage in an idea-based EFP. Sovereignty concerns and domestic constituency concerns are reasons for this, but they are unlikely to be the only reasons'. A 'realist' (i.e. national interest) description of EFP is sometimes valid 'even for states whose international identity and power may be based very much on a rejection of the realist worldview'.

Tàbara's case study in Chapter 7 examines what he calls the 'failure' of GCC foreign policy in Spain, and its domestic and international causes and consequences. As with other case studies, Tàbara finds that a number of major variables at different levels of analysis are germane, with domestic politics being paramount. Unlike other chapters, however, Chapter 7 finds that state variables are almost determinative. In particular, institutional changes that resulted from the advent of a new Leftist government, as well as intra-state relations between Spain's central government and its highly autonomous regions, need to be considered when trying to understand this European country's GCC policies. Within this process one must factor in Spain's political culture, which is 'characterized by the strength and predominance of the state over regional governments, individual actors and private interests'. When combined with the fact that environmental issues have low salience in the Spanish public's mind (relative to other European countries), a political culture that 'confers a stronger authority to the state and to the community above the individual', including in the environmental issue area, results in little demand from citizens for environmental protection to be reflected in foreign policies. However, when a government more concerned about GCC came to power, this inordinate focus on the state allowed for significant changes at the policy level.

Chapter 8, by Tjernshaugen and Lee, focuses on strategies used by environmental NGOs (ENGOs) to influence the Norwegian government's responses to global climate change. The strategies of ENGOs to influence policy are characterized as 'shaming' and 'framing': 'Shaming the Norwegian government into changing its policies, and framing the domestic debate over international commitments in terms that underpinned [the NGO's preferred] policies'. As with most of the other case studies, processes at more than one level were important. In this case, while the NGOs were directing their attention at domestic actors, the Norwegian public in particular, they operated internationally on the sidelines of the GCC conferences, often in concert with multinational NGOs (e.g., the Climate Action Network). 'For our understanding of EFP', Tjernshaugen and Lee write, one lesson is that ENGOs 'engage in influencing foreign policy and could potentially be one of the actors shaping actual policies. In Norway, ENGO's opinions strongly influence public *debate* on EFP, and it is not unreasonable to assume that this could have some bearing on foreign policy *decisions*. By harshly criticizing Norwegian governments, ENGOs tried to make it politically costly to pursue policies [the ENGOs] disagreed with'. Importantly, however, Chapter 8 notes that the ENGO's resources were 'mostly symbolic rather than material', with one important 'currency' of the ENGOs being legitimacy, which helped them shame the government into a more pro-environmental set of GCC policies. Given these efforts by NGOs, Tjernshaugen and Lee conclude that a '*societal* understanding of EFP processes is in fact crucial to understand NGO influence even in international negotiations'.

In Chapter 9, Lacasta et al. find explanatory power in all levels of analysis, although they suggest that state-centric EFP approaches have the least to say about the EU's position on climate change: 'State-centric theory currently has limited applications in the EU case due to the diversity of actors and the absence of a hierarchical structure headed by a single leader'. Having said that, they argue that behavior of larger member states, whereby they 'try to impose their own priorities, or to broker agreement with other bigger states or influential smaller states', is an example of state-centric factors at work (in terms of viewing the EU as a sort of state; otherwise, this can be more traditionally viewed as reflecting the exercise of international power). Indeed, larger member states often tried to push their weight around during internal EU negotiations, and to wield their power in international

GCC negotiations, as when France tried to push its own agenda while EU president and when the UK attempted to unilaterally broker a deal with the US. These actions by powerful states, along with continued efforts by the EU to exert its 'soft' power and leadership in this issue area, reflect forces emphasized in systemic approaches to EFP. Lacasta et al. also see societal-level forces at work, reflected in the 'fact that NGOs, business, the media and the public are in different ways involved in the GCC negotiations' and in some smaller member states (e.g., Belgium) taking a 'consensus-based approach internally and [relying] considerably on the assistance of the Commission as well as on a division of labor between member states, especially at the technical level'.

Much like in Chapter 9, in Chapter 10 Jung et al. deploy several theoretical approaches in their case study of the EU's policy on carbon sinks. They argue that their 'three-level approach' encompasses 'influences of domestic actors on country positions (societal theory) as well as the influences of structural characteristics of the EU and the international regime on the negotiation process (systemic theory)'. According to these contributors, EU policy on carbon sinks has involved a prolonged co-ordination process among member states and has been 'subject to the pressure of various domestic interest groups', including environmental NGOs, research organizations and industry. It has also been influenced by the 'power constellation in the international negotiations', including pressure from a variety of international stakeholders, as well as efforts by other states to shape the outcome of international negotiations on carbon sinks. Ideas – knowledge – was an important variable insofar as interest groups used it to push their policy preferences on member states so that those preferences would be reflected in international negotiations and outcomes (although we might characterize this as falling under the interests-in-society approach to EFP, without discounting the central role played by knowledge).

More so than the other case studies in this volume, in Chapter 11 Oberthür and Tänzler assign major explanatory value to the influence of the international system, particularly with regard to the variables of interests and, albeit less so, ideas. They simultaneously deploy constructivist and rationalist approaches to EFP, referring to 'both national interests and ideas/learning, paying particular attention to states' interests as influenced by their perceptions and ideas'. Chapter 11 indicates that the international GCC regime and the Kyoto Protocol in particular drove the diffusion of GCC policy instruments in the EU. Oberthür and Tänzler's case study begins with the assumption that international institutions 'can play a substantial role as drivers and triggers of processes of policy diffusion' by providing states with 'special incentives to implement particular innovative policy instruments and/or put diffuse pressure on state governments to take action in order to be able to comply with their international commitments. [International institutions] may also serve to induce learning processes at the national level. Thus, the unspecific "compliance pull" of the Kyoto Protocol or particular incentives provided by it may increase the willingness of EU governments to implement innovative policy instruments, as may redefinitions of their interests induced by the Kyoto process'. Oberthür and Tänzler treat EU states largely as unitary actors, thus discounting societal and state/governmental variables, which they nevertheless acknowledge would be important to consider for a full explanation of GCC policy-instrument diffusion in the EU.

Bang, Vevatne and Twena focus on several variables that they find determined the choice of GCC policy instruments in Germany and Norway. They use systemic, societal and state levels of analysis while, like Oberthür and Tänzler, employing an integrated approach drawn from rationalist and constructivist theory. They begin by using a social constructivist approach to explain norms and values shaping German and Norwegian environmental foreign policies. According to their case study, these two countries: 'developed key objectives – environmental leadership, respect for the precautionary and polluter pays principles and a determination to protect their competitive position within the internal market – on the basis of essentially socially constructed norms. These normatively oriented objectives cannot be explained and understood purely on rational grounds since there is little economic gain from adopting the precautionary principle or being an environmental leader. Rather these objectives are perceived to be important as they give "soft power", prestige, respect among European and international peers, and reinforce domestic values'. In other words, 'key norm-based objectives helped guide perceptions and evaluations' of policy instruments.

Chapter 12's case study also focuses on power relationships in German and Norwegian societies and in their governments, thereby 'demonstrating how ideas are introduced and incorporated into the national policy program' as societal elites set agendas and politicians, bureaucrats and stakeholders wield power to push aside alternative policies. Within the state, the case study shows that 'executive-legislative relations are central for defining the policy agenda. In parliamentary systems it matters for policy outcome what the constellation of the parliamentary majority looks like, and whether it shares the same policy objectives and conceptions as the executive/prime minister'. Finally, Bang, Vevatne and Twena focus on interests at the systemic and societal levels, in the former instance showing that national interests were constrained by the EU's common GCC policy, and in the latter instance showing how implementing policy instruments was significantly a result of bargaining among stakeholder groups in these societies. According to the authors, their case study: 'demonstrates how Barkdull and Harris's typology of approaches to environmental foreign policy can be applied empirically, and illustrates how "the line between one approach and another can blur" (Barkdull and Harris 2002: 68). ... Overall, we find that a social constructivist approach provides the most convincing explanation of how norms and values shape EFP objectives, at least in the cases examined here, while rational choice-based theories give a better understanding of how interests are pursued once they have been defined'. Massai's case study on EU enlargement arrives at a decidedly systemic explanation for the GCC policies examined.

In Chapter 13 Massai finds that requirements of the Framework Convention on Climate Change and its Kyoto Protocol 'contributed to the development of an advanced European GCC strategy and policy'. Negotiations between the EU and central and eastern European states related to climate change prior to their entry into the union, along with the formulation of broader EU GCC policies, were in Massai's reckoning significantly a byproduct of the climate change regime. International rules established under the rubric of the Kyoto Protocol, and the means for implementing those rules (the so-called flexible mechanisms, such as emissions trading) were key variables shaping European GCC policies and associated legislation. His account of EU enlargement suggests that: 'systemic approaches to environment and foreign policy may be the most suitable perspective from which to analyze European foreign policy on global climate change. The relationship between the international and regional systems (the latter being the EU), and the relationship between the EU and the national systems of member states (and member states-to-be), play a central role in defining roles and identities, as well as policies and measures at the local level, in terms of addressing the problem of global climate change'.

The case studies by Jaggard (Chapter 14) and Harris (Chapter 15) focus on the role of ideas in European climate change policies. Jaggard's chapter shows how ideas in society, 'being expressed through public opinion, the media and social movements' are important for explaining Germany's GCC policies and in turn their 'reflexive' impact on broader EU and even international (beyond the EU) negotiations and policies on climate change. Jaggard argues that 'ideas have the ability to permeate' domestic, European and international GCC discussions. While she acknowledges that power and interest-group lobbying remain significant in shaping GCC policy, she finds that ideas are 'paramount' even if we deploy analytical approaches at all levels of analysis – systemic, societal and state. Similarly, Harris focuses on ideas in his case study, namely ideas identified as what he calls international environmental equity (IEE). He argues that considerations of this nature have 'started to permeate the GCC process and shape European behavior', with the EU and several member states actually fostering principles associated with IEE in their own EFP. In so doing, they have increased the degree to which IEE is 'embedded' in the broader GCC regime. Europe is unusual in this respect: 'International environmental equity has contributed more to the GCC policies of European states and the EU than to the policies of other countries. The EU is in turn driving the construction of IEE more broadly as it becomes a European *interest* and a shaper of its internationally relevant domestic policies (e.g., on energy use and conservation) and foreign policies on GCC'. Thus, within the EFP typology explicated by Barkdull and Harris, Chapter 15's case study points to constructivist theory, 'particularly those theories arguing that objectives of foreign policy and even national interests are in fact constructed from powerful ideas'.

These explanations for European and EU GCC policies are roughly depicted in Figure 16.1. Other scholars may interpret the theoretical implications of the case studies differently; how the chapters are read may send different messages and thereby possibly highlight additional or alternative cells in the Figure 16.1 matrix (or, indeed, those not found in it at all). Some theoretical purists might not like the mixed bag of findings emanating from the collection of case studies in this volume. However, this variety is precisely the point: no one approach to foreign policy or international relations can adequately explain European and EU responses to global climate change. To assert otherwise is simply not supported by our research.

CONCLUSION

Where our explanations fall within the Figure 16.1 matrix is useful for what it says and for what it does not say. Ideas can drive European GCC policies to a substantial degree, showing that *ideational* variables are often very important or even central to understanding EFP and international environmental relations. Certainly complete explanations for GCC policy will normally have to consider ideational forces. This challenges notions of foreign policy and international relations that discount the importance of ideas, particularly principled ideas that seem to be playing quite a big part in European (and, arguably, broader international) responses to GCC. That this is true for the EU should not come as a surprise because it is an organization that has built its collective identity on upholding, and expressing internationally (for example, through its relatively generous development aid), ideas that underpin its own domestic societies, and violation of which grates on national and common identities to an extent where these ideas do influence policy – if seldom entirely overwhelming more short-term, particularistic and traditional national interests.

	Systemic (International System)	Societal (Domestic Politics)	State-Centric (Government)
Power	Kanie (4) Bokwa (5) Lacasta (9) Jung (10)	Bang (12)	Hatch (2) Tàbara (7) Lacasta (9) Bang (12)
Interests	Marshall (6) Oberthür (11) Bang (12) Massai (13)	Hatch (2) Kanie (4) Marshall (6) Tjernshaugen (8) Lacasta (9) Jung (10) Bang (12)	Hatch (2) Cass (3)
Ideas	Cass (3) Oberthür (11) Bang (12) Massai (13) Jaggard (14) Harris (15)	Cass (3) Kanie (4) Bokwa (5) Marshall (6) Bang (12) Jaggard (14) Harris (15)	Jaggard (14) Harris (15)

Note: Cells in the matrix show examples from the chapter case studies, identified here by first authors (as listed in bylines), with chapter numbers in parentheses.

Source: Adapted from Barkdull and Harris (2002: 69).

Figure 16.1 Theoretical explanations for Europe's responses to global climate change

Our case studies validate some theories of international relations and foreign policy, including EFP, that point to the importance of *interests* at all levels: interests shaping and sustaining international environmental regimes, lobbying by interest groups in societies, and interests operating within government agencies. These interests often are not amenable to aggressive action on climate change, although as the importance of responding to this problem (and even the idea of the responsibility of doing so) starts to permeate state bureaucracies, as it has started to do within some of the agencies of the European Commission, this is changing. Interests matter, to be sure, but states do not always act on what is clearly their short-term (e.g., economic) interests. When we realize how those interests are formed, sometimes with principled ideas helping to shape them, it becomes clearer that longer-term interests, including those related to global climate change, are being acted upon more and more (if not to the extent that the science tells us is necessary).

Where our case studies' theoretical messages do not fall in the matrix says something, too. For example, the importance of *power* in this issue area is not quite as important as some may expect. Great powers cannot push their weight around very easily unless they are backed by commonly accepted conceptions – ideas – of what is appropriate policy, as shown by the inability of the US to derail the Kyoto Protocol process and the EU's contrasting ability to keep it alive because doing so was 'the right thing to do' in the eyes of most actors. This does not mean that power is unimportant in the GCC debate; powerful states in the traditional sense matter greatly even if they cannot dictate as many outcomes as they might want. Power also matters for states that do not possess traditional power resources: small and 'middle power' states in Europe, backed by solid science and principled policy objectives, were able to influence EU and international GCC policies, in the latter case by leveraging EU 'environmental' power at the global level.

An important message about theory emanating from the case studies is that foreign policy, defined as the crossover and interaction between the domestic and international policy arenas, can be a key to understanding GCC policies at all levels. As Gerner has described it: 'Although no subfield in political science is completely self-contained, the study of foreign policy is somewhat unusual in that it deals with both domestic and international arenas, jumping from individual to state to systemic levels of analysis, and attempts to integrate all of these aspects into a coherent whole' (Gerner 1995: 17, quoted in Harris 2005: 18). Indeed, most of the contributors to this volume have deployed multi-level analyses, or otherwise invoked the importance of it when focusing on variables and processes at one level. To borrow a line from Bang, Vevatne and Twena's chapter, 'rather than seeing different theoretical approaches to environmental foreign policy as competing alternatives, they are best viewed as complementary tools that can help us to paint a more comprehensive picture and allow us to simultaneously utilize concepts such as norms, power and interests within a multi-level context'. This suggests that when we seek to understand European and EU policies on GCC we cannot get a true picture of reality without explicitly considering (or deploying) to some degree the 'foreign policy level' of analysis, which is by definition amidst and between the domestic (including society and the state) and the international (including the systemic) levels of analysis.

This is especially true for Europe – more so than for, say, the US – because the EU organization and the interactions that membership in it engenders are often a place or venue for crossover and interplay between the domestic and the international. Put another way, EU member states (and arguably most other European states, given their extensive connections with their neighbors and with the EU and other European institutions) possess less autonomy than 'normal' countries to do what they want domestically, and their international interactions with states beyond Europe and with the international system are mediated or at least greatly influenced by the EU, *especially* in this issue area (despite what some member states may say or want). As Ott aptly puts it, 'Climate policy thus equals foreign policy. It is for the European Union to decide whether to take up that challenge' (Ott 2001: 296). I think it has, certainly more so than any other state or group of states. We can at least say that Europe has started down the right path and is now the leader on the path to addressing global climate change. It is a slow, faltering leader, with several member states clearly trailing behind, but it is a leader nonetheless. It looks set to continue down that path, with the increasing willingness of its members and other European states. Our case studies show that many actors, forces and institutions will be involved in this process, so it will never be smooth and very likely will not go far enough to 'prevent dangerous anthropogenic interference with the climate system' (UNFCCC 1992: Article 2) for which the climate change regime was created. But it is a start.

REFERENCES

Barkdull, John and Paul G Harris (2002), 'Environmental change and foreign policy: a survey of theory', *Global Environmental Politics* **2**(2) (May): 63–91.

Finnemore, Martha (1996), *National Interests in International Society*, Ithaca, NY: Cornell University Press.

Gerner, Deborah (1995), 'The evolution of the study of foreign policy', in Laura Neack, Jeanne Hey, and Patrick Haney (eds), *Foreign Policy Analysis: Continuity and Change in Its Second Generation*, Englewood Cliffs, NJ: Prentice Hall.

Harris, Paul G (2005), 'Environmental politics and foreign policy in East Asia: a survey of China and Japan', in Paul G Harris (ed.), *Confronting Environmental Change in East and Southeast Asia: Eco-Politics, Foreign Policy, and Sustainable Development*, London: Earthscan/United Nations University Press.

Ott, Hermann E (2001), 'Climate change: an important foreign policy issue', *International Affairs* **77**(2): 277–96.

United Nations Framework Convention on Climate Change (UNFCCC) (1992), *ILM* **31**, 849, Article 2, available at http://unfccc.int/essential_background/convention/items/2627.php.

Index

Acquis Communautaire 29–30, 258, 305–10, 319–20
activists 5, 18, 148, 152, 188–92, 194–205, 248
see also citizens, civil society, nongovernmental organizations
activities implemented jointly (AIJ) 234, 239, 247
see also joint implementation
'actorness' of EU 23–33, 212, 239–40
Ad hoc Group on the Berlin Mandate (AGBM) 11, 78, 85, 98, 102, 105, 109, 222
adaptation 3–4, 6, 10, 13, 15, 33, 171, 175, 244, 267, 323, 351, 353, 359–60, 364, 370–72
Adaptation Fund 13, 15, 359
afforestation *see* forests
agenda setting 94, 281, 341
agriculture 6, 8, 163, 165, 169, 245–6, 288, 355, 372
Alliance of Small Island States (AOSIS) 15, 223, 242, 244, 334–5, 341, 345, 462, 381
see also small island states
allocation of emissions 55, 57, 59, 154–5, 162, 168 170–71, 173, 178, 180, 201, 220, 285, 291–2, 296–7, 315, 337, 355, 358
alternative energy 148–50
see also energy, renewable energy
Annex I countries 11, 102, 105, 130, 165, 167, 218, 235, 242, 244, 248, 256, 313, 335, 361
see also developed countries, industrialized countries
autonomy *see* policy autonomy

Berlin Mandate 11–12, 14–15, 49–50, 222, 357, 362
see also Ad hoc Group on the Berlin Mandate, common but differentiated responsibility
biodiversity 131–2, 233, 245
see also species
Blair, Tony 64, 80–82, 219–20
Britain *see* United Kingdom
Brundtland Commission *see* World Commission on Environment and Development
bubble, EU emissions 12, 23, 30, 147, 170, 223–4, 226, 301, 312, 318, 367, 370, 381
Buenos Aires Plan of Action 11–12
burden sharing agreement (BSA) 78, 100, 217, 224–5, 301, 312, 318
see also bubble
burden sharing 27, 30, 78, 96–8, 100, 102, 195, 224–6, 260, 262, 349–50, 370, 377–8
see also burden sharing agreement, bubble, equity, justice
bureaucracy 19, 21, 66, 116
Bush, George H.W. 47, 70, 75
Bush, George W. 12, 193, 228, 267, 336, 365, 369, 377
business as usual 30, 48–9, 241, 243, 316, 369, 372
business interests 27, 67, 91, 124–5, 169, 186, 196–7, 200, 202–3, 219–21, 237, 246, 265, 290, 295, 299, 311, 328, 339, 400
see also industry

capacity building 308, 312, 317–18, 364
carbon sinks 12–13, 15, 28, 53–4, 79–81, 120, 175, 193, 211, 233–8, 240–41, 243–4, 246–8, 251, 336, 344, 364, 371, 381, 400
carbon taxes 74, 146, 261, 267–70, 291, 294, 301
see also energy taxes, green taxes

carbon trading *see* emissions trading
categorical imperative 324
Central and Eastern European Countries (CEEC) 305–20
see also Central European countries, Poland
Central European countries 131–2
see also Central and Eastern European Countries
certified emission reductions (CERs) 314–15
China 13–15, 21, 80, 193, 219, 242, 244, 249, 352, 358, 361, 380–81
chlorofluorocarbons (CFCs) 6, 75, 332
citizens 14, 117, 152, 172, 194, 218, 323, 325, 338, 343, 353, 399
civil society 53, 173, 177
see also activists, citizens, nongovernmental organizations
Clean Development Mechanism (CDM) 12, 15, 53–5, 60, 128, 155, 235–7, 241–5, 247, 249, 285, 292, 296–7, 301, 310–11, 314–15, 336–7, 362, 370, 381
climate change *see specific aspects*
climate protection plans 258–63, 269, 272
Clinton, William J. 12, 81
CO_2 taxes *see* carbon taxes
coal 6, 25, 43–4, 51, 57, 70, 73–4, 76, 116, 123–4, 129, 133, 151, 269, 288, 299, 328, 368, 370, 381, 395–6
see also fossil fuel, natural gas, oil
coastal areas 8–9
see also coastal countries
coastal countries 15, 345, 356, 360
see also Alliance of Small Island States, small island states
Cold War 114
collective action 71, 178, 375
combined heat and power (CHP) 78, 255, 311, 332
commitments, Kyoto *see* Kyoto Protocol
common but differentiated responsibility 14, 318, 334, 352–3, 357, 361, 364, 377
common foreign policy 4, 63, 65, 77, 82–4, 351
see also common position
common policy 233–51 *passim*
common position, European 50, 99, 214–17, 219, 223, 227, 244, 250, 314, 319
communism 25, 128, 397
compensation 122, 357, 360
competence, European 22, 27, 45, 64, 68, 74, 212, 216, 219, 226, 240, 306
competitiveness 46, 48, 55, 58, 67–8, 96, 104, 162, 174, 284, 289–90, 293, 298
compliance 13, 15, 49, 75, 77, 165, 173, 178, 243, 246, 256, 263, 267, 270, 285, 292, 296–7, 307–8, 312–14, 317, 364, 376, 397, 401
Conferences of the Parties to the FCCC (COP)10–16, 194–7, 199, 203, 205, 220–21, 313, 335, 342, 358
COP1 48–9, 105, 110, 190, 194, 198, 216, 222–3, 234, 335, 362
COP2 97–8, 191, 223
COP 3 41, 49–50, 76, 91, 97, 105, 130, 192–3, 197–8, 211, 221, 243, 335–6, 351, 363
COP4 80, 351
COP5 130
COP6 52, 54, 80–81, 108, 201, 212, 217, 221, 237–8, 242–3, 247, 313, 336, 364
COP7 29, 54, 130, 193, 212, 217, 264, 279
COP9 176, 221, 236, 246, 249, 251
COP10 176, 249–50
COP11 109, 130, 176, 193, 250, 312, 365
consensus 23, 27, 30, 32, 42, 45, 57–9, 74, 77–8, 91, 94, 97–8, 107, 289, 211–29 *passim*, 247, 294–5, 298, 313, 324–6, 334, 396, 400
constructivism 29, 350
cooperation, international 5, 17, 20, 30, 42, 88–9, 120, 128, 131–2, 139, 144–5, 168, 178, 186, 188–9, 193, 198, 212, 229, 351
regional 4, 22, 31, 371–2

deforestation *see* forests
democracy 21, 123, 133–4, 180, 306, 333, 358

developed countries 10–15, 31, 102, 110, 205, 222–5, 228, 242, 335, 351–3, 355–6, 360, 364, 368–9, 370–74, 376–7, 381
see also Annex I countries, industrialized countries
developing countries 3, 10–15, 27, 31, 49, 53, 105, 130, 150, 189, 202, 222, 225–6, 229, 242–3, 245, 248, 264, 332, 335, 337, 340, 352–3, 357–65, 371–5, 378–9
see also least developed countries, less developed countries, poor countries
development assistance, overseas (ODA) 343
development *see* sustainable development
diffusion of policy *see* policy diffusion
diplomacy 3, 5, 10–16, 24–6, 31, 88, 97, 106, 140, 143–5, 153, 156, 176, 180, 369, 398
discourse 30, 168, 178, 180, 279, 282, 295, 298, 300, 324–7, 334–7, 341–2, 344, 353
distribution of emissions 163, 165, 262, 313, 316, 353, 357–8
distributive justice *see* equity, fairness, justice
domestic politics 5, 10, 16–19, 31, 42, 59, 143, 187, 215, 327, 333, 335, 343–4, 396–8, 404
domestic stakeholders *see* stakeholders
drought 5–8, 10, 355, 379

Earth Summit 11, 132, 169, 190, 298, 330, 335, 361, 338
ecological taxes 50–51, 57–9, 267, 270, 294, 331
see also carbon taxes, energy taxes, green taxes
economic development 14, 122–3, 218, 340, 351
see also sustainable development
economies in transition (EIT) 130, 211, 218, 242
electricity 51–2, 57, 72–3, 100, 121, 123, 126, 129, 149, 171, 175, 246, 287–8, 310, 331, 333, 381
elites, influence of 19, 281, 291, 294–5, 297–8, 300–301, 394, 401
emissions reduction 13, 15, 29, 49, 54, 63, 65, 72, 74–80, 82, 87, 95, 97, 99–100, 109, 124, 126, 130, 150–52, 176, 189, 198–9, 201, 205–6, 211, 225, 236, 244, 262, 264, 270, 279, 284–5, 288, 290, 292–4, 299, 301, 313, 357, 360, 370, 381
see also distribution of emissions, emissions targets, fugitive emissions, historical emissions
emissions targets 29, 41, 73, 165, 182, 192, 235, 262–4, 267, 313, 362
see also targets and timetables
emissions trading (ET) 12–13, 15, 28–9, 33–4, 53, 55, 57–9, 79, 139, 170, 173, 179–80, 189, 197, 200–201, 211, 255, 257, 261, 264–7, 270–72, 279, 282–5, 289–97, 310, 313–18, 368, 402
see also emissions trading directive, emissions trading system
emissions trading directive 219, 255, 266
Emissions Trading Scheme (ETS) 154, 220, 241, 255, 264–6, 271, 280, 282–5
energy *see* energy taxes, nuclear energy, oil, renewable energy
energy taxes 46–8, 58, 77, 86, 213, 222, 255, 257, 267–73, 365, 367
see also carbon taxes, green taxes
enforcement 172, 307–8
enlargement of the EU 29–30, 153, 258, 269, 305–19 *passim*, 402
Enquete Commission 43, 45
entry into force of the Kyoto Protocol *see* Kyoto Protocol
environmental foreign policy (EFP) 4, 16, 18–19, 21, 23, 25–7, 29, 33, 41, 42, 59, 63, 65, 67, 69, 83–4, 88, 91, 106–9, 113, 119–20, 132, 140–42, 145, 156, 161–3, 171–2, 187, 203–4, 215, 221, 250, 255, 257, 270–72, 280, 289, 298, 301, 305, 314, 330, 334, 342–3, 350, 366, 374, 393–406

environmental nongovernmental organizations (ENGOs) *see* nongovernmental organizations
environmental security *see* security
equity 31, 35, 92, 163, 177, 179, 228, 334, 349–81, 403
  see also fairness, justice
EU as an actor *see* 'actorness'
EU Bubble *see* bubble
EU Burden Sharing Agreement *see* burden sharing agreement
European Climate Change Program (ECCP) 255, 367–81
European Commission 14, 20, 27, 68–9, 83, 102, 153, 214, 219, 255, 266, 269, 284, 306–8, 311–12, 315, 329, 342, 353, 361, 364, 366, 368, 371, 375, 405
European Community (EC) 41, 69, 96, 147, 162, 213–14, 228, 239, 260, 288, 305–9, 312–13, 316, 318, 329, 379
European Council 98, 215, 225, 227–8, 269, 284, 306, 329, 366, 369
European Parliament 215, 283, 310–11, 314, 317, 379
European Union (EU) *see specific issues and* European Community
executive-legislative relations 19, 281, 394, 401

fairness 21, 325, 349–50, 353–7, 360, 364, 369, 377–8, 380
  see also equity, justice
finance 359–60
foreign aid *see* development assistance
foreign policy analysis (FPA) 4–5, 17–18, 21, 257, 344
foreign policy 4–5, 16–21, 32, 41–2, 59, 63, 65, 82–4, 87–90, 106–9, 119, 128, 132–4, 140–45, 153, 155–7, 161–2, 171, 173, 176, 178–80, 185–7, 203–4, 211–12, 214–15, 221, 226–7, 233, 236, 250, 255, 257, 280, 289, 301, 305, 313, 319, 323, 326, 328, 330, 333–4, 337, 342–4, 350–51, 353, 374–8, 393–9, 402
  see also common foreign policy, environmental foreign policy
forests 7, 9, 12, 28, 54, 67, 72, 92, 116, 163, 175, 193, 197, 233–9, 241–51, 328, 330, 344, 364, 372
formulation of policy *see* policy formulation
fossil fuel 6–8, 44, 72, 233, 244, 269, 299, 339
  see also coal, natural gas, oil
Framework Convention on Climate Change (FCCC) 11, 13, 15, 23, 27–30, 33–4, 37, 41–2, 48–9, 64, 71, 75–6, 80, 82, 87, 91, 96–7, 125, 130–31, 154, 162–3, 165, 169, 170–71, 176, 185, 189–90, 193, 197, 199–211, 214, 216, 220–23, 233–4, 242, 255–6, 262–4, 279, 306, 312, 316, 334–7, 342, 349–71, 402
framing 27, 145–6, 185–205 *passim*, 399
fugitive emissions 165, 288

G77 *see* Group of 77 developing countries
Germany 12, 23, 29–30, 33, 41–60, 68–9, 71, 73, 77–8, 81, 83, 101, 120, 167, 212, 217–18, 223–5, 236, 238, 247, 259, 268, 270, 279–301, 323–44 *passim*, 362, 367–8, 373, 381, 393, 396, 401–2
glaciers 6, 8–9
global climate change *see specific aspects*
Global Environment Facility (GEF) 342, 345
global warming *see specific aspects*
globalization 16, 124, 339–40, 343, 353
green taxes 291–2
  see also carbon taxes, ecological taxes
greenhouse gases (GHGs) *see specific gases*
Greenpeace 99, 107, 185, 193–4, 197, 200, 203–4, 244
Group of 77 developing countries (G77) 14, 105, 193, 242, 244, 361

Hague Summit on the Atmosphere 70
health, human 9, 16, 66–8, 131, 308, 370

hegemony 88, 267
historical emissions 279, 357
see also emissions
historical responsibility *see* responsibility
human health *see* health
human rights 306, 340, 358, 370
hurricanes 377

ideas 18–21, 24, 29–31, 65, 83, 87–8, 90, 106–9, 113, 116, 118–19, 125, 133, 140–44, 153, 187, 196, 234, 248, 250, 256–7, 271–2, 280–81, 290, 298, 323–45, 349–50, 374–5, 377, 393–405
impacts of climate change 4, 6, 8–10, 15, 18, 32, 218, 236, 245, 352
see also specific impacts
implementation *see* activities implemented jointly, joint implementation, Kyoto Protocol
India 13, 80, 219, 244, 352, 358
industrialization 6, 113–15, 117, 119, 323
industrialized countries 11, 23, 32, 41, 94, 106, 185, 191, 193, 200, 223, 235, 243, 264, 271, 284, 332, 335, 343, 352–3, 358–9, 362, 364–5, 371, 379–80
see also developed countries
industry 6, 9, 14, 27, 42–9, 52, 54–5, 57, 59, 66–9, 72, 74, 91, 95–6, 104, 107, 114–16, 119, 121, 124, 129, 131, 146, 148–9, 152, 169, 171, 173, 177–8, 197, 200–202, 219–20, 236–7, 240–41, 246, 255, 265, 281, 284, 286–7, 289–97, 299, 311, 317, 327–8, 331, 337, 363, 370, 394, 396, 398, 400
see also business
injustice *see* justice
institutionalism 20, 57, 335
see also international institutions
institutions *see* international institutions
interest groups 14, 19, 42, 45, 51, 57, 200, 240, 250, 291, 296, 335, 338, 400, 405
intergenerational justice *see* equity, fairness, justice
Intergovernmental Negotiating Committee (INC) 11, 75–6, 334–5
intergovernmental organizations *see* international organizations
Intergovernmental Panel on Climate Change (IPCC) 6–11, 32, 45, 71–3, 75, 95, 146, 161, 235, 246, 330, 354–5, 380
international cooperation *see* cooperation
international emissions trading (IET) *see* emissions trading
international environmental agreements 22, 26, 179, 205, 239, 337
see also specific agreements
international environmental negotiations 14, 27, 30–31, 41, 46, 50, 69–70, 84, 96, 103, 108, 185–8, 194, 196, 198, 201–4, 214, 216–17, 224–6, 228, 233, 239–40, 250, 262, 342–3, 349, 357, 360–61, 365, 369, 381, 399–400
international institutions 28, 64, 109, 187, 256–8, 270–71, 400–401
see also international organizations, regimes
international negotiations 14, 30, 46, 50, 70, 84, 108, 185–8, 196, 198, 202–4, 214, 216–17, 219, 226, 228, 239–40, 250, 262, 342–3, 349, 399–400
see also international environmental negotiations
international organizations 29, 214, 245, 350, 395
see also specific organizations
international politics 4–5, 10, 31, 143, 156, 326, 333–4, 375, 393
international regimes *see* regimes
international relations 4, 16, 18, 32, 59, 161, 186, 215, 257, 375, 380, 396, 403–5
international system 19–21, 133, 143–4, 187, 266, 313–14, 350, 394, 396, 398, 400, 404, 406
irrigation 9

Japan 12, 53, 72, 78, 105–6, 130, 167, 193, 211, 219, 224, 229, 235, 241–2, 248, 251, 255, 335–6, 341, 374

joint implementation (JI) 12, 53, 79, 96, 110, 128, 154, 211, 244, 285, 310, 360, 366
see also activities implemented jointly
justice 30, 323–5, 327, 333, 335, 338, 341, 349–79
see also equity, fairness

knowledge 106, 128, 141–2, 187, 197, 220–21, 250, 295, 375, 395, 400
see also science
Kyoto Protocol (KP) 11–15, 23, 28, 30, 41–2, 53–5, 79–80, 106–9, 140, 145, 147, 150, 153–5, 169, 170, 177, 192–3, 199–200, 202, 211–12, 214, 216, 219, 221–4, 229, 235, 238, 242, 250, 255–6, 258, 260, 262–72, 282, 284–5, 290–91, 293, 295, 300, 305–6, 310, 312–14, 318, 332, 336, 351–2, 360–61, 363, 365, 370–71, 377, 400–402, 405
commitments 23, 41, 49, 58–60, 109, 165, 192, 195, 235, 256, 262–3, 266, 270, 279–301, 312, 314, 332, 335–6, 370, 373
entry into force 127, 154–5, 258, 335
implementation 12, 15, 30, 64, 131, 161, 202, 206, 235, 238, 247, 267, 269, 271, 311, 313, 317, 351, 359, 373
negotiations 12–13, 23–4, 49–50, 52, 64, 87, 91, 98, 102, 107, 109, 151, 224, 238, 258, 269, 271, 301, 357, 363
ratification 13, 50, 193, 216, 247, 249–50, 260, 287, 309, 336, 365
see also post-Kyoto negotiations

laggards 22–3, 140–41, 150, 157, 297, 367, 369
see also lead states
land use *see* land-use, land-use change and forestry
land-use, land-use change and forestry (LULUCF) 7, 28, 163, 233–51 *passim*
Large Combustion Plant Directive (LCPD) 69, 73
leadership 20, 22–7, 29, 31, 33, 47, 50, 53, 65, 70, 72, 74, 76–7, 80, 87–109, 140–41, 143–6, 150–52, 155–7, 195, 211, 216–18, 226–8, 257–8, 262, 265, 271–2, 280, 282, 284, 289–90, 293, 295, 298, 329, 342, 350–53, 357, 361–9 *passim*, 372–3, 378, 397, 399–401, 406
see also laggards
less-developed countries 162, 352
see also developing countries, least-developed countries, poor countries
least-developed countries (LDCs) 372
see also developing countries, less-developed countries, poor countries
Least Developed Countries Fund 13, 242
lobbying 26, 47, 99, 103, 186, 195–7, 203, 205, 240, 246, 250, 291, 297, 403, 405

Maastricht Treaty 22, 336
Major, John 64, 74–5, 219
markets 54, 168, 202, 265, 375, 381
Marrakech Accords 11, 13, 15, 130, 212, 310, 359
media 19, 27, 200, 203, 205, 220–21, 326, 400, 402
Meeting of the Parties to the Kyoto Protocol (MOP) 15, 109, 130, 235, 250, 312, 365
mitigation 28, 54, 171, 175, 189, 228, 233, 236, 244, 250, 290, 311, 313, 352–3, 358, 371–2
Montreal Protocol on Substances that Deplete the Ozone Layer 33, 129–30
multilateralism 20, 24, 69–70, 76, 83, 87, 89–91, 97, 104, 106, 108, 397

national communications 170, 256, 312
national interests 16–18, 21, 25, 130, 219, 257, 281, 298–300, 356, 374, 395–6, 400, 402–3, 405
National Programs for Adoption of the Acquis (NPAA) 307–9
national security *see* security

natural gas 51, 73–4, 76, 190, 256, 288–9, 367, 381
  see also coal, fossil fuel, oil
Netherlands 23–4, 55, 70–71, 73, 77, 87–109, 120, 150, 167, 224–6, 229, 239, 265, 330, 367–8, 393, 397–8
non-Annex I countries 218, 312, 364
  see also developing countries
nongovernmental organizations (NGOs) 3, 24, 63, 91, 185–205, 220, 237–8, 244, 248–9, 340, 394–5, 399
norms, international 16, 188, 376
Norway 22–3, 26–9, 70, 89, 97, 108, 127, 167, 185–205, 229, 242–3, 279–301 *passim*, 360, 393, 399, 401
nuclear energy 74, 147–8, 150, 174, 332, 340

oil 6, 14, 51, 92, 147, 190, 195, 199, 269, 285–9, 291, 293, 314, 328, 334, 365
Organization for Economic Cooperation and Development (OECD) 47, 121, 129, 225–6, 245
Organization of Petroleum Exporting Countries (OPEC) 14–15, 242, 244, 341
ozone depletion *see* stratospheric ozone depletion
ozone-depleting substances (ODSs) 129–30

Paris Club 127
perceptions 20, 144, 174, 178, 195, 204, 257, 272, 281, 290, 300, 376, 400–401
petroleum *see* oil
Poland 23, 25, 29, 113–34, 146, 167, 218, 305, 309, 317, 319, 393, 397–8
policies and measures 30, 77, 100, 105, 163, 222–3, 244, 256, 259–60, 267, 309, 311, 319, 364, 367, 402
policy autonomy 22, 152
  see also subsidiarity
policy diffusion 28, 256–63
policy formulation 279, 396
policy instruments 28–9, 54, 179, 189, 237, 255–8, 263, 271–2, 279, 290–91, 295–300, 400–402
politics *see* domestic politics, international politics
polluter-pays principle 280, 289, 298, 327, 370, 401
poor countries 5, 13, 352, 358–9, 371, 380–81
  see also developing countries, least developed countries
post-Kyoto negotiations 80, 173, 200, 246–50, 318
poverty 6, 10, 14, 340, 357, 359, 371–2
precautionary principle 68, 83, 280, 282, 289, 298, 334, 336, 366, 401
presidency of the European Union 97–102, 105, 110, 169–70, 180, 214, 216–19, 224, 227, 238–40, 329, 337
prestige 17, 119, 281, 401
psychology 19, 156, 395
public health *see* health
public opinion 19, 117, 144, 200, 204, 221, 293, 298, 326, 395, 402

ratification of the Kyoto Protocol *see* Kyoto Protocol
rationality 324, 327, 334
realism 19–20, 376, 394
reforestation *see* forests
regimes 16–20, 28–9, 31, 88, 187, 219, 227, 300, 335, 350, 394, 397, 405
regional actors 26, 162–3, 177, 180
regional cooperation *see* cooperation
regulation 17, 64, 66–9, 73, 77, 124–5, 129, 145, 165, 172–4, 178, 195, 200, 212, 214, 226–7, 256, 262, 264, 267, 269, 271, 284, 290, 294, 300, 306, 327–9, 356, 361, 363
reinsurance *see* insurance
renewable energy 50–52, 54, 57–9, 78, 237, 331–3, 340–43, 366–7
  see also alternative energy, energy
research organizations 237, 240, 245–6
responsibility 11, 14, 53, 200, 338–9, 353, 355–7, 359–61, 363, 369, 373, 375–8, 405
  see also common but differentiated responsibility

risk 9, 178, 238, 247, 282, 293, 340
Russia 13, 131, 153–4, 167, 189, 193, 227, 229, 241–3, 247–9, 251, 336

science, role of 10, 14, 57, 67, 72, 92, 100, 128, 134, 175, 220–21, 377, 395, 405
  see also knowledge
sea-level rise 4, 6–9, 15, 244, 334, 356, 362, 378
security 172, 227
  energy 46, 133
  global 228
  human 5
  national 5, 116
  social 51
shaming 27, 185–205 *passim*, 399
shared competence *see* competence
sinks *see* carbon sinks
small island states 4, 242, 334
  see also Alliance of Small Island States
social justice *see* equity, fairness, justice
socialism 114, 130, 133
societal theory 215, 218, 221, 250, 400
society, role of 19–21, 32, 42, 53, 91, 99, 104, 106–7, 116, 118, 132–3, 144, 172–3, 177, 187, 325–6, 338, 344, 394–8, 400, 402, 406
sovereignty 26, 69–70, 140, 144–5, 150–53, 155–6, 213, 244, 398
Soviet Union 70, 72, 113–15, 120, 123, 397
  see also Russia
Spain 23, 26, 33, 50, 52, 55, 77, 101, 109, 161–81, 224–5, 260, 269, 367, 381, 393, 398
species, other 9, 355, 380
  see also biodiversity
stakeholders 18, 28, 51, 58–9, 91, 175, 220, 233, 236, 238, 247, 250, 281, 311, 316, 318, 400–401
Stalinism 114–16, 118, 133, 397
state, role of the 18–20, 42, 84, 91–2, 113–14, 116–19, 124, 133, 144, 147, 152, 165–6, 172, 175, 178, 180, 194–5, 215–16, 257, 280–81, 286, 326, 329, 334, 338, 344, 394–5, 397–9, 401, 404, 406
Stockholm Conference on the Human Environment *see* United Nations Conference on the Human Environment
stratospheric ozone depletion 6, 75, 83, 92, 129, 219, 329
  see also ozone-depleting substances
subsidiarity 22, 163, 179, 213
  see also policy autonomy
Sulfur Protocol 120, 128
sustainable development 93, 98, 121, 123–4, 128, 153, 245, 328, 339–40, 342–3, 371
Sweden 23, 25–6, 33, 50, 77, 81, 89, 94, 97, 101, 109, 120, 127, 139–57, 167, 193, 216, 225, 238, 265, 367, 393, 398
systemic theory 18–19, 26, 32, 65, 83, 113, 116, 118–20, 130–33, 140–41, 143–4, 152, 215, 241, 250, 257, 280–81, 293, 305, 313–14, 318–19, 326, 334, 338, 344, 394, 398, 400–406

targets and timetables 15, 74, 76, 223–4, 282, 284, 362, 381
targets, emissions *see* emissions targets, targets and timetables
taxes *see* carbon taxes, energy taxes, green taxes
technology transfer 12, 130, 333, 359–60, 364
temperature, changes in 6–10, 366, 368
theory 16–21, 133, 156, 374–7, 393–406
  see also specific theories
timetables *see* targets and timetables
trade *see* emissions trading
transport 6, 52, 57, 59, 73–4, 79, 95, 115–16, 123, 129, 162–3, 165, 169, 173, 190, 201, 287–8, 293, 311, 339, 370
Treaty of Rome 22, 379
trees *see* forests
Triptych approach 100–101, 105–6
two-level game 18, 140, 188, 203

Umbrella Group 193, 228–9, 242–3, 249
uncertainty, scientific 14, 18, 53, 142, 200, 247

Union of Soviet Socialist Republics *see* Soviet Union
United Kingdom (UK) 23–4, 33, 63–84, 101, 105, 120, 154, 167, 211, 216–17, 219, 224–5, 227, 229, 237, 247, 265, 268, 319, 367–8, 393, 396–7, 400
United Nations Conference on Environment and Development (UNCED) *see* Earth Summit
United Nations Conference on the Human Environment (Stockholm Conference) 118, 145
United Nations Environment Program (UNEP) 71, 245, 341
United Nations Framework Convention on Climate Change (UNFCCC) *see* Framework Convention on Climate Change
United Nations General Assembly 11, 71, 334
United States (US) 3–4, 12, 15, 19, 21, 23–4, 27, 31–2, 47, 52–4, 57, 63–5, 69, 70–84, 106, 108–9, 127, 130, 148–50, 167, 189, 191, 193, 201, 211–12, 217, 219, 222–5, 228–9, 234, 241–2, 247–8, 250, 264, 267, 272, 282, 284–5, 289, 291, 297, 334–8, 341, 350, 352, 355, 361–5, 371, 373–9, 381, 400, 405–6

values and principles 16, 280, 289–90, 293–4, 375
Vienna Convention on the Protection of the Stratospheric Ozone Layer 75, 129
voluntary agreements 49, 52, 55, 57–9, 255, 290–91, 299
vulnerability 6, 132, 244, 290, 345

waste 69, 92, 95, 121, 265, 275, 288, 308–10, 319, 329, 340
water 5, 8–9, 98, 127, 176, 233, 308–10
weather 6–7, 221
welfare 116, 370, 375–6
World Climate Conference, First 10
Second 11
World Commission on Environment and Development (WCED) ('Brundtland Commission') 93, 189–90, 289–90
World Summit on Sustainable Development (WSSD) 336, 341